Soil and Soil Fertility Management Research in Sub-Saharan Africa

T0174291

Judicious soil fertility management is crucial for sustainable crop production and food security in sub-Saharan Africa (SSA). This book describes the various concepts and approaches underlying soil and soil fertility management research in SSA over the last fifty years. It provides examples of important innovations generated and assesses the position of research within the research-to-development continuum, including how innovations have been validated with the intended beneficiaries.

Using the experience of the International Institute of Tropical Agriculture (IITA) as a case study, the authors analyse how processes, partnerships and other factors have affected research priorities, the delivery of outputs, and their uptake by farming communities in SSA. They evaluate both successes and failures of past investments in soil fertility research and important lessons learnt which provide crucial information for national and international scientists currently engaged in this research area. The book is organised in a number of chapters each covering a chronological period characterised by its primary research content and approaches and by the dominant research paradigms and delivery models.

Henk Mutsaers is a Farming Systems Agronomist based in the Netherlands.

Danny Coyne is a Soil Health Scientist at the International Institute of Tropical Agriculture (IITA), based in Nairobi, Kenya.

Stefan Hauser is a Root and Tuber Systems Agronomist at IITA, based in Ibadan, Nigeria.

Jeroen Huising is a Soil Scientist at IITA, based in Ibadan, Nigeria.

Alpha Kamara is a Savannah Systems Agronomist at IITA, based in Ibadan, Nigeria.

Generose Nziguheba is a Soil Ecologist at IITA, based in Nairobi, Kenya.

Pieter Pypers is a Soil Fertility Specialist at IITA, based in Nairobi, Kenya.

Godfrey Taulya is a System Agronomist at IITA, based in Kampala, Uganda.

Piet van Asten is a Systems Agronomist at IITA, based in Kampala, Uganda.

Bernard Vanlauwe is R4D Director for Central Africa and Natural Resource Management at IITA, based in Nairobi, Kenya.

Earthscan Studies in Natural Resource Management

For more information on books in the Earthscan Studies in Natural Resource Management series, please visit the series page on the Routledge website: www.routledge.com/books/series/ECNRM/

Soil and Soil Fertility Management Research in Sub-Saharan Africa

Fifty Years of Shifting Visions and Chequered Achievements

Henk Mutsaers, Danny Coyne,
Stefan Hauser, Jeroen Huising,
Alpha Kamara, Generose Nziguheba,
Pieter Pypers, Godfrey Taulya,
Piet van Asten and Bernard Vanlauwe

Routledge
Taylor & Francis Group
LONDON AND NEW YORK

earthscan
from Routledge

First published 2017
by Routledge

2 Park Square, Milton Park, Abingdon, Oxfordshire OX14 4RN
52 Vanderbilt Avenue, New York, NY 10017

Routledge is an imprint of the Taylor & Francis Group, an informa business

First issued in paperback 2019

British Library Cataloguing-in-Publication Data
A catalogue record for this book is available from the British Library

Library of Congress Cataloging in Publication Data
Names: Mutsaers, H. J. W., 1940–
Title: Soil and Soil Fertility Management Research in Sub-Saharan
Africa : Fifty Years of Shifting Visions and Chequered Achievements
/ Henk Mutsaers, Danny Coyne, Stefan Hauser, Jeroen Huising,
Alpha Kamara, Generose Nziguheba, Pieter Pypers, Godfrey
Taulya, Piet van Asten, and Bernard Vanlauwe.
Description: New York : Routledge, 2017. |
Series: Earthscan studies in natural resource management |
Includes bibliographical references and index.
Identifiers: LCCN 2016044663| ISBN 9781138698512 (hbk) |
ISBN 9781315518855 (ebk)
Subjects: LCSH: Soils--Research--Africa, Sub-Saharan. | Soil
fertility--Management--Research--Africa, Sub-Saharan.
Classification: LCC S599.5.A1 M88 2017 | DDC 631.4/220967--
dc23
LC record available at https://lccn.loc.gov/2016044663

ISBN: 978-1-138-69851-2 (hbk)
ISBN: 978-0-367-33513-7 (pbk)

Typeset in Bembo
by HWA Text and Data Management, London

Contents

Figures

Acronyms

AARP	Applied Agricultural Research program
ACIA	African Cassava Agronomy Initiative
ACIAR	Australian Centre for International Agricultural Research
ADP	Agricultural Development Projects
AEZ	Agro-ecological zone
AfNet	Africa Network for Soil Biology and Fertility
AFNETA	Alley Farming Network for Tropical Africa
AfSIS	Africa Soil Information System
AGRA	Alliance for a Green Revolution in Africa
AMF	Arbuscular mycorrhizal fungi
ARC	Advanced Research Centres
ARI	Advanced Research Institute
ASB	Alternatives to Slash and Burn
ATER	Area-time equivalent ratio
BADP	Bida Agricultural Development Program
BMGF	Bill & Melinda Gates Foundation
BNMS	Balanced Nutrient Management System
CAADP	Comprehensive Africa Agricultural Development Programme
CAN	Calcium ammonium nitrate
CARDER	*Centre d'Action Régionale pour le Développement Rural*
CB	Climbing bean
CEC	Cation exchange capacity
CEF	*Conseil à l'Exploitation Familiale*
CERES	Crop Environment Resource Synthesis (model)
CGIAR	Consultative Group for International Agricultural Research
CIALCA	Consortium for Improving Agriculture-based Livelihoods in Central Africa
CIAT	*Centro Internacional de Agricultura Tropical*
CIDA	Canadian International Development Agency
CIEPCA	*Centre d'Information et d'Echanges sur les Plantes de Couverture en Afrique*

CIMMYT	International Maize and Wheat Improvement Center (*Centro Internacional de Mejoramiento de Maíz y Trigo*)
CND	Compositional nutrient diagnosis
COMBS	Collaborative Group on Maize-Based Systems
COMPRO	Commercial products
CORTIS	Collaborative Group for Root and Tuber Crops Improvement and Systems
COSCA	Collaborative Study for Cassava in Africa
CRC	Catholic Resource Centre
CRP	Consortium Research Programs
CSIRO	Commonwealth Scientific and Industrial Research Organisation
DANIDA	Danish International Development Agency
DAP	Di-ammonium phosphate
DfID	Department for International Development
DM	Dry matter
DR	Congo Democratic Republic of Congo
DRIS	Diagnosis and recommendation integrated system
DS	Derived savannah
DSS	Decision support system
DST	Decision support tools
ECEC	Effective cation exchange capacity
EPHTA	Ecoregional Program for the Humid and Sub-humid Tropics of Sub-Saharan Africa
EPMR	External Program and Management Review
FAO	Food and Agriculture Organization of the United Nations
FARA	Forum for Agricultural Research in Africa
FARMSIM	Farm-scale Resource Management SIMulator
FCC	Fertility Capability Classification
FF	Ford Foundation
FFS	Farmer field schools
FSP	Farming Systems Program
FSR	Farming Systems Research
FT-NIR	Fourier-Transform - Near Infrared Reflectance
FUE	Fertilizer Use Efficiency
FYM	Farmyard manure
G × E	Genotype by environment
GIS	Geographic information system
GTZ	German Technical Cooperation Agency
HFS	Humid Forest Station
IAR	Institute for Agricultural Research
IARC	International Agricultural Research Centers
IAR4D	Integrated Agricultural Research for Development
IBSNAT	International Benchmark Sites Network for Agrotechnology Transfer

ICRAF	World Agroforesty Centre (formerly International Centre for Research in Agroforestry)
ICRISAT	International Crops Research Institute for the Semi-Arid Tropics
ICT	Information Communication Technology
IDRC	International Development Research Centre
IFAD	International Fund for Agricultural Development
IFDC	International Fertilizer Development Center
IITA	International Institute of Tropical Agriculture
ILCA	International Livestock Center for Africa
ILRI	International Livestock Research Institute
INRM	Integrated natural resource management
IPG	International public goods
IPNI	International Plant Nutrition Institute
ISPC	Independent Science and Partnership Council
IRAD	*Institut de Recherche Agricole pour le Développement*
ISFM	Integrated soil fertility management
ISNAR	International Service for National Agricultural Research
ISRIC	International Soil Reference and Information Centre
KIT	Royal Tropical Institute, Amsterdam
LEXSYS	Legume Expert System
LINTUL	Light Interception and Utilisation (model)
LSD	Least Significant Difference
M&E	Monitoring and Evaluation
mATER	modified Area-Time Equivalent Ratio
MIDAS	Managed Inputs and Delivery of Agricultural Services
MPT	Multi-purpose trees
NA	Not applicable
NAERLS	National Agricultural Extension and Research Liaison Services
NAFPP	Nationally Accelerated Food Production Program
NARO	National Agricultural Research Organisation
NARS	National Agricultural Research Systems
NCRI	National Cereals Research Institute
NE	Nutrient Expert
NERICA	New Rice for Africa
NifTal	Nitrogen Fixation by Tropical Agricultural Legumes
NIRS	Near-infrared reflectance spectroscopy
NGO	Non-governmental organisation
NGS	Northern Guinea savannah
NRM	Natural Resource Management
NUE	Nutrient Use Efficiency
ODA	Overseas Development Administration
OFR	On-farm research

PAPR25	Partially acidulated phosphate rock (25%)
PAPR50	Partially acidulated phosphate rock (50%)
POM	Particulate organic matter
PR	Phosphate Rock
PRA	Participatory Rural Appraisal
PROSAB	Promoting Sustainable Agriculture in Borno State
QUEFTS	Quantitative Evaluation of the Fertility of Tropical Soils
RAE	Relative Agronomic Efficiency
RAMR	*Recherche Appliquée en Milieu Réel*
RCMD	Resource and Crop Management Division
RCMP	Resource and Crop Management Program
RF	Rockefeller Foundation
R-for-D	Research-for-Development
R-in-D	Research-in-Development
RIS	Resource information system
RP	Rock phosphate
RSE	Relative symbiotic efficiency
RUSLE	Revised universal soil loss equation
RYT	Relative yield total
SAFGRAD	Semi-Arid Food Grains Research and Development
SE	Standard error
SED	Standard error of the difference
SEM	Standard error of the mean
SG2000	Sasakawa Global 2000
SGS	Southern Guinea savannah
SOC	Soil organic carbon
SOM	Soil organic matter
SPOT	*Satellite pour l'Observation de la Terre*
SRF	Strategic and Results Framework
SSA	Sub-Saharan Africa
SSA-CP	Sub-Saharan Africa Challenge Program
SSP	Single super phosphate
STCP	Sustainable Tree Crops Program
TAC	Technical Advisory Committee
TAMASA	Taking Maize Agronomy to Scale in Africa
T&V	Training and visit
TSBF	Tropical Soil Biology and Fertility
TSP	Triple super phosphate
T-XRF	Total X-ray fluorescence
UNESCO	United Nations Educational, Scientific and Cultural Organization
USA	United States of America
USAID	United States Agency for International Development
USD	United States Dollar

USDA	United States Department of Agriculture
USLE	Universal Soil Loss Equation
WAFSRN	West African Farming Systems Research Network
WARDA	West African Rice Development Agency (now called AfricaRice)
WRB	World Reference Base
WURP	Wetlands Utilisation Research Project
XRPD	X-ray power diffraction

Introduction and justification

Preamble

Agricultural research for sub-Saharan Africa (SSA) has thoroughly changed over the past half century, a change that was driven by its own experiences while seeking solutions to the constraints of Africa's agricultural production, as well as by external factors, including priorities of the donor community and African countries themselves, climate change, urbanisation and food price surges. Worldwide views on likely or desirable pathways for African agricultural development have also changed significantly and the calls for research to become more relevant to development and show impact on the livelihoods of smallholder farmers have grown much louder.

The International Institute of Tropical Agriculture (IITA) celebrates its 50th anniversary in 2017 and has been a major player in agricultural research for the sub-humid and humid tropics of Africa. This book uses IITA's research on soils, land use and soil fertility as a case study to analyse how the position of research in the research-to-development continuum has changed and how this change has affected the generation, validation and uptake of technologies by smallholder farmers in SSA. We have used the summary term 'soil and land use research' for the work of the first two decades, which, as will be seen, aimed at a fairly drastic overhaul of agricultural land use and crop production systems. For the following decades, as the emphasis shifted to providing tools to farmers for intensified production within their existing systems, we use the term 'soil and soil fertility (management) research'.

Despite significant progress in several research areas, particularly in crop breeding and pest and disease control, many of the problems facing Africa's smallholder farming have remained as severe today as they were 50 years ago. IITA's founding fathers' concerns about low and declining land productivity in the face of growing populations and the inadequacy of traditional farming methods to curb this trend remain as valid as ever, or even more so, with growing populations and ever-declining availability of good agricultural land.

In its first 25 years, IITA remained very much the institute as originally conceived with problems of food production tackled through basic and applied research, and its results and products, once considered ready for dissemination, handed over to extension services. Today, IITA has become an entirely different institute, in the methods and content of its soil and soil fertility research, as well as its physical set-up, and would probably be barely recognised by its founders. It has moved much closer to the farm and now carries out most of its work in the context of the development of that farm. How this happened and what can be learned from its successes and failures for today's and future development-driven soil and soil fertility research form the main topic of this book.

The conclusions and lessons learned have validity beyond IITA and can enrich future soil and soil fertility research for smallholder farming, in particular in SSA. Furthermore, the book serves as a repository for the research that IITA has carried out over the years, allowing today's scientists to look back and make their own judgement.

Fifty years of soil, land use and soil fertility research in sub-Saharan Africa

Deterioration of soil resources in SSA through deforestation, nutrient mining, loss of organic matter, and erosion and inadequate fertility restoration by traditional means was a major concern 50 years ago and continues to be so, while many of the potential solutions offered by research have remained just that: potential. Today's heightened concerns about the continuing degradation of natural resources and its effect on climate, agriculture's role in the global carbon economy, and the recent food price hikes have brought the soil back into the forefront of the international debate.

Non-adoption by farmers of many of their soil-related technologies has led research organisations in the past decades to rethink their research methods, stimulated by donors' increasing impatience with the lack of impact. As a result, research has become much more outward-looking. Whereas, in the early years, much attention was focused on soil and production characterisation and designing alternative land and soil management systems and practices, the attention has now shifted towards providing practical solutions for farmers that fit within their systems and take their local soil conditions into account. Attitudes have changed, with researchers now offering their services to farmers and farming communities, rather than handing out ready-made solutions. Whilst this will likely enhance the chances of success and avoid the dead ends of the past, a participatory attitude must be married to technical and managerial innovations if research is to have real impact. So, the ultimate challenge remains: what technical solutions can be offered to address farmers' concerns, and if none are available, how can they be developed? In other words: what can research put into the 'baskets of technologies' from which its clients and partners may choose to satisfy their needs? An integrated approach is now thought to be needed, involving various stakeholders in finding ways to raise productivity, while acknowledging the farming conditions of the African smallholder and taking into account the absence, in many cases, of functional extension, input delivery and marketing services.

Research often tries to tackle unanswered challenges with 'proven' concepts, whose logic is compelling, even if they did not work in the past. A striking example in agriculture are 'auxiliary' legumes of all kinds, which have been hailed by successive generations of development workers and scientists, starting early in the twentieth century, as a panacea for many ills, but which have run up to reluctance by farmers to grow them beyond a certain minimum acreage. Other examples are various biological weed control and soil protection methods and fertiliser recommendations which performed well on-station but hardly reached the real farm. The failure of these technologies in the past has been explained by the absence of proper targeting to specific conditions, their testing without adequate involvement of the intended beneficiaries and their promotion without due attention to factors outside the immediate farm affecting uptake, such as the functioning of agricultural input and produce markets. Today, the need to pay attention to such factors drives most development-related research, and soil and soil fertility research is more and more embedded in programmes with broader agricultural development goals that create conditions facilitating the uptake of research results. In support of this reorientation, this study looks back and presents an inventory of technologies and approaches for soil and soil fertility management, developed and tested in the past, using IITA as a case study, and examines their adoption and non-adoption record.

The question of adoption is of course crucial and has often been insufficiently addressed or even neglected, until it was taken up as a priority concern in the recent past. Whether or not a technology will be taken up by farmers depends on a number of equally important factors:

1   Does the technology solve an important technical constraint?
2   Is the technology robust or does it have to be fine-tuned to different ecological conditions and farm types?
3   Is the constraint considered important by the farmers themselves (relative to all other constraints that the farming household is facing), and is the solution economically attractive and socially acceptable?
4   Can the technology be incorporated into the farmer's current system or is a major system change required?
5   Is the farmers' production environment (e.g. input and output markets) conducive to adoption and what additional support is needed for effective dissemination?
6   Do the expected benefits justify the risk associated with the required investment?

Early research essentially addressed the first factor and to some extent the second and third, but over the years the importance of the other factors has come to the forefront. Today scientists are aware that research cannot stop at the gate of their station, nor address technical problems in isolation. This review therefore examines both the technical results and the research approaches used by IITA and their changes over time, in particular the changing balance between controlled station research on one hand and 'adaptive' research with and by farmers on the other hand. By identifying the factors that affected adoption or non-adoption, today's scientists may retain what remains valuable from earlier research while avoiding its pitfalls and dead alleys of the past.

Although IITA is used as case study, most of the issues touched upon in this study do apply more widely. The Institute is part of an international research community with shared opinions, motivations and methods, and it is exposed to and responds to trends and concerns in the world at large. The lessons drawn from this case study should therefore have validity across the research-to-development continuum, and across countries and ecologies, especially on the African continent.

The chapters and the story-lines

There are different ways in which a story covering five decades of research on soils, land use and soil fertility can be told. One is by identifying a number of major themes and presenting them diachronically; another is a chronological account with different themes recurring in each episode until they are completed, or abandoned. We have opted for the second approach which has the merit of showing the evolution in thinking, both in a scientific sense and in relation to the role of research in the context of agricultural development where its results must find application.

The book has seven chapters, each covering a specific episode from 1967 till today. In order not to lose sight of the grand research themes, we have traced those themes through successive episodes, as 'story-lines' or 'red threads' running through the book. They correspond broadly with each chapter's sections, and deal with goals and concepts, technical content, dissemination and adoption by farmers, as well as the factors and drivers affecting them. The last section of each chapter presents an analysis of the findings of the episode as well as external developments which influenced the views about the desirable direction of future research. The first section of each following chapter then outlines how these views were translated into implementation in the research programmes. Thus, the first and last sections of each chapter are more reflective on the role of research, its content and the progress made, while the other sections present extensive, though non-exhaustive, accounts of the actual research carried out. Readers thus can make their own choice on how to access the material and messages presented in this book.

Figure 0.1 shows major events in the world of international and African agricultural research and development which occurred during the period covered in this book and influenced the visibility, position, approaches and methods of soil, land use and soil fertility management research.

The first two chapters cover the period 1967–1982 when exploratory research on soils and land use dominated, thus Chapter 1 mostly deals with the formative years of resource characterisation, and Chapter 2 treats the early phases of the experimental work. This was a time of great expectations, raised by the early

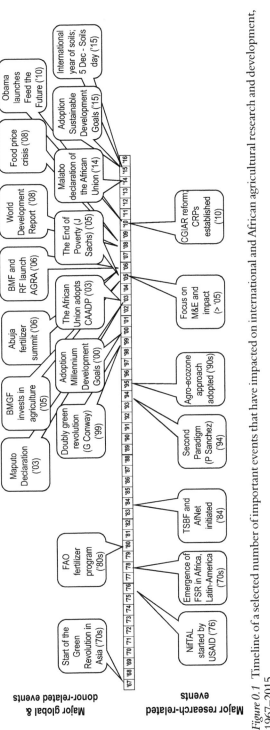

Figure 0.1 Timeline of a selected number of important events that have impacted on international and African agricultural research and development, 1967–2015

The abbreviations are spelled out in full as: AfNet: African Network for Tropical Soil Biology and Fertility; AGRA: Alliance for a Green Revolution in Africa; BMGF: Bill & Melinda Gates Foundation; CRP: CGIAR Research Program; CAADP: Comprehensive Africa Agriculture Development Programme; FAO: Food and Agriculture Organization of the United Nations; FSR: Farming Systems Research; M&E: Monitoring and Evaluation; NifTAL: Nitrogen Fixation by Tropical Agricultural Legumes; RF: Rockefeller Foundation; TSBF: Tropical Soil Biology and Fertility; USAID: United States Agency for International Development.

successes of the Asian Green Revolution (Figure 0.1) and IITA had been created to bring the blessings of modern agricultural science to the African farmer. The integrity of the fragile soil and the improvement and maintenance of its fertility was considered paramount to that goal and the work during these years was expected to lay the foundation for new land use and crop production technology to help African farmers break out of their age-long circle of long fallows and small yields.

By 1982, it was felt the time had come to move closer to the farmer through the initiation of farming systems research. During the period 1983–1988, covered in Chapter 3, Farming Systems Research (FSR) concepts, developed elsewhere in Africa and in Latin America (Figure 0.1), were hesitatingly introduced, initially more as a tool to take technologies out of the research station than as a way to associate farmers with the technology development process itself. Meanwhile the more basic research on soil and soil fertility management continued mostly in controlled research plots. This was also the time novel land use technologies like mechanised land clearing, zero tillage and alley cropping emerged, however, without a clearly defined target farmer.

The period 1989–1995, covered in Chapter 4, was characterised by the exploration of constraints and leverage points in smallholder agriculture and by integrated natural resource management programs seeking for alternative solutions that are less dependent on agro-inputs, combining such inputs with farm-produced organic material. More and more of the work was carried out on the farm, especially in the moist savannah areas.

Eco-regional programs, initiated in the period 1996–2001, are covered in Chapter 5. These programs intended to bring research to the African farm in a structured manner, by way of Benchmarks, representing the conditions in the major Agro-Ecological Zones (AEZ; Figure 0.1), where the technology development and validation would be carried out, for subsequent extension or 'scaling' to the wider eco-regional zone for which the benchmarks stood model. In the course of the 1990s, the USDA Soil Taxonomy classification, which was used as a standard by the Institute during the first twenty-odd years, was replaced by the World Reference Base for Soil Resources (WRB)[1].

The period 2002–2011, covered in Chapter 6, saw the establishment in several countries across the continent of integrated programs aimed at strengthening livelihoods of farming families. They cut across all research areas, including agronomy, breeding and pest and disease management. This wide web of inter-related projects resulted in no small manner from the growing influence of a variety of important donors during these years (Figure 0.1).

During the last period – from 2012 until today, covered in Chapter 7 – integrated systems research programs were initiated in specific locations, with soil and soil fertility management research again occupying a more prominent, but this time fully integrated position. Investments in research on soil and soil fertility management increased, driven by the recognition that soil fertility is an essential piece of the agricultural intensification puzzle. Certainly the reform within the CGIAR and the establishment of CGIAR Research Programs (CRPs) influenced these dynamics.

The final chapter brings together the story-lines, as a summary of the findings, results and lessons learned and presents the authors' reflections on the way forward for soil and soil fertility research for smallholder farming in SSA and beyond.

We have also added a list of abbreviations for easy reference[2], an Annexe with soil profile data from the earlier years and a tentative correlation the Soil Taxonomy classifications and the WRB classification systems since both are used in the original documents used for the book.

This book is published at a time of great changes, promising greater relevance for international agricultural research by associating itself closely with development. Whether these promises are going to be fulfilled will depend on many factors, important among them being that research heeds the lessons of the past, uses its achievements and avoids its errors and pitfalls. It is hoped that this book can contribute to that goal.

Notes

1　A tentative correlation between the earlier (benchmark soil) Taxonomy classifications and the corresponding WRB classifications is given in Annexe II.

2　Abbreviations and acronyms tend to be used abundantly in a scientific book dealing with international agricultural research.

1   The exploratory years

1967–1976

1.1 Scope, approaches and partnerships

IITA's original mandate was formulated by the institute's management and Board of Trustees on the basis of a 1967 document written by Will Myers, its first director and later Board Chairman, which stated that research should:

> focus primarily on problems of improving food crop production in the humid tropics and on the soil and crop management requirements for developing a stable, permanent agriculture in which food crops occupy a central position.

During a series of international seminars in 1970/71 the mandate was sharpened, whereby soils acquired a dominant position because of the obvious lack of 'techniques for raising or maintaining soil fertility to allow a safe transformation from shifting to settled agriculture' (IITA, 1992). A 'broadly based Farming Systems' research approach was adopted for the Institute in order to:

> devise production and management practices that will permit farmers to intensify their farming operations without causing loss of the soil due to erosion, loss of nutrients due to leaching, or deterioration of the soil's physical structure.
>
> (Annual Report, 1972/73; IITA, 1974)

More specifically, the development of more productive alternatives to shifting cultivation or bush fallow-based farming in the humid lowland tropics became an explicit goal, as it was thought that the then current system would not be able to satisfy the increasing demand for food. According to that view, traditional systems would eventually have to be replaced by more productive ones which research was going to design. Research on soils and land use was organised as a separate program, called the Farming Systems Program (FSP), whose operational objective was stated as:

> the rapid development of permanent, productive farming systems for staple food production in the humid tropics...

Although not explicitly stated, this objective implies the criterion of sustainability, since permanent systems will only be possible if their sustainability is assured. The research agenda was essentially driven by what the research team, supported by management and Board of Trustees, thought were the things to do, without systematic interaction with stakeholders in the target countries, except through several Board members from those countries.

Three crop improvement programs, for cereals, grain legumes and root and tuber crops, were to develop suitable varieties for crops currently grown in, or with potential for, the humid zone. The FSP would integrate such crops into the farming systems and explore the potential of other 'auxiliary' species, which could enhance or stabilise production, such as mulch and cover crops and tree species. The FSP would thus be expected to play a pivotal role, integrating the various components into stable, productive systems.

In the first two decades, the FSP was the Institute's largest program, dominated by soil scientists of different denominations, which reflected the importance attached to the soil. The emphasis during the early years was on understanding and characterising the ecological and economic setting of agriculture in humid and sub-humid sub-Saharan Africa (SSA) and developing innovative production technology for the African smallholder, to be incorporated into existing systems or assembled into new more efficient systems in replacement of existing ones (e.g. IITA, 1976). The process to develop and disseminate such technology followed the then widely-accepted linear model, shown in Figure 1.1. According to this model, the conception and development of technology and its testing under controlled conditions were the task of research in collaboration with National Agricultural Research Systems (NARS). At completion of this process,

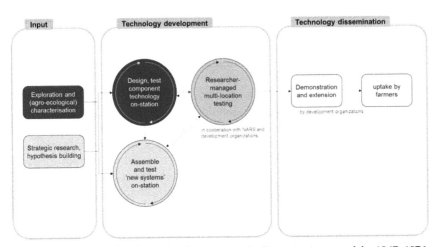

Figure 1.1 IITA's technology development and dissemination model, 1967–1976: emphasis on characterisation and component technologies

Shading indicates relative emphasis on the various components with darker ones indicating a relatively higher emphasis

extension and development organisations would take over to demonstrate and disseminate the technology to the farmers. Partnerships with NARS in Nigeria were informal and subsidiary, rather than systematic. Partnerships with development and extension organisations were expected to be taken care of by the NARS, although some *ad-hoc* contacts with Institute scientists existed.

Five grand orientations can be perceived in the Institute's early soil and land use-related research, which followed directly from the goals formulated at the start. These were:

1   Understanding the nature and potential of the soils, current land use and farming practices and their limitations;
2   Developing sustainable, intensive land management practices as alternatives for shifting cultivation and the fallow-based systems derived from it;
3   Designing improved crop production practices through soil amendments, judicious crop combinations and management and small scale mechanisation;
4   Developing methods and tools for the exploitation of under-utilised soil resources, in particular inland valley bottoms; and
5   Understanding farmers' constraints and potentials and conceiving effective approaches for the dissemination of improved technology among the intended 'clients'.

Although there was a general awareness that innovations would have to fit into the farmers' systems, it was only at the end of the technology development process that they would be exposed to the conditions of the real farm. Separate studies of farmers' productive practices and their constraints were carried out by the social scientists, but there was little or no involvement of the soils group, which concentrated on characterisation of the soil resources and development of fertility-enhancing technologies, mostly in the south-west and south-east of Nigeria, the Institute's country of residence, harbouring a wide range of the conditions found in the humid and sub-humid zone.

During the initial years, and as a consequence of the linear dissemination model, partnerships with other stakeholder groups were not seen as critical, rather *ad-hoc*, and mostly based on personal relationships whereby both sides benefited professionally. There was no mention of development, extension or policy partners, and even partnerships with national agricultural research institutes in Nigeria were scanty and subsidiary rather than thought of as critical. Several international organisations were associated with the soil characterisation activities (see Section 1.2)[1]. The soils group also collaborated with NARS in Nigeria in some Benchmark Soil activities and other soils-related research, whereby some of the multi-locational trials were hosted at the national research stations or sub-stations. Surprisingly, there was no formal linkage with the Benchmark Soils project, supported by the United States Administration for International Development (USAID), whose objectives were very similar to those of the Institute's project of the same name (see Section 1.2).

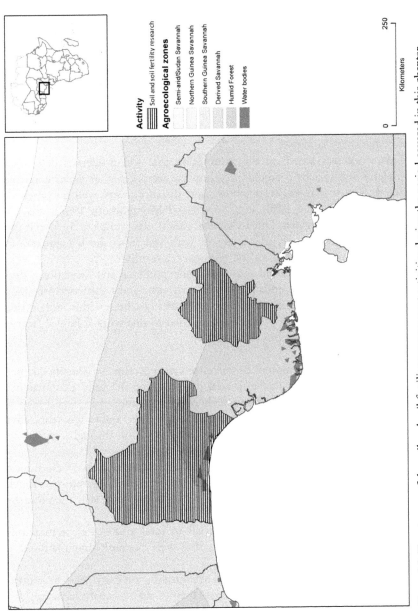

Figure 1.2 Target areas of the soil and soil fertility management activities during the period covered in this chapter

1.2 Characterisation of soils, farms and farming systems

In the 1970s, soil characterisation and mapping, based on the tradition of soil survey and land evaluation in the western world, was being transferred to developing countries. The original aim of the approach was to assess the actual or likely performance of land under different land use systems and was therefore associated with land use planning. In Africa, however, planning the use of land, already occupied by farmers' traditional systems was not a relevant option and the characterisation work adopted the more modest aim of explaining current land use and developing innovations consistent with the conditions of the soil. Initially it meant evaluation of soil resources only with little attention to other factors that shape land use.

This work was based on the assumption that soil properties determine crop yields and that by correcting for limiting factors, higher yields could be obtained. The spatial soil variation at short distances, very well documented by soil research in the 1950s (e.g. Smyth and Montgomery, 1962), was not being further investigated. Once characterisation was complete, field and pot experiments could be carried out in, or with soil from the benchmarks to investigate nutrient deficiencies and responses to fertiliser.

Characterisation of the soil resources, their potential and limitations, thus dominated the research agenda during the early years and was carried out along disciplinary and to some extent inter-disciplinary lines, both within and outside the Institute's station. The importance attached to this work followed from a number of initial observations (IITA, 1974):

1   Existing soil data could not be correlated with productive potential, hence exploratory work would be needed to determine the soils' potential and deficiencies, as a basis for future technology targeting;
2   In existing farming systems, crop yields decline after only a few years, after which the land is returned to fallow; intensification through continuous cropping will require measures to prevent this decline;
3   Exposing tropical soils to the impact of the torrential rains of the humid tropics will quickly lead to major losses of topsoil, the repository of much of the soils' fertility; a protective cover of some kind will therefore be essential; and
4   Inland valleys, though relatively small in total area, 'offer tremendous potential' for major crop production; they were currently under-utilised.

The research undertaken by the soil scientists in the early years to address these issues was clustered around five broad themes, viz. soil characterisation, soil erosion control, soil fertility management, inland valley development and cropping systems improvement. A number of representative benchmark soils would have to be identified and characterised across West and Central Africa, where soil fertility research would be carried out, the results of which could hopefully be extrapolated to the zones for which the benchmark soils were representative.

An ambitious characterisation project was therefore started in the early years, in collaboration with several foreign laboratories, to collect and analyse information on soils and land use across humid West Africa. This was needed 'in order to relate work at IITA to other parts of the humid tropics' and build a basis for targeting technologies to different environments' (IITA, 1976). The project did not include systematic agronomic surveying of farmers' soil management practices and crop yields as baseline information, probably because such information would be collected in the experimental phase of the project.

Regional small-scale soil maps had been available for many areas, but their agronomic interpretation was problematic. The project would therefore collect local soils information in well-chosen locations and assess their production capability. The two levels – regional soil maps and on-the-ground capability assessment – would then be linked for technology targeting. This would evolve into the Benchmark Soils approach, which was initiated around 1976. It was inspired by the Benchmark Soils Project coordinated by the Universities of Hawaii and Puerto Rico, which started in 1974 in Asia and Latin America (Beinroth, 1984), with one West African country taking part, viz. Cameroon. The objectives were very similar (Swindale, 1984):

> to correlate food-crop yields with soil properties and soil-use practices on a network of benchmark soils [...] to determine scientifically the transferability of agro-production technology among tropical countries, to develop methodologies, and to create the required infrastructure for successful agro–technology transfer.

There was no formal collaboration between the two projects, however.

1.2.1 Soil mapping

In the early 1970s, two major standards for regional soil mapping became available: the Soil Taxonomy classification, promoted by the United States Department for Agriculture (USDA) and the Soil Classification system, promoted by the Food and Agriculture Organization of the United Nations (FAO), with comparable hierarchical structure. In Soil Taxonomy, the classification hierarchy up to the local level was:

Order > Sub-order > Great Group > Sub-group > Family > Series

The nomenclature in both systems uses 'self-explanatory' names, except at the lowest level where soil series are named according to local convention, often pre-dating the USDA and FAO classifications.

The Institute's soil scientists opted for Soil Taxonomy as the standard for their Benchmark Soils project. The dominant Soil Taxonomy Orders found in humid and sub-humid West Africa with their major properties and some common sub-groups are shown in Table 1.1 (Juo, 1980; Juo and Moormann,

1981; Moormann et al. 1975). The names of the series in south-west Nigeria in the fourth column of Table 1.1 were those given by Smyth and Montgomery (1962) in their comprehensive soil and land use study carried out in the 1950s. They were brought into correspondence with Soil Taxonomy sub-groups by the Institute's pedologist.

Table 1.1 Soil types according to the Soil Taxonomy classification[a]

Soil orders	Major properties	Common sub-groups	Soil series in south-west and south-east Nigeria, belonging to this sub-group
Alfisols	Clayey B horizon and base saturation >50%; savannah and drier forest zone	Oxic Paleustalf	Egbeda, Iwo, Ibadan, Alagba
		Oxic Haplustalf	Iregun
		Udic Haplustalf	Iregun
		Rhodic Plinthoxeralf	–
		Typic Plinthustalf	Gambari
		Aeric Tropaqualf	Adio
Ultisols	Clayey B horizon and base saturation <50%; acidic leached soils from humid tropics and sub-tropics	Plinthic Paleustult	–
		Plinthic Paleudult	Uyanga toposequence
		Oxic Paleudult	Onne
		Typic Paleudult	Uyo
		Orthoxic Tropohumult	–
		Orthoxic Tropudult	Uyanga toposequence
Oxisols	Strongly weathered acidic soils of uniform texture; rainforest and savannah	Tropeptic Haplorthox	–
Entisols	Mostly alluvial soils with little horizon development; e.g. lower slope, valley bottoms	Psammentic Tropaquent	Uyanga toposequence
		Psammentic Ustorthent	Apomu[b]
		Aeric Tropaquent	Matako
Inceptisols	Young soils with limited profile development, mostly formed on colluvial and alluvial materials	None of the Benchmark soils were classified as Inceptisols	–

Notes:

a A description of the Soil Taxonomy nomenclature is given in Juo (1980); for a recent account see Soil Survey Staff (2015).
b Later reclassified as Alfic Ustipsamment.

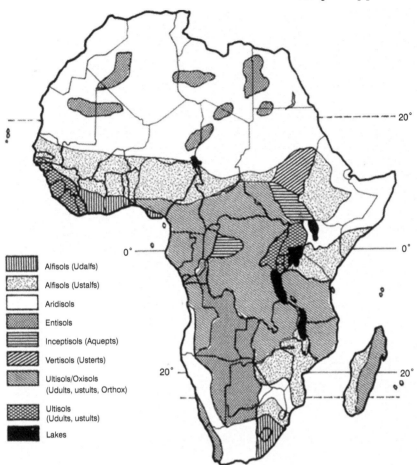

Figure 1.3 Soils of tropical Africa (adapted from Aubert and Tavernier, 1974, by Kang and Osiname, 1979)

At least two soil orders are usually found even within a small toposequence, e.g. sedentary Alfisols or Ultisols in upper and middle slope positions, and Entisols in the valley bottoms. Existing regional and continental maps only showed the distribution of the dominant soil orders at a small scale[2] (e.g. the soil map of Africa shown in Figure 1.3). It is clear that the Ultisols, Oxisols and Alfisols are by far the most prevalent soils in humid Africa, but their association with other orders along the toposequences results in considerable soil variation even at small distances.

1.2.2 Capability assessment

For the field level benchmark studies, representative toposequences were chosen, with their succession of land types and soil series between crest and valley bottom[3]. In a rolling topography, the land types were designated as

(i) upper and middle slope, (ii) lower slope and (iii) valley bottom, further sub-divided as needed. Representative profiles at different positions along a toposequence were then classified according to Soil Taxonomy and, if possible, identified with soil series according to local nomenclature. The samples were analysed for physical and chemical properties, which were translated into a semi-quantitative assessment of the soil's productive capability. As an example, a detailed description of the soils of the Institute's Ibadan station was made, consisting of (i) a land type map, (ii) drawings of typical toposequences (Figure 1.4) with their soil series, (iii) detailed descriptions of representative profiles with physical and chemical analyses and finally (iv) a 'tentative' (very simplistic) land evaluation for the land types of the map units (Moormann et al., 1975). Similar studies were conducted in the following years in many toposequences across humid and sub-humid West and Central Africa, most of them (14) in south-west and south-east Nigeria (Figure 1.5). The latter cover a range of neutral to slightly acidic Alfisols in the south-west and acidic Ultisols in the south-east of the country, with Entisols in the valley bottoms (IITA, 1976; Juo and Moormann, 1981). They were presumed to 'contain a range of soil profiles that are representative of the landscapes and the soils of West Africa and other parts of the humid tropics' (Moormann, 1981). Chemical analyses of two upland soils each from south-west and south-east Nigeria are shown in Annexe I.

1.2.3 Physical and chemical properties of benchmark soils

Chemical and physical soil properties of the benchmark soils were measured and their limitations for cropping and the need for amendments were determined on the basis of their mineralogy, the nature and saturation of the adsorption complex, the mineral reserves, soil acidity and P sorption. Some preliminary results are presented here, to be extended in later chapters.

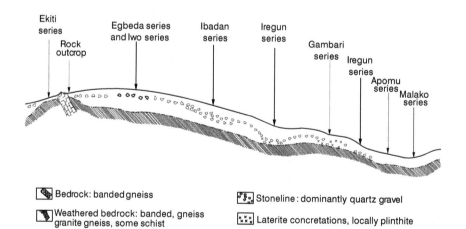

Figure 1.4 Schematic cross-section of a toposequence at IITA (adapted from Moormann et al., 1975)

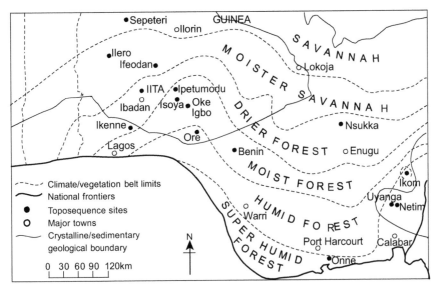

Figure 1.5 Location of benchmark toposequences in Southern Nigeria (adapted from Moormann, 1981)

As an example for the range of soil conditions occurring in the humid and sub-humid zone, the texture and exchangeable base saturation of three typical profiles (an Alfisol, Ultisol and Oxisol) are shown in Figure 1.6, illustrating the very different nature of the major soil groups of these zones.

The potassium status of Alfisols and Ultisols expressed by their K quantity/ intensity relationship[4] was found to be very different indicating that no immediate K deficiency would be expected in the Alfisols, contrary to the Ultisols, in particular those derived from sandstone and sandy coastal sediments. K-deficiency would be expected to develop in all soils under intensive cropping, the earliest in the Ultisols and Oxisols and in soils derived from coastal sedimentary rock, because of their low amounts of weatherable minerals (Juo and Grimme, 1980). Pot studies with maize (*Zea mays*) on five soils derived from coastal sedimentary rock and six soils from basement complex rock showed a lower K-status of the maize plants in the former soils (IITA, 1977; soil classification not given). This corresponded with an average (adequate) available K-content of 0.214 cmol/kg in the basement complex soils and a low 0.103 cmol/kg for the soils derived from sedimentary rock (IITA, 1977). These were just indicative results, since data from such small numbers of soils is very little to draw general conclusions about soils where large variability might be expected.

Phosphorus status was characterised by total P content of the soils, percentage in the organic matter and the P-sorption capacity, which measures the degree of P-immobilisation and has a strong influence on the P-fertiliser rate needed for maximum crop production. P-immobilisation is strongly connected with the soils' parent material and particularly high in soils derived from volcanic rocks, which are of minor importance in West Africa. Acid soils in the tropics

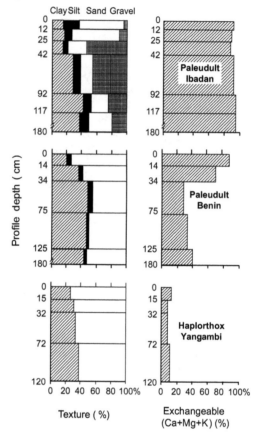

Figure 1.6 Texture and percentage base saturation of three upland soil profiles (adapted from IITA, 1977)

also have higher P-sorption than non-acid soils, because in the former P is mainly present in the form of highly insoluble iron and aluminium phosphates. Ultisols therefore have higher P-requirements than Alfisols, because applied P is converted into less soluble forms (Juo and Fox, 1977; Kang, 1980).

Most Alfisols on basement complex rock are highly sensitive to erosion, especially under large-scale farming, due to the combination of a sandy top soil and high gravel content of the upper B-horizon.

In order to validate the interpretation of these assessments in terms of production capacity and limitations, field trials would be needed and a network of testing sites may initially have been envisaged.

1.3 Technology development

A start was made during this episode with field and laboratory research which should result in component technologies for stable and permanent food crop production systems for humid and sub-humid Africa. A wide range of themes was thereby tackled, such as erosion control under intensive cropping, nutrient responses and response to liming in acid soils, mixed cropping and crop rotation with leguminous crops. Most of this work was carried out at the Institute's own experimental station in Ibadan in the Alfisol zone and Onne in the Ultisol zone, with some involvement of individual national researchers.

1.3.1 Soil and soil fertility management

According to the benchmark concept, research on soil management had to be carried out in representative sites or with soil from such sites, in order to assess nutrient limitations, soil acidity and Al-toxicity, as well as physical problems such as erosion hazards. The findings would then form the basis for the development and testing of corrective and yield enhancing technologies. In order to carry out this work systematically a testing network would have to be set up covering the chosen benchmarks, which would require intensive collaboration with NARS. The capacity of the NARS, however, was not adequate for such programs, nor was the economic condition of agriculture conducive to the set-up of such an intensive scheme (e.g. Juo, 1980). Systematic trials were therefore only conducted at a few of the benchmark sites, viz. at or near the main station in Ibadan and sub-stations in Ikenne (south-west Nigeria, medium rainfall, Alfisol area) and Onne (south-east Nigeria, high rainfall, Ultisol area). Some preliminary results are reported in the following sections, while more comprehensive summaries will be given in following chapters.

Soil erosion control

The risk of massive soil losses upon exposure of bare soil to torrential tropical rains was one of the great concerns from the outset. This was not so much based on actual erosion problems in traditional farming systems in the humid zone of West Africa, where it was of minor importance, but rather on problems to be expected under more intensive semi-mechanised land use, a major (implicit) goal of research at the time.

In order to measure the losses under different conditions, a long-term trial with a set of large run-off plots was established in 1971 at the Ibadan station to compare the effect of minimum tillage, mulches and conventional management. This was one of the first evaluations of minimum/no-till agriculture in SSA, an important component of what would become 'Conservation Agriculture'.

The slopes in the run-off plots ranged from 1 to 15% and the slope length was 25 m, except where the effect of slope length was tested (Lal, 1976). After 3 years, it became already clear that soil losses under continuous cropping (2 crops/year) could be practically eliminated and yields maintained by leaving maize stover and weeds, killed by herbicide, on the surface as mulch and planting through the rubble. Tests with different amounts of rice straw mulch showed that 4–6 t/ha of straw was sufficient to reduce soil loss to negligible amounts. Apart from reducing erosion, the soil cover dampened the high temperatures in the surface soil, which are harmful to germinating seeds and young seedlings. Also, moisture loss by evaporation and weed growth were reduced and soil structure was improved, all of which contributed to higher moisture storage (IITA, 1976; Lal, 1983). The *in situ* production of 4 to 6 tons of straw was beyond farmers' current production levels, but was thought to be attainable under the more intensive system envisaged for the future.

Several studies were initiated on the mechanisms of soil erosion and the influence of rain intensity and amount, soil cover, soil texture, aggregate stability etc. The aim was to measure or estimate the different factors in the USDA Revised Universal Soil Loss Equation (RUSLE) for soils and rainfall regimes in the humid and sub-humid tropics. Eventually this should allow the mapping of erosion risks for the humid zone. In the RUSLE formalism soil loss due to a rainstorm is proportional to both erosivity of the rain (R) and erodibility of the soil (K), whereby R is a function of the kinetic energy and the amount of rain in a storm[5]. Lal (1976) found that for a tropical storm the erosivity is better represented by the product of its total rainfall (a) and its maximum intensity i_m and calculated a total annual rainfall erosivity index as:

$$A_m = \sum_{1}^{12}\left[\sum_{1}^{31} ai_m\right]/100 \tag{1.1}$$

It had a correlation coefficient of 0.69 with soil loss from a bare fallow Alfisol at the Ibadan station with a slope of 10% (which is actually not very high, explaining less than 50% of the variation). It is a convenient index for mapping and a hypothetical *iso-erodent* map (showing lines of equal erosivity of the rainfall regime) was prepared for West and part of Central Africa (Figure 1.7).

In subsequent years, soil erosion remained a priority research area, which was one of the characteristic features of the Institute's soil research of the 1970s and 1980s. Much attention was going to be given to the different factors in

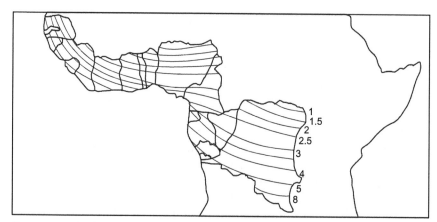

Figure 1.7 Hypothetical iso-erodent map of West Africa (adapted from Lal, 1976)

The figures express AI_m values as per equation (1.1) and have as units cm²hr⁻¹·

the soil-loss equation, including the erodibility K of the soil and the effect of crop management and cover represented by the factor *C*. The unfolding erosion story will be further reviewed in the next chapter.

Soil fertility management

One of the first priorities of soil fertility research was to close the perceived shortage in experimental information on the effect of fertilizer under continuous cropping, both in the Alfisol and the Ultisol zone (Kang and Balasubramanian, 1990). Soil fertility in the Alfisol zone was relatively favourable, with a pH around 6.5 and medium to high base saturation, but nutrient deficiencies were to be expected, especially when high crop yields were the aim. Soil fertility problems are much more severe in the high rainfall Ultisol zone, because of low pH, low mineral reserves and medium to high Al saturation of the adsorption complex. These problems are often concealed by the farmers' cropping–fallow cycle, but they become more pronounced when the cycle is broken in (semi-) permanent cropping. Such cropping was expected to require expensive correction measures through liming and balanced fertiliser dressings, including micro-nutrients, combined with biological N-fixation and enhanced P-nutrition through mycorrhiza (IITA, 1977), and possibly improved fallow or agroforestry-type systems (e.g. Kang, 1980). Some work was started during this episode and was going to be expanded in the next to address these issues.

It is worth repeating here that much of the work on nutrient responses in the early years was carried out under high inputs of the non-treatment nutrients and 'optimum management', including clean weeding, insect control and sometimes even supplementary irrigation, resulting in yields far exceeding those of farmers. Similarly, crop breeders would conduct much of their selection work under equally unrepresentative conditions.

Since P-deficiency was expected to be the first (besides N) to appear in many soils a trial was set up in 1971 at the Ibadan station (Egbeda soil) and at Ikenne (Alagba soil on coastal sediment) on P-response of sole maize on newly cleared land with all vegetation as well as crop residues removed. Fertiliser-P was applied once as single super phosphate (SSP) followed by two maize crops annually in Ibadan and one in Ikenne for three years, each crop amply supplied with other nutrients (Kang and Osiname, 1979). Figure 1.8 shows significant responses in the first two maize crops to P applied in the first year at the IITA station, up to 26 kg P/ha, but in the third year the response levelled off, suggesting depletion at the lower rates. Note the (very) high maize yields in this trial, due to application of presumably optimal rates of other nutrients and effective management. At Ikenne, P-response was significant up to 52 kg P/ha for a single annual maize crop, at an overall yield level of only about half that at Ibadan. It was thought that the early P-response observed in these trials was due to complete removal of crop residues.

There was good relationship between maize yield and the Bray-1 P-test for the Egbeda soil in this trial, with a critical level (95% of maximum yield) of 14 mg P/kg (Figure 1.9).

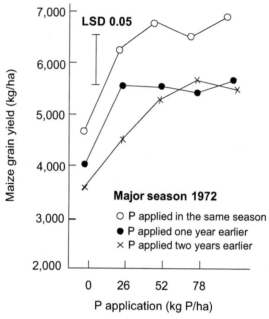

Figure 1.8 Main season maize yield in 1972 as a function of P applied in same season (o) or one (•) or two years earlier (×) (adapted from Kang and Osiname, 1979)

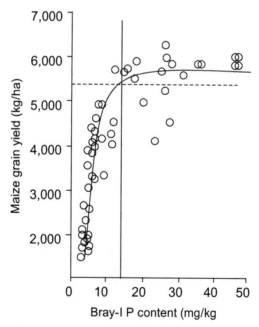

Figure 1.9 Maize yield as a function of Bray-I P, Egbeda soil, Ibadan (adapted from Kang and Osiname, 1979)

A more comprehensive long-term exploratory trial with N, P, K, S, Mg, Fe and Zn combinations on a maize/sweet potato (*Ipomoea batata*) relay crop was started in 1972 on Egbeda and Apomu soil at the Ibadan station. Crop residues were returned in all cases except in two extra treatments, one with high and one with low fertiliser rates where residues were removed. Fertiliser rates were sometimes modified in the course of the trial in adjustment to the observed effects. The sweet potato relay crop was rather odd, since it is hardly grown as a component of the cropping systems in the Alfisol zone. In 1976, it was replaced by cowpea (*Vigna unguiculata*) planted in the second season after first season maize. Responses to N and P were registered on both soils from the start, but no response to K nor to the secondary and micronutrients so far. Full results for this trial will be reviewed in the next chapter.

A greenhouse study was conducted on the response to sulphur by paddy grown in valley bottom soils collected from 11 locations in central and southern Nigeria. Total S-content of the soils ranged from 46 to 487 mg/kg. In 50% of the soils rice grown under flooded conditions responded to S-application, while only 20% showed S response under non-flooded conditions. No further information seems to have been published by the Institute on the S-status of inland valleys.

Iron toxicity was found frequently in paddy grown in inland valleys, in particular in the Ultisol zone, due to interflow from surrounding uplands (IITA, 1977). This topic will come to the foreground in inland valley research of the following years.

A long-term liming trial was started on an Oxic Paleudult at the Onne sub-station in 1976, on a field which had been cleared and burned the previous year and uniformly cropped with maize. The objectives were to study (i) retention and movement of applied Ca, (ii) leaching losses of fertiliser nutrients and (iii) effective lime rates for sustained optimum crop yields. The trial consisted of an annual rotation of maize and cowpeas at 4 liming levels of 0.5, 1, 2 and 4 t/ha $Ca(OH)_2$ applied in the first year, in all cases with zero tillage, surface mulch of crop residues and 'adequate annual application' (IITA, 1977) of all necessary nutrients (N, P, K, Mg, S, Zn, Mo). The trial yielded important results right from the start. There was slower than expected downward movement of Ca, maize yield only responded to liming when Al-saturation was above 45% (pH between 4.5 and 5) while the threshold for cowpea was 20% (pH between 5 and 5.5). The full results of the trial which ran from 1976–1981 will be discussed in the next chapter.

Magnesium deficiency was observed in some plots elsewhere at the Onne station where 1-2 t/ha of lime had been applied, and pot trials showed that zinc deficiency could be induced by high rates of P and lime. These results pointed to the need for well-balanced fertiliser applications and liming in these highly weathered acid soils.

In other preliminary tests and observations a high tolerance of cassava (*Manihot esculenta*) for soil acidity and a limited response to liming was observed (IITA, 1977).

1.3.2 Cropping systems

Elimination of the need for fallowing being one of the major objectives of these early years, several lines of research were initiated during this episode to study the evolution of soil conditions under permanent cropping and to develop alternatives to fallow which could prevent degradation, as fallowing did. Although the ultimate aim was to design systems which would be adoptable by farmers, in this phase the research was 'biophysical' and exploratory, emphasising the effects on soil conditions, without much concern about adoptability. That was apparently thought to be of later concern, to be addressed once the technology was mature.

Continuous cropping, mixed cropping and crop rotation

A trial was started in 1972 at the Ibadan station on Egbeda soil to monitor the evolution of soil properties under continuous cropping, compared with fallow (IITA, 1974; Juo and Lal, 1977). The trial field was cleared from secondary forest and four cropping and four fallow treatments were installed. The cropping treatments were continuous maize with residue either returned or removed, continuous soybean (*Glycine max*) and maize+cassava, the latter two with residues returned. The fallow treatments, with which continuous maize growing was compared, were bush regrowth, Guinea grass (*Panicum maximum*), *Leucaena leucocephala*[6] and pigeon peas (*Cajanus cajun*). These species covered the range of plant types considered candidates for artificial fallow. All crops were given 'adequate fertiliser and plant protection' and were grown under minimum tillage. The trial would eventually run until 1982, with interesting results in respect of maize yield and changes in soil properties, which will be reviewed in the next chapter.

Once seen as archaic by many agronomists, mixed cropping was becoming a favourite topic for scientific study in the 1970s in many parts of the world. At the Ibadan station, various crop combinations were tested expecting that intercropping species with complementary properties would be more productive than sole crops, and perhaps incorporate some of the functions played by fallow. Several large screening trials were set up about compatibility and productivity of a range of crop combinations and arrangements, including maize, cassava, yams, melon and various grain legumes (cowpea, groundnuts (*Arachis hypogaea*), soybean, french beans (*Phaseolus vulgaris*), lima bean (*Phaseolus lunatus*) and pigeon pea). The trials usually confirmed the logic farmers' mixed cropping practices reflected in high land use efficiency (Land Equivalent Ratios) of many mixtures. Mixtures involving a long-duration crop, such as cassava, relayed into a short-season crop, such as maize, were particularly effective in exploiting the extended growing season of the humid and sub-humid tropics. They are precisely the kind of patterns used by the farmers in those zones. Such tests would continue for a number of years and we will look into the findings in following chapters.

Integration of herbaceous legumes

The incorporation of herbaceous legumes as a planted fallow in annual cropping systems had been tested in Nigeria as early as the 1930s, showing that legumes such as *Mucuna pruriens* alone could not maintain organic matter (Vine, 1953). In their summary of this early work, Smyth and Montgomery (1962) concluded that grassland-legume leys would probably be the best option for a planted fallow, combined with balanced fertiliser in case of combination with livestock.

At the Ibadan station renewed studies were started in the early 1970s with herbaceous legumes as short-term cover crops. The first trials tested planting methods with winged bean (*Psophocarpus tetragonolobus*), *Pueraria phaseoloides* and *Mucuna* followed by crops without removing the residue. The best method was as follows: begin with killing the fallow vegetation by herbicide, followed by planting the legume through the rubble, then, in the next season kill the legume by herbicide and plant the crop straight through the mulch in a minimum-till system (IITA, 1976).

There is also first mention in the 1975 Annual Report of live mulch, i.e. a permanent 'non-competitive' live legume ground cover, with a crop planted through it. Tests were being carried out with *Desmodium triflorium* and *Arachis prostrata*, and the potential of the system was described in glowing terms: 'The significant aspect of this method of crop culture is the potential it holds for a really low-input and 'minimum-toilage'[7] farming system. There is no tillage or weeding necessary'. Live mulch was going to gain prominence in the FSP in following years.

An exploratory greenhouse test about tolerance of five legumes to soil acidity was carried out with soils from Onne at lime rates of 0-5 t/ha. There were considerable differences in tolerance between the species, expressed in terms of percentage of maximum yield at 0.5 t lime/ha, in this order:

soybean<lima bean<pigeon pea<cowpea<winged bean.

Several of the trials and tests with legumes were backed up by studies on biological N-fixation by legumes (and grasses) and Rhizobium inoculation and on the effect of the legumes on nematodes.

1.3.3 Technology validation by farmers

No significant on-farm research was carried out yet during this episode, as it was felt that more time would be needed in technology development before it could be confidently exposed to farmers' conditions, in line with the model presented in Figure 1.1.

1.4 Technology delivery and dissemination

A first analytical summary of the work carried out thus far on soil erosion (Section 1.3.1) was published in 1976, with an extensive review of international research in this field (Lal, 1976).

The Benchmark Soils project was building its soils database during these years, but with the exception of a monograph on the soils at IITA (Moormann et al., 1975) no consolidated results were published yet.

Neither were any significant technology dissemination activities conducted during this episode. The opportunities opened by Nigeria's Nationally Accelerated Food Production Program (NAFPP) were not exploited by the FSP, presumably because it was felt that no definite recommendations could yet be made.

1.5 Outputs, impact and emerging trends

1.5.1 Highlights of research outputs

The Benchmark Soils project led to the following preliminary conclusions and observations:

- Poor fertility of many soils is related to the lack of weatherable minerals and the dominance of kaolinites and Fe-oxides in the clay fraction;
- Ultisols (and Oxisols) may have good physical structure due to the large specific surface area of Fe and Al oxides, but due to low nutrient stock, high acidity and Al saturation, they have particularly low fertility;
- Successful food crop production on Ultisols (and Oxisols) will require lowering the (toxic) level of Al and applying practically all essential nutrients;
- Many soils are low in available P and their P-requirement for maximum yield can be estimated from the P-sorption isotherm. It is highest in Ultisols derived from volcanic rocks;
- K-content of soils from sedimentary materials in the high rainfall zone is extremely low and K-deficiency is to be expected; and
- N being considered as a limiting nutrient for high yield in all soils, no specific attention was paid to this element until later.

Studies at the Ibadan station showed that soil erosion by runoff could be largely prevented by keeping the soil covered by a live vegetation or by a mulch layer of 4–5 t/ha of crop residues. The combination of zero or minimum tillage with crop residues left on the surface so far allowed continuous cropping for three years of a fertilized maize and cowpea rotation without yield decline.

Some results from crop production trials, all carried out on experimental farms, were:

- The Al-saturation threshold for a response to liming in the acidic Ultisols at the Onne station was 45% for maize and 20% for cowpeas;
- Ultisols are very sensitive to unbalanced nutrient applications and liming; excess P and lime may result in Mg and Zn deficiencies;
- The Bray 1 P-test appears most suitable to measure P-sufficiency with a threshold of 14 mg/kg for maize;

- A response of flooded rice to S was found in greenhouse studies in 50% of valley bottom soils from central and southern Nigeria, while only 20% showed S response under non-flooded conditions; and
- Research with legumes was intensified and tolerance to soil acidity of five legumes was found to increase in the order: soybean<lima bean<pigeon pea<cowpea<winged bean.

1.5.2 Uptake and impact

In this and the following chapters by 'uptake' we mean genuine technology adoption, independent of further intervention by the researchers, and by 'impact' we mean that a significant effect on people's livelihood has been made. Besides the contribution to the 'body of knowledge' and the inspiration to collaborators in national institutes, no direct uptake or impact was apparent during this exploratory episode.

1.6 Emerging trends

The inventory of soil conditions in the Institute's mandated zone was seen as an essential task in support of future technology targeting. Soil Taxonomy was found to be a suitable system for the characterisation of the soils of tropical Africa and it was going to be used as the major tool in the years to come.

Minimum tillage and mulch management were beginning to emerge as important elements for future sustainable intensive cropping systems, as were herbaceous legumes as cover crops for soil improvement and weed suppression. The question remained however, as to how these practices would fit within the system practiced by farmers. This could only be answered by testing them on-farm, but such testing was not yet taking place. There was an awareness that the technology might turn out to be most suitable for medium- or large-scale mechanised farming, as such farming was regularly mentioned in the annual reports.

The rotation of maize and early maturing determinate cowpea was most often used in the experiments, with chemical pest control and sometimes even supplementary irrigation. Perhaps justified as a test combination, it was rather artificial from an extension point of view, as maize-cowpea rotation was rarely practised by farmers in the humid zone, who almost invariably intercropped maize with other crops, such as cassava in SW Nigeria.

Appropriate fertiliser formulations and rates, would eventually have to be defined for specific cropping systems and soil types, bringing together the results of soil characterisation, fertiliser trials and the farming system studies carried out by the social scientists. By the end of this episode that had not happened yet. Many of the plant nutrition trials of this and following episodes, however, were carried out at yield levels far exceeding those customary in farmers' fields, thereby raising questions about their validity for smallholder conditions.

In 1976 an opportunity presented itself to extend some of the research results to the farmers through the country-wide Nationally Accelerated Food Production

Program (NAFPP), launched by the Nigerian Government. The Cereals Improvement Program hired several scientists to work with the project, while all crop improvement programs supplied crop varieties for the production 'mini-kits', i.e. demonstration packages, consisting of planting material and other inputs with instructions for their use. 'Pre-mini-kit' trials were conducted by NARS in different areas on fertiliser rates and minimum tillage, which were to provide an experimental basis for the mini-kits, possibly with advice from the Institute's soil scientists. The FSP did appoint a 'planning economist', but no record has been found of active involvement of the soils group, nor of the results of the pre-mini-kit trials. It was probably considered too early to confidently make recommendations for the widely differing conditions of the various zones in the country.

In spite of the intended role of the FSP as 'integrator' of different components into novel or improved productive systems, the linkage with the breeding departments was weak. The soil scientists and agronomists did use IITA's crop varieties in their trials, but otherwise cooperation was incidental. This was symptomatic for the monodisciplinary orientation of much agricultural research worldwide at this time. For example, the work on 'promiscuous' soybean varieties, capable of nodulating with native rhizobia in African soils, which was started by the breeding program in the late 1970s, would not impact significantly on the cropping systems and soil fertility work until decades later. The call for more inter-disciplinarity was, however, already beginning to be heard.

Notes

1   The Overseas Development Administration – Land Resources Division, Reading and Leuven University, the Rothamsted Experiment Station and the International Soil Museum, Wageningen.
2   The scale is the ratio between the distance on the map and the distance on the ground.
3   Smyth and Montgomery use the term 'soil associations' (named after the dominant soil series) instead of toposequences.
4   This relationship is a measure for the buffering capacity for K of a soil.
5   The full RUSLE is: $A=R*K*LS*C*P$, where R refers to the erosivity of the rain and K to the erodibility of the soil, S and L are slope steepness and length factors, C is a cover management factor, and P reflects the effect of conservation farming (e.g. contour ploughing). For an explanation of the logic of the RUSLE, see USDA (1997).
6   This is the first reference to *Leucaena leucocephala* which would become one of the most widely promoted 'auxiliary' species in West Africa for many years to come. Its area of origin is Central America, but it has long been used in South-East Asia for shade, contour planting and many other purposes (Jones et al., 1997).
7   This is not wrongly spelled but refers to a comment made by an agronomist when referring to live mulch systems.

References

Beinroth, F.H., 1984. A synopsis of the Benchmark Soils Project. In: ICRISAT (International Crops Research Institute for the Semi-Arid Tropics). *Proceedings of the International Symposium on Minimum Data Sets for Agrotechnology Transfer*, 21–26 March 1983, ICRISAT Center, India.

CGIAR, 2014. Guidelines for SRT 2 activities within the humid tropics CGIAR Research Program (working document).

IITA, 1974. *Report 1972–73*. IITA, Ibadan.

IITA, 1976. *Annual Report 1975*. IITA, Ibadan.

IITA, 1977. *Annual Report 1976*. IITA, Ibadan.

IITA, 1992. *Sustainable Food Production in Sub-Saharan Africa 1. IITA's Contributions*. IITA, Ibadan.

Jones, R.J., J.L. Brewbaker and C.T. Sorensson, 1997. *Leucaena*. In: I. Faridah Harun and L.J.G. van der Maesen (Editors), *Plant Resources for South-East Asia No. 11. Auxilliary Plants*. Backhuys Publishers, Leiden, The Netherlands.

Juo, A.S.R., 1980. *Characterization of Tropical Soils in Relation to Crop Production*. University of Nairobi/IFDC/IITA Training course on Fertilizer Efficiency Research in the Tropics.

Juo, A.S.R. and R.L. Fox, 1977. Phosphate sorption characteristics of some benchmark soils of West Africa. *Soil Science* 124, 567–584.

Juo, A.S.R. and H. Grimme, 1980. Potassium status in major soils of tropical Africa with special reference to potassium availability. *Proceedings of the Potassium Workshop*, IITA and International Potash Institute, October 8–10, 1980.

Juo, A.S.R. and R. Lal, 1977. The effect of fallow and continuous cultivation on the chemical and physical properties of an Alfisol in western Nigeria. *Plant and Soil* 45, 567–584.

Juo, A.S.R. and F.R. Moormann, 1981. Characteristics of two soil toposequences in south-eastern Nigeria and their relation to potential agricultural land use. *Nigerian Journal of Soil Science*, 1, 47–61.

Kang, B.T., 1980. *Soil Fertility Management of Tropical Soils*. University of Nairobi/IFDC/IITA Training course on Fertilizer Efficiency Research in the Tropics.

Kang, B.T. and V. Balasubramanian, 1990. Long term fertilizer trials on Alfisols in West Africa. *Transactions 14th International Congress of Soil Science*, Kyoto, Japan, August 1990, Volume IV, 20–25

Kang, B.T. and O.A. Osiname, 1979. Phosphorus response of maize grown on Alfisols of Southern Nigeria. *Agronomy Journal*, 71, 873–877

Lal, R., 1976. *Soil Erosion Problems on an Alfisol in Western Nigeria and their Control*. IITA Monograph no. 1, IITA, Ibadan.

Lal, R., 1983. *No-till Farming. Soil and Water Conservation and Management in the Humid and Subhumid Tropics*. Monograph no. 2, IITA, Ibadan.

Moorman, F.R., 1981. Representative toposequences of soils in southern Nigeria, and their pedology. In: D.J. Greenland (Ed.). *Characterization of Soils in Relation to their Classification and Management for Crop Production: Examples from Some Areas of the Humid Tropics*. Clarendon Press, Oxford.

Moormann, F.R., R. Lal and A.S.R. Juo, 1975. *The Soils of IITA. A Detailed Description of Eight Soils near Ibadan, Nigeria with Special Reference to their Agricultural Use*. IIITA Technical Bulletin No. 3. IITA, Ibadan.

Smyth, A.J. and R.F. Montgomery, 1962. *Soils and Land Use in Central Western Nigeria*. Ministry of Agriculture and Natural Resources, Government Printer, Ibadan, Nigeria

Soil Survey Staff. 2015. *Illustrated Guide to Soil Taxonomy, Version 2*. U.S. Department of Agriculture, Natural Resources Conservation Service, National Soil Survey Center, Lincoln, Nebraska.

Swindale, L.D., 1984. Keynote Address. In: ICRISAT (International Crops Research Institute for the Semi-Arid Tropics). *Proceedings of the International Symposium on*

Minimum Data Sets for Agrotechnology Transfer, 21–26 March 1983, ICRISAT Center, India.

USDA, 1997. *Predicting Soil Erosion by Water: A Guide to Conservation Planning with the Revised Universal Soil Loss equation (RUSLE).* Agricultural Handbook Nr. 703. Agricultural Research Service, United States Department of Agriculture, 384 pp.

Vine, H., 1953. Experiments on the maintenance of soil fertility at Ibadan, Nigeria, Part I. *Empire Journal of Experimental Agriculture,* 21, 65–85.

2 More exploration, lagging integration, weak impact
1977–1982

2.1 Scope, approaches and partnerships

Although IITA was conceived as an institute where scientists of different disciplines would work together for the common goal of systems intensification, the practice of the first decade had been one of increasing 'compartmentalisation', both across programs and within the Farming Systems Program (FSP). This is explained by the perception that the knowledge base had to be built first, without which integration would be meaningless.

The 1977 Quinquennial Review stressed the need for integration of research results for the attainment of one of the Institute's major goals: developing alternatives to shifting cultivation and 'to give priority to the validation and off-site evaluation of systems presently being developed at IITA for the sub-humid zone' (TAC, 1978).

After the Quinquennial Review, farming systems research was somewhat reorganised into five activity areas: *Regional Analysis, Crop Production, Land Management, Energy Management* and *Technology Evaluation*. This was a rather inconsequential reshuffle, except for the last activity area: *Technology Evaluation*. Its aim was (IITA, 1978):

> to develop, evaluate and adapt appropriate systems of crop management and land use for different ecologies, drawing on the findings of Farming Systems Research and the Institute's crop improvement programs.

It foreshadows the struggle in the following decades to find new ways and methods to reach the farmer. In 1977 and the following five years, however, apart from a few attempts at real on-farm testing, the conventional linear research–extension pathway to the dissemination of technology continued to be followed. But doubts about its effectiveness were growing.

Meanwhile, the objectives and content of soils-related research continued to be decided by the soil scientists themselves and their leadership. The emphasis remained on characterisation of the soil resource base (through the Benchmark Soils project), testing of conventional technology (e.g. fertiliser) and development of novel land management ideas (e.g. planted fallows), mostly on-station and with only occasional interaction with extension or development organisations: the technologies were not thought to be ready yet for transfer to the farm. The characterisation work, of course, took the researchers to the other side of the research station gate, and studies of existing farming systems were carried out under the *Regional Analysis* umbrella by the economists, but most of the experimental work was carried out in long-term trials at the experimental farms and at testing sites in Nigeria controlled by collaborating institutions and projects. In most of the long-term trials there was a degree of collaboration among FSP scientists according to their discipline and some involvement of scientists in national (Nigerian) institutes, mostly on a personal basis.

Although the period of 1977–1982 was transitional in some respects, most technical research continued along the lines set out during the previous period, seeking refinement rather than integration. A few attempts were made during this episode to expose IITA's soil and soil fertility management technologies to the real farm, but there was no systematic approach to reach the farmer directly and the assumption continued to be that technology would find its way through the existing channels of the conventional research–extension pathway depicted in the chart of Figure 2.1. It was only around 1980, when it had become clear that the conventional linear model of technology transfer was not working in the African smallholder context, that the new *Farming Systems (Approach to) Research* concepts which had been pioneered elsewhere[1] were hesitantly embraced by some in the Institute management.

Research continued around the same five themes as before, viz. soil characterisation, control of soil erosion and structural degradation, soil fertility

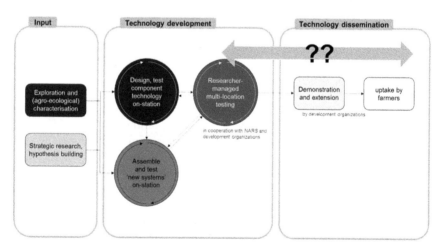

Figure 2.1 IITA's technology development and dissemination model, 1977–1982, demonstrating some systems assembly with little effort at dissemination

Shading indicates relative emphasis on the various components with darker ones indicating a relatively higher emphasis. The question marks indicate that it was not clear whether and how the transfer process would be facilitated at that time

management, wetlands development and cropping systems. It was becoming abundantly clear that fertility issues of the major soil taxa (Alfisols, Ultisols, Oxisols[2] and hydromorphic soils of various classifications) were entirely different and had to be addressed for their own sake.

Partnerships at this time continued to be mostly with research institutions, based on personal relationships. Several cooperative arrangements with advanced institutions[3] were set up during this episode, in most cases with funding from the institutes' home countries. Some of these were a continuation of partnerships established earlier, while others were new. The Wetlands Utilization Project, for example, was initiated on the initiative of a Dutch member of the Institute's Board who brought together the Dutch partners as well as facilitating the funding from the Netherlands. They were all engaged in the project because of a shared desire by IITA and the partner institutes to contribute to the advancement of agriculture in sub-Saharan Africa (SSA), and essentially on the Institute's own initiative.

In Ghana, Cameroon, Tanzania and Zaire (now Democratic Republic of Congo), projects funded by the World Bank and the United States Agency for International Development (USAID) with National Agricultural Research Systems (NARS) were initiated on the donors' initiative and engaged IITA for technical assistance. The intention was to assist the NARS in performing their task of generating and testing production-enhancing technology which was to be delivered to smallholder farmers through the countries' extension services. In Tanzania and Zaire, they initially dealt mainly with improved crop varieties, while in Ghana and Cameroon some farming systems surveying and soil fertility research was undertaken as well. In Nigeria, there were several personalised cooperative arrangements with scientists in NARS, some of them hosting 'benchmark trials' within their stations or substations (Ikenne, Mokwa). Furthermore, there was some, more or less formal, cooperation with The National Accelerated Food Production Project (NAFPP), mostly by the Crop Improvement Programs, while some individual FSP scientists cooperated with the World-Bank-funded Agricultural Development Projects (ADPs).

The areas where soil and land use research as well as some farming systems research took place during this episode are shown in Figure 2.2.

2.2 Characterisation of soils, farms and farming systems

In eastern Nigeria, broad characterisation studies of farming systems, mostly by the Institute economists (Diehl, 1982; Lagemann et al. 1975), had little influence on the FSP's technology development process. In Cameroon (Atayi and Knipscheer, 1980) and Ghana (Balasubramanian et al., 1982; USAID/ IITA, 1983), scientists working in collaborative programs conducted diagnostic studies, whereby attention was paid both to the bio-physical and the socio-economic aspects of farming environments, to provide a basis for technology targeting and dissemination. By combining system diagnosis with on-farm technology testing, the projects may be seen as precursors of the later Farming Systems Research (FSR) approach, initiated at the Institute in the 1980s.

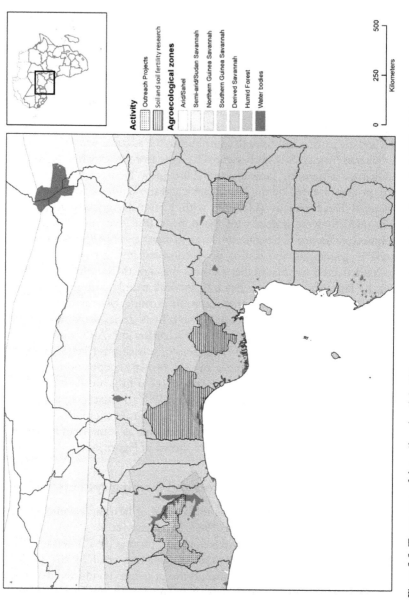

Activity

Outreach Projects

Soil and soil fertility research

Agroecological zones

Arid/Sahel

Semi-arid/Sudan Savannah

Northern Guinea Savannah

Southern Guinea Savannah

Derived Savannah

Humid Forest

Water bodies

0   250   500

Kilometers

Figure 2.2 Target area of the soil and soil fertility management program during the period 1977–1982

Detailed characterisation of soils continued during this episode, coupled with an assessment of their fertility status and its likely implication for nutrient responses, whereby the marked variation in soils and soil fertility was apparent. These studies were mostly conducted in south-west and south-east Nigeria and by the Managed Inputs and Delivery of Agricultural Services (MIDAS) project in Ghana, where considerable soil data were collected, using both the USAID and the FAO classification system. The Fertility Capability Classification (FCC), developed in the USA was adapted to the African context and applied to the benchmark soils. Characterisation also expanded to wetlands as an important under-utilised resource.

2.2.1 Benchmark Soils

The Benchmark Soils project was the flagship project of soils research of the 1970s. Its objective was quite practical, as briefly described in Chapter 1: provide a handle on the great variation in soil conditions, to be used by applied research and extension for targeting technology. The eventual outcome of the project would be a combination of soil maps showing the major soil formations, combined with physical and chemical analyses of representative soils (the Benchmark Soils) and an analysis of their capability and likely nutrient responses.

The project[4] extended its activities beyond Nigeria during this period and undertook chemical and mineralogical characterisation of soils in several countries in West Africa and tropical America or obtained data already available from those countries. From the accumulated information, it was becoming clear that the Alfisols, Ultisols and Oxisols of humid- and sub-humid Africa could be meaningfully subdivided on the basis of their geological origin (basement complex or coastal sediments), resulting in the listing of the most important Great Groups in the region (Table 2.1; IITA,1980). It should be remembered that these soils always occur in association with Entisols and Inceptisols along the toposequences, which will have similar mineralogy and chemical limitations (IITA, 1978).

Yet another grouping was proposed in 1980, based on the mineralogy of the clay and silt fractions and the 'base' status of the soil (Table 2.2), which strongly affect their expected response to agricultural land use (IITA, 1981).

2.2.2 Characterisation of hydromorphic soils

In the early 1980s, soil characterisation activities turned more and more to the hydromorphic soils of West Africa, which were considered as a large under-utilised resource for year-round cropping, including wet season rice. A preliminary inventory of wetland areas in the forest, forest–savannah transition and savannah zones was carried out in 1980 (IITA, 1981). It showed that 80% of the soils in the forest zones of southern Nigeria, Sierra Leone and Liberia (coastal and inland swamps and small river valleys) were physically and chemically very poor, requiring high fertiliser inputs and water control, if they were to be put to agricultural use (see Juo and Lowe, 1986 and Windmeijer and Andriesse, 1993

Table 2.1 'Great Groups' of soils with common occurrence in humid and sub-humid Africa and their limitations according to the Fertility Capability Classification (FCC) system (IITA, 1980)

Soils (great group)	Parent material or rocks	Land form or topography	Textural types	FCC condition modifiers[a]
Alfisols				
Paleustalfs Plinthustalfs	Granitic gneisses Granites, quartz-schists	Rolling to undulating	LL or LLR	e t* m* r w d
Paleustalfs	Coastal sediments sandstones	Flat to undulating	LL	e t* w d
Ultisols				
Paleudults Tropudults	Granites, quartz-schists, acid gneisses	Rolling	LL, SL, LLR	e h k a* t* r
Tropohumults Tropudults	Basalts, diabases	Rolling to hilly	LC, CC	e k I h*
Paleustults Plinthustults	Coastal sediments sandstones, shales	Undulating to rolling	LL or SL	e h k w d t* r
Paleudults	Coastal sediments	Flat to undulating	LL or SL	e h a* k t*
Oxisols				
Tropohumox	Amphibolites	Rolling to hilly	LC, CC	e k I h*
Haplorthox	Old alluvium	Level river terraces; small upland plateaux	LL, CC	e k h a* t*
Hydromorphic soils				
Aquolls Aqualfs	Colluvium or alluvium	Inland valleys	SLR, LLR, LL	q f
Aquults Aquox	Colluvium or alluvium	Inland valleys or lower; river terrace	LL, LC, CC	q h f e
Aquents Aquepts	Alluvium	Inland valleys and swamps, river deltas and flood plains, mangrove swamps	LL, SL, LC, LLR, SLR, CC	q, other modifiers vary widely

Note:

a S=Sandy, L=Loamy, C=Clayey, with the first capital referring to the topsoil (0-20 cm), the second to the subsoil (20-50 cm) and the 'R' to a hard pan or gravelly layer occurring in the subsoil. For a detailed legend of the textural types and the FCC condition modifiers see Table 2.3.

Table 2.2 A sub-division of the soils in humid and sub-humid Africa on the basis of the mineralogy of their clay and silt fractions and their base status and the respective Fertility Capability Classification (FCC) modifiers (IITA, 1981)

Clay mineralogy	Sub-group (base status)	FCC condition modifiers[b]	
		Physical	*Chemical*
Kaolinitic soils	Eutric[a]	w, r, c	t*, (m*)
	Dystric[a]	w, r, c	t, k, a, (m)
Siliceous soils	Eutric	w, c	t*, k*, (m*)
	Dystric	w, c	t, k, a, (m)
Oxidic soils	Eutric	w	i
	Dystric	w	I, t, k, a
Allophanic soils	Eutric	–	i
	Dystric	–	I, t, k

Note:
a Eutric = high exchangeable base status; dystric = low exchangeable base status.
b For a legend of the FCC condition modifiers see Table 2.3.

for extensive analytical data). The more favourable soils with good agricultural development potential comprised only 15–20%.

In the forest–savannah transition zone in Nigeria, small inland valleys occupying 15–20% of the total land surface were being used but little for cropping. Land quality was assessed on the basis of the groundwater regime, N-deficiency and Fe-toxicity. Chemical quality varied from poor to reasonable (Annexe I), with invariably low P-status and proposals for year around utilisation were made. In the savannah zone of central and northern Nigeria, extensive wetlands are found in river valleys, flood plains and inland depressions, with a wide variation in physical and chemical properties, in some areas with high activity (montmorillonitic) clays. Nutrient status was generally low and development for rice cultivation would require fertilisation and effective water management.

More detailed characterisation was carried out in 1982 in three locations in Nigeria: (i) the *fadamas*[5] in the Bida area of central Nigeria with coarse-textured soils derived from Cretaceous sandstones, (ii) the Anambra floodplains and terraces near Adani in the south-east with medium- to fine-textured soils derived from Tertiary shales and (iii) the inland valleys or swamps near Bende in the humid south-east, with fine-textured soils derived from Tertiary shales. The poorest soils were found in the Bida *fadamas*, with low pH and cation exchange capacity and very low nutrient status. The most suitable for lowland rice production were the wetlands in the Bende area, but measures would have to be taken in 5–10% of them to keep out iron-rich seepage water from the surrounding uplands.

2.2.3 Capability assessment

It was clear that the methods used in industrialised countries to correlate laboratory-measured soil parameters to production capacity and nutrient requirements was not suitable for Africa, at least in the medium term, because of the lack of research infrastructure as well as the cost involved. Recourse was therefore made to a methodology for soil capability classification developed at North Carolina State University (Buol and Couto, 1981), called the FCC, which uses 'condition modifiers', i.e. tags attached to a morphologically classified soil to indicate its fertility characteristics[6]. Table 2.3 shows the legend for the FCC tags. The Institute's soil scientists adjusted these modifiers for African conditions, based on previous characterisation and field testing, resulting in the tags of Tables 2.1 and 2.2. The groupings of these two tables with their FCC modifiers

Table 2.3 Legend for the Fertility Capability Classification textural types and modifiers (IITA, 1980)

Textural types (0–20 and 20–50 cm)	
S	Sandy, > 50 % sand
L	Loamy texture, <35% clay
C	Clayey texture, >35% clay
R	Gravels or other hard layer in 20–50 cm depth
Condition modifiers (0–50 cm)	
w	Low available water reserve (i.e. < 50 mm)
q	Wet soil moisture regime, profile saturated most of growing season
d	Dry, ustic, or xeric environment, soil remains dry > 60 consecutive days/year within 20–60 cm depth
w	Low available water reserve (<50 mm)
r	High erosion hazard, unsuitable for large-scale food crop farming
c	High soil compaction hazard
e	Low effective cation exchange capacity (<4 cmol^c/kg soil)
h	Acidic; exchangeable Al saturation 10–45%, pH(H_2O) < 5.5
a	Al-toxicity for most legume crops; exch. Al saturation >45%
k	K-deficient, exchangeable K <0.15 cmol^c/kg, < 10% weatherable minerals in silt and sand fractions
i	High P-fixation, standard P-requirement at 0.2 mg/kg in solution > 350 mg/kg
m	Mn-toxicity for most legume crops, soil pH < 5.0 for soils derived from high Mn-containing rocks
t	Secondary and micronutrient deficiencies (e.g. Mg, S, Zn) and/or imbalances
f	Fe-toxicity in wetland rice
(*)	Potential soil fertility constraints resulting from decline of organic matter under continuous cropping, and from moderate to high rates of chemical fertilisation

can be seen as a preliminary summary of the characterisation and classification work carried out by the Benchmark Soils project[7]. The soils were classified by both the USDA (United States Department of Agriculture) and the FAO (Food and Agriculture Organization of the United Nations) taxonomy, combined with tentative FCC modifiers, to hypothesise likely fertility constraints (USAID/IITA, 1983). Unfortunately, this project was terminated after two years for political reasons.

The fertility modifiers are the essence of the FCC system and their quantification for African soil conditions determines their usefulness. According to the 1980 Annual Report, 'tentative quantitative limits for the modifiers had been defined and would be further refined on the basis of further research'. Some threshold values for available nutrients published in various IITA publications are shown in Table 2.4. The figures are only meant to be indicative since they do not discriminate for soil type and interaction with other soil parameters nor for differences among crops. The table has been somewhat expanded using later results by the Institute and others. Obviously, 'maximum yield' is a very remote concept for most African smallholders, so the threshold figures are only a warning signal to be alert for possible deficiencies.

Sulphur deficiency had long been known to be prevalent in African savannah soils, but the extent of its occurrence in West Africa had not been systematically investigated. Kang (1980) prepared a tentative map showing locations where S deficiency had been found (Figure 2.3).

In a greenhouse study of Nigerian soils, clear differences were found in total S content of the surface soil from forest, Derived savannah and Guinea savannah areas[8], with averages of 273, 183 and 96 mg/kg respectively, as well as a high correlation between total S and total N and organic P content. There was no apparent relationship with parent material or soil type (Kang et al., 1981a). The differences between savannah and forest were explained by lower soil organic matter (SOM) content of the savannah soils and the lower nutrient recycling capacity of the shallow-rooted fallow vegetation, especially in the Guinea savannah. Maize (*Zea mays*) grown in the greenhouse showed significant S responses for all soil samples from the Guinea savannah, but for only one sample from the Derived savannah and the forest. The correlation between total S content and dry matter yield was relatively weak, and much better for heat-soluble and $Ca(H_2PO_4)_2$-extractable sulphur.

A very preliminary map of micronutrient deficiencies in tropical Africa was published later by Kang and Osiname (1985), based on their own results and published sources (Figure 2.4)

2.3 Technology development

Technology development and testing during this episode took place under controlled conditions, mostly at IITA stations and sub-stations of national research partners. The Benchmark Soils concept can be recognised in the scatter of research locations, but the number of sites was too limited to really qualify as a

Table 2.4 Some indicative toxicity and sufficiency threshold values of plant nutrients in the soil; sufficiency is for 90% of maximum yield, yield levels for soybean as indicated

| Crop | pH | Toxicity threshold | | Sufficiency threshold | | | | | |
| | | Al saturation (%) | Mn toxicity (mg/kg) | P | | K (cmol$_c$/kg) | Mg (cmol$_c$/kg) | Zn | |
				Bray-1 (mg/kg)	Soil solution (mg/L)			(mg/kg)	Soil solution (µg/L)
Maize[a]	4.7	30–45	100–120[e]	14	–	0.15–0.18	0.50	2	35
Cassava[a]	–	80	–	4–9	0.04	–	0.20	–	–
Cowpea[a]	–	25–60[b]	–	7	–	–	–	–	–
Soybean	–	–	–	–	–	–	–	–	–
55%[c]	–	–	–	10	–	–	–	–	–
80%[c]	–	–	–	32	–	–	–	–	–
90%[d]	–	–	–	15	–	–	–	–	–

Notes:

a Sources: IITA 1977, 1978, 1981; Friesen et al, 1982; Juo and van der Meersch, 1983; Kang, 1980, 1984; Howeler, 2002; RCMD, 1992.
b There are strong varietal differences in tolerance.
c BNMS-II, 2007.
d Kang, 1980; obviously, these are mutually incompatible results with those of BNMS-II, 2007.
e KCl-extractable, estimate (see Chapter 3).

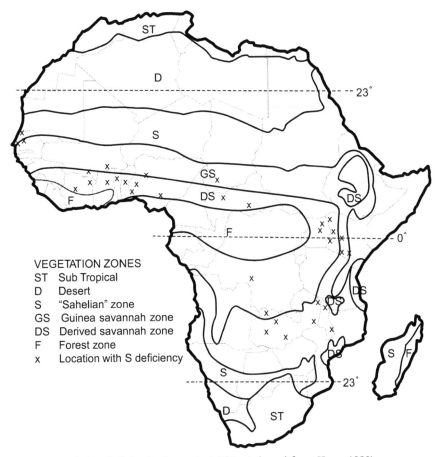

Figure 2.3 Sulphur deficiencies in tropical Africa (adapted from Kang, 1980)

network of testing sites associated with benchmark soils, possibly because of the limited staff strength for a job of this magnitude. Some off-campus technology validation was carried out, but always under full researcher management.

2.3.1 Soil and soil fertility management

Strategic research continued to dominate this period, whereby the question for whom technologies like erosion control, land clearing and 'auxiliary' legumes were intended was not explicitly raised. Soil tillage appeared on the agenda as a result of the internal dynamics of the work on soil management, whereby the soil erosion theme called up questions about the effect of different tillage methods. This was especially important in view of the reigning belief that current farming systems would need to be replaced by commercial, mechanised farming that would require land clearing. Due to the recognised prevalence and

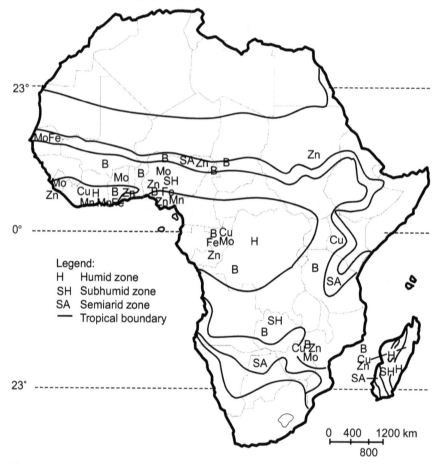

Figure 2.4 Micronutrient deficiencies in tropical Africa (adapted from Kang and Osiname, 1985)

diversity of nutrients deficiencies, fertilisers were integrated in all studies and nutrient response studies were usually carried out on at least two types of soils.

Soil erosion

Soil erosion is a function of the *erosivity* of the rainfall, the *erodibility* of the soil, the *slope* of the land and the way the land is managed, in particular *soil cover* (see Chapter 1). Although it was clear that erosion could be prevented to a large extent by mulching and minimum tillage, soil physics research was looking for a comprehensive understanding of the mechanisms involved, by measuring or estimating the factors in the USDA Revised Universal Soil Loss Equation (RUSLE, USDA, 1997). A key factor is soil erodibility K. This is most reliably measured in long-term runoff plots, but establishing such plots across the zone would not be feasible. So erodibility had to be estimated from runoff experiments using field and laboratory rainfall simulators, and from measured soil characteristics. This work, together with the erosivity of the rainfall, would eventually allow mapping of erosion hazards and form a useful addition to the soil characterisation work of the benchmark soils project.

Obtaining reliable data for the RUSLE factors, especially K, turned out to be a major challenge and could not be completed during this episode. In a review paper published in 1984, the Institute's soil physicist presented an overview of the status of erosion research for the tropics in general and the Institute's findings to date (Lal, 1984). Some of the important findings were that soil loss increases exponentially with increase in slope *for a given slope length*, and the counter-intuitive result that runoff per unit area *decreases* with slope length, while soil loss does not follow any definite trend with slope length.

According to the 1979 Annual Report, an erosion hazards map had been prepared for different regions in Zaire (now Democratic Republic of Congo), but so far no record of this work has been found. There has been no further mention of erosion hazard mapping, but the work on quantification of erosion phenomena continued in the following episode.

Physical soil management

TILLAGE PRACTICES

Minimum tillage with retention of plant residue and the use of various kinds of mulch was being shown to stimulate soil health, prevent erosion and improve yields. Special measures would be needed, however, for crop establishment and maintenance in mulch-covered land. Herbicide use was unavoidable to control weeds and special tools were needed to seed crops through the rubble (see also the sections on mulching below). The most significant tools developed during the 1970s were a hand-operated fertiliser applicator and a rolling injection planter (IITA, 1978), for which hand-pushed and tractor-mounted versions were built.

The advantages of zero-tillage were elaborated in a synthetic monograph in 1983 which outlined the conditions that had to be satisfied for zero-tillage to be suitable[9], as well as the remaining issues to be studied (Lal, 1983).

At the Ibadan station, zero-tillage was now applied in most trials, either as standard operation or as a treatment, based on earlier results showing very low erosion and stable maize yields in zero-tillage combined with at least 4 t/ha of maize stover as mulch (Chapter 2). Subsequent experiences brought out a number of problems with the technology, however:

*   There was up to 60% volatilization of urea but only 5% for 15:15:15 and 13:52:0 compound fertiliser *placed on the surface* (IITA, 1979, p. 94);
*   The herbicides, necessarily used for weed control, were partly intercepted by the mulch layer, leading to higher cost (IITA, 1982, p. 39);

- Seedling emergence in the plant residue mulch was often found to be reduced (IITA 1982, p. 34);
- Rodent damage could also lead to stand reduction in zero-till plots (IITA, 1983, p. 126); and
- In less fertile soil with high bulk density, tilled maize produced higher yields than no-till and the latter required more N (IITA, 1980, p. 12).

LAND CLEARING AND LAND MANAGEMENT

During this period, soil physics research took a new turn: to land clearing and post-clearing land management. A large project was started at IITA in 1979, entitled the 'Hydrology and Watershed Management Project'. The aim was to investigate the effect of different methods of land clearing and subsequent soil management on soil properties and crop production. The hypothesis was that the deleterious effect of mechanised clearing methods could be corrected by subsequent tillage and land use methods. The outcomes of this work could provide guidelines to projects considering medium- to large-scale land clearing. Some of the large World-Bank-funded ADPs, which were being started in Nigeria, used heavy equipment for land clearing, making them immediate potential 'clients' for the results of this research.

A large land clearing-cum-management trial was carried out on land which had been under secondary forest fallow. Plot sizes were approximately 3 ha with two replications, and the trial had the following treatments (IITA, 1982):

1   *Traditional methods*: hand clearing, trees left standing, traditional cropping without 'agricultural chemicals';
2   *Manual clearing, zero-tillage*: complete clearing, burning the dried vegetation, no terracing, no tillage, keeping residue of previous crop as mulch;
3   *Manual clearing, conventional tillage*: burning the dried vegetation, ploughing and harrowing, terracing and grass waterways;
4   *Shear blade, zero-tillage*: clearing by tractor-mounted, vegetation in 50 m spaced windrows, no tillage, keeping residue of previous crop as mulch;
5   *Tree pusher-root rake, zero-tillage*: same as 4, but using tree pusher plus root rake instead of shear blade; and
6   *Tree pusher-root rake, conventional tillage*: same clearing as 5, same land management as 3.

Physical data collected in this complex trial were expected to explain differences in soil loss, moisture conditions and crop yield as a result of different land clearing-cum-land use methods.

In the first season after land clearing, maize and cassava (*Manihot esculenta*) were planted and an extensive program of measurements of parameters associated with moisture infiltration and storage and soil erodibility was carried out. Some of the results through 1981, the third year after clearing, are shown in Table 2.5. The top soil (0–5 cm) bulk density had increased steeply from a pre-clearing value of

Table 2.5 Effect of clearing method on soil erosion (Lal, 1984) and bulk density, penetrometer resistance and cassava yield (IITA, 1981, 1984). No statistics are provided

Treatment	Soil erosion[a] (t/ha)	Bulk density[b] (g/cm³)	Penetrometer resistance[b] (kg/cm²)	Maize yield[c] in 1979 (t/ha)	Cassava yield[c] in 1980 (t/ha)
Traditional farming	0.02	1.27	1.32	0.50	7.7
Manual clearing/no tillage	0.4	1.40	1.19	1.56	15.0
Manual clearing/disc ploughing	9.8	1.38	NA	1.59	11.7
Shear blade/ no tillage	4.8	1.38	2.19	1.98	14.1
Tree pusher/ no tillage	15.7	1.47	1.23	1.36	20.2
Tree pusher/disc ploughing	24.3	1.37	NA	1.75	17.5

Notes:
a Total 1979–1981; no statistics provided.
b Measured at the start of the first season after clearing, 2 weeks after ploughing the conventionally tilled plots.
c There were considerable differences in management, so yield differences between land clearing treatments have little meaning.

0.66 g/cm³ to an average of 1.37 g/cm³ at the start of the second season of 1981. It was highest in the no-till plots, but the differences among the treatments had become small at the time of harvest and in the next year (IITA, 1982, 1983). The only striking result appears to be the low erosion loss in the plots cleared with shear blade and the very high losses in the tree pusher plots. The maize yields in all treatments were quite low while cassava yields were higher in tree pusher plots than in other plots, but assigning the differences to the land clearing treatments seems dubious because of differences in management. After the cassava harvest, all plots were planted to a maize–cowpea (*Vigna unguiculata*) rotation until 1982. As from that year, mechanised no-till operations were used in all plots. The trial will be further discussed in the next chapter.

Soil fertility management

Studies on soil fertility management continued to be a major line of work during this episode, in both the Alfisol and the Ultisol Zone. At this time, the emphasis was still on fertiliser as the principal factor for soil fertility improvement, but it was already becoming clear that fertiliser alone could not maintain soil productivity indefinitely. In addition to research on nutrient responses, crop residue management and SOM maintenance were therefore also addressed. In the Ultisols in the high rainfall zone, soil acidity, Al toxicity and nutrient imbalances had been identified as overriding problems, which are aggravated by intensive cropping and fertiliser. Research in the Ultisol zone was therefore expanded around 1977, mainly at the Onne substation, to study potential solutions for these problems.

The trials, which were all carried out in research plots, just looked at the 'physical' effects of soil amendments without considering the economic aspects, as it was thought that the processes of nutrient requirements and uptake had to be well understood first. In many cases, trials were replicated on the major soil types occurring even within short distances along a toposequence, such as Egbeda and Apomu in south-west Nigeria.

PLANT NUTRIENT RESPONSES

A large experiment to study the long-term effect of fertiliser on maize and second season crops had been started in 1972 (Chapter 1) on two soils (Egbeda and Apomu series[10]) at the Ibadan station. The second season crop initially was sweet potatoes, after 1974 replaced by cowpeas. The treatments formed a partial factorial trial with combinations of N, P, K, S, Mg and Zn. Stover was retained except in two extra treatments at the highest and zero fertiliser rates. Retention of stover was standard practice in most trials, for soil protection and SOM maintenance, except where its effect was an explicit objective of studies. The trial was concluded in 1981. Figure 2.5 shows that over a period of ten years maize yields declined slowly at the highest NPK rates with stover retained, from close to 7 to around 5.5 t/ha on both soils[11]. The figure shows further that on the Egbeda soil maximum maize yield was obtained at an annual application of 60 kg N/ha plus 30 kg P/ha. A K effect became visible only after eight years (not shown). On the Apomu soil, however, a maintenance dressing of 150 kg N/ha and 60 kg P/ha was needed and after 2–4 years 40 kg K/ha had to be applied for maximum yield as well. Since the yields at zero-N (with adequate P and K) fluctuated around a high 4.5 t/ha for Egbeda and 2 t/ha for Apomu, the agronomic efficiency for applied N at the optimum rates of 60 and 150 kg N/ha respectively was about 25 kg grain/kg applied N for Egbeda and 23 kg/kg for Apomu, about equal to the best N efficiency in farmers' fields in the area for a maize+cassava crop, where part of the N is taken up by the cassava (Mutsaers and Walker, 1990).

The removal of maize stover had a depressing effect on maize yield after seven years in the Egbeda soil, and after only 3–4 years in Apomu soil. No Mg, S and Zn effects were observed at any time on either soil (IITA, 1981, 1982). It was concluded that high maize (and cowpea) yields could be maintained for many years in Egbeda soil with modest fertiliser, while Apomu soil required much higher rates after only a few years and should probably be left to fallow after 3–4 years.

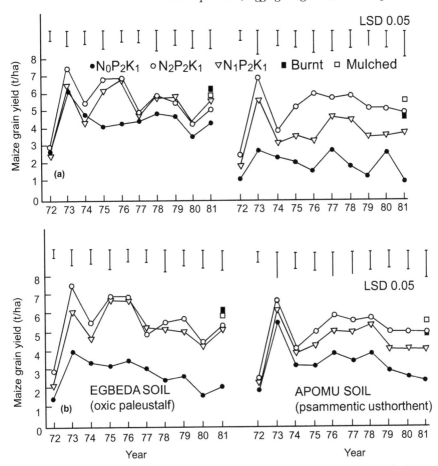

Figure 2.5 Maize response to (a) N and (b) P at IITA, Ibadan; N-rates 0, 60, 120 kg/ha on Egbeda soil and 0, 75, 150 kg/ha on Apomu as from 1975; P-rates 0, 30, 60 kg/ha (adapted from IITA, 1982)

These results illustrate the considerable differences in nutrient response of two soil types within short distances and the need for specific soil information for appropriate plant nutrient recommendations. They also reinforce the importance of information on the intrinsic soil properties and their expected nutrient response patterns, such as the Soil Benchmark Project originally intended to provide. The combination of the two, the distribution of soil types and the results of fertility amendment trials on representative soils would form the basis for future recommendations.

Fertiliser trials under controlled conditions were conducted with maize at three sites in Central Ghana in the first season of 1981 and the second season of 1982. Four levels of K and Mg and three levels of Zn were applied, in view of expected deficiencies of these elements in the country's Southern Guinea savannah. All treatments were also given 120 kg N/ha and 26.2 kg P/ha. The

Table 2.6 Maize response to Zn, averaged over other nutrients in two sites in C. Ghana (USAID/IITA, 1982)

Zn (kg/ha)	Maize grain yield (t/ha)				
	Owowam (Haplustalf)		Jatozongo (Rhodustult)		Beks[a]
	1981	1982	1981	1982	1981
0 (2)[b]	1.72	5.46	3.48	1.15	2.86
5 (7)	1.74	5.10	4.11	1.32	3.83
10 (1)	2.19	6.08	5.09	1.48	4.50

Notes:
a   Classification not mentioned.
b   Between brackets is the number of treatments used for averaging.

nutrient effects were erratic, with only a significant Zn effect found in all three sites in both seasons[12] (Table 2.6). No convincing explanation was provided for the large differences in yield between sites and seasons, which was attributed to dry spells.

SOIL ACIDITY, AL-SATURATION AND LIMING

In the long-term liming trial (1976–1981) on a Typic Paleudult at Onne[13] with 'adequate fertilisation' (see Chapter 1), lime $(Ca(OH)_2)$ was applied once in the first trial year at 0, 0.5, 1, 2 and 4 t/ha. A lime rate of 0.5 t/ha maintained maize yield at near maximum level for two years[14] (Figure 2.6), the maximum varying from year to year, from 3.5 to 4.6 t/ha (Friesen et al., 1982). The authors commented that this corresponds with the shifting cultivator's experience that the liming effect of burned fallow vegetation lasts for 1–2 years. Liming at a rate of 2 t/ha sustained maize yield in the trial at more than 85% of the maximum for five years. In the non-limed plots yield declined to 55% of the maximum in year 2 and went further down slowly thereafter. The trial did not include an unfertilised control, so the effects of fertiliser and liming could not be separated. When the yields of all treatments and all years were plotted against pH and exchangeable Al, fairly clear response patterns emerged with maximum yield obtained from a pH of around 5.5[15] and 85% of the maximum yield when exchangeable Al did not exceed 40% (Figure 2.7a and b). From the combined graphs, it follows that Al saturation must have been practically zero at a pH of 5.5. In a soil incubation experiment by Friesen et al. (1980a) with the same soil, exchangeable Al also declined steeply with pH to practically zero at pH 5.5.

The response to liming by cowpeas, grown in rotation with the maize, was erratic and complicated by varietal differences in Al tolerance and by heavy insect attacks in the crucial second and third year. Another complication was the effect of the heavy fertiliser regime, which caused an increase in the Al

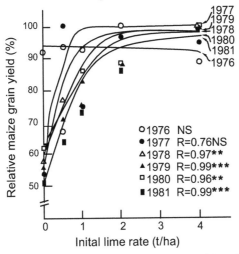

Figure 2.6 Response of maize to lime applied in 1976 (adapted from Friesen et al., 1982)

Note: ** , *** probably indicate significant differences at P<0.05 and P<0.01 respectively

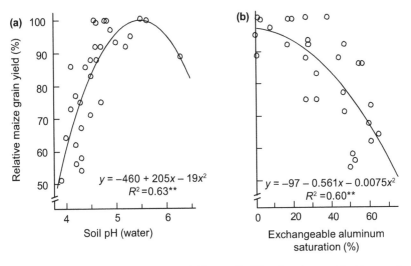

Figure 2.7 Relationship of maize grain yields to soil pH and exchangeable Al saturation (adapted from Friesen et al., 1982)

Note: ** probably indicates significant differences at P<0.05

concentration in the soil solution, with a negative effect on nodulation. The only conclusion which can be drawn from the yield data is that the Al-tolerant cowpea variety Vita-1 responded similarly to liming as the maize did, while the susceptible variety Vita-4 responded up to a lime rate of 2 t/ha.

There was considerable downward movement of Ca as from the second trial year, especially at the higher lime rates, without an appreciable effect on the acidity and exchangeable Al-saturation in the subsoil. High lime rates were therefore not effective and the authors recommended an initial lime application of about 1 t/ha in this soil, to reduce the exchangeable Al-concentration of the topsoil to 20%, followed by annual maintenance dressings of 100–300 kg/ha. Recycling leached nutrients from the subsoil would require deep-rooting perennials tolerant to high Al levels.

Greenhouse maize trials with two Ultisols (a Paleudult from Onne and a Paleustult from Nkpologu) showed strong phosphorus–lime–zinc interaction. The Paleustult, with low P status (Bray-1 < 5 mg/kg) only responded to liming if combined with P application, contrary to the Onne soil (Bray-1 of 35 mg/kg) where liming alone doubled maize dry matter yield. As to Zn, dry matter yield was depressed if the Zn concentration of the soil solution fell below 35 μg/kg This would happen when the Nkpologu soil was limed and given P fertiliser, without Zn application, due to immobilisation of Zn at pH above 5 (IITA, 1978; Friesen et al., 1980b).

CROP RESIDUE MANAGEMENT

There appears to have been a controversy about the effect of residue burning vs mulching, especially in the Ultisol zone. A long-term test was therefore started in 1978 in Onne on a Typic Paleustult and in Ikenne on a slightly acidic Oxic Paleustalf, comparing residue mulching or burning, with a maize–cowpea test rotation, at various fertiliser treatments. Maize yield figures for a selection of years are presented in Table 2.7. For Onne, no yield data were given after 1981, but it was mentioned that the yield of the control had declined in 1982 to 'low levels', while a yield of over 3 t/ha was still maintained with the full nutrient package (Kang, 1983). At Ikenne, there was little difference over the years in maize yield between burning and mulching, except for the higher yield as from 1979 for the residue burning in the plots without P (NK), which was attributed to the release of P by the burned residue (Kang, 1983). No significant differences in cowpea yield were observed. At Onne, the results were erratic, showing no consistent advantage for either mulching or burning. The high fertiliser rates may have prevented the role of residue management from being expressed in both sites. In any case, in the unfertilised control mulching mostly gave higher maize grain yield than burning, in spite of better initial growth in the burned plots in Onne, possibly due to higher leaching losses in the burned plots.

MINERALISATION OF SOIL ORGANIC MATTER AND THE EARLY SEASON N-FLUX

Early planting is thought to be advantageous because of an N flux from decomposing organic matter in the soil triggered by the early rains. The occurrence of this effect was monitored in a 0.25 ha Alfisol plot (Egbeda) at the Ibadan station after clearing from secondary forest. Table 2.8 shows the decrease in total organic C, N and P after one year due to SOM mineralisation and Figure 2.8 shows the change in nitrate- and ammonia-N content of the topsoil in the

Table 2.7 The effect of residue mulching or burning on maize yield in an Alfisol and an Ultisol at different fertiliser treatments; data from IITA (1979, 1982, 1984)

Fertiliser treatments[a]	Maize grain yield (kg/ha)									
	Ikenne						Onne			
	1978		1981		1983		1978		1981	
	Mulch	Burn	Mulch	Burn	Mulch	Burn	Mulch	Burn	Mulch	Burn
Control	4,228	3,706	3,257	3,613	2,468	2,055	1,740	1,190	2,493	2,162
PK	5,069	4,830	4,880	5,042	3,273	3,895	–	–	–	–
NK	4,939	4,812	3,825	4,503	2,678	3,254	–	–	–	–
NP	5,205	4,720	4,894	5,196	3,487	3,410	–	–	–	–
NPK	5,327	5,275	5,308	5,835	3,365	4,229	3,412	2,727	3,700	3,552
NPK Mg Zn	5,838	5,824	5,486	5,592	4,267	4,669	2,878	2,802	3,643	4,034
Above+lime	–	–	–	–	–	–	3,495	3,008	4,432	4,422
LSD[c] (5%) I[b]	822	467	490	671	396					
LSD[c](5%) II[b]	738	741	762	679	908					

Notes:
a Nutrient rates: N 200, P 40, K (missing in source), Mg 30, Zn 5, lime 1000 kg/ha.
b I = within mulch management; II = across mulch management.
c 'LSD' = 'least significant difference'.

Table 2.8 Changes in soil properties after forest clearing (0–10 cm), Egbeda soil, Ibadan (IITA, 1981). Values between brackets are standard errors

Soil property	After clearing	After 1 year
Bulk density (g/cm³)	1.04 (0.05)	1.22 (0.02)
Organic C (g/kg)	15.5 (0.7)	12.4 (0.6)
Total N (g/kg)	1.71 (0.07)	1.46 (0.07)
Organic P (mg/kg)[a]	194 (7)	160 (7)

Note:
a   Crop residue excluded.

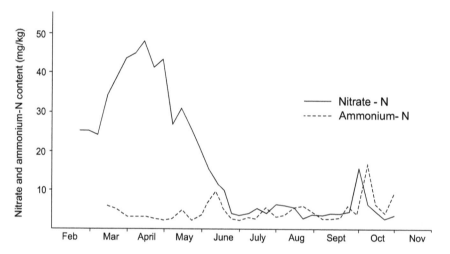

Figure 2.8 Seasonal fluctuation of mineral-N in a newly cleared forest Alfisol (0–10 cm) (adapted from IITA, 1981)

course of the season after land clearing. These data were a clear confirmation of the early rainy season N flux. It is a strong argument for early planting in spite of drought risk, in order to capture the N before it is washed down. It also shows that the N condition of the top soil is less favourable for a non-leguminous crop in the second season, necessitating a higher N application rate for the same yield.

2.3.2 Cropping systems

During this episode further studies were conducted on the effect of continuous cropping on soil conditions and experiments were carried out on mixed cropping and crop rotation as possible approaches to alleviate some of the negative effects of intensive crop production.

It was becoming increasingly clear that continuous cropping would require measures to maintain SOM, replenish nutrient stocks and prevent nutrient losses and pH decline, especially in the high rainfall Ultisol (and Oxisol) area. Legumes were coming to be seen as a necessary component of a (semi-) continuous cropping system, as a source of biologically fixed nitrogen. Nutrients leached from the topsoil would have to be recovered to reduce losses, and weeds being a major concern in continuous cropping, (cost-)effective weed control would be imperative. In traditional farming, all these roles were achieved by fallowing, albeit at the cost of a low cropping intensity taken over a full crop–fallow cycle.

There was growing interest in the potential of legume and non-legume 'auxiliary' species to play the same roles in intensified cropping systems as fallowing did in traditional systems. Tests and observations were therefore initiated at Ibadan and later at Onne on:

- Leguminous and non-leguminous herbaceous cover crops for soil improvement and control of speargrass (*Imperata cylindrica*);
- Perennial legumes and non-legumes as fallow substitutes; and
- Improving the efficiency of biological N fixation in legumes.

The first two lines of work evolved in a few years' time into three long-term technology development paths: (i) herbaceous legumes for *in-situ* mulch production and short fallow, (ii) herbaceous legumes as live mulch and (iii) leguminous or non-leguminous perennials for planted fallow or alley cropping. The question of how those technologies could fit into the existing farmers' cropping system was hardly raised at the time, probably because a thorough overhaul of cropping practices was deemed necessary anyway to be realised by an as yet imaginary new type of modern farmer.

Continuous cropping, mixed cropping and crop rotation

CONTINUOUS CROPPING

The evolution of soil properties under continuous maize was monitored in a ten-year trial which started in 1972 at the Ibadan station on Egbeda soil (see Chapter 1). Continuous zero-till maize, with stover removed or retained, was compared with maize/cassava, soybeans and four different fallow types. The trial produced an unambiguous picture of soil degradation under intensive maize cropping and the effect of retaining stover as mulch. The yield of no-till maize with stover removed and given 'adequate fertiliser and plant protection', declined from the start from an initial 6.5 t/ha (two crops/year) to 2.5 t/ha after ten years. With stover retained the yield remained high for four years at about 8 t/ha and then also declined to 3.5 t/ha. The trend over time is shown in Figure 2.9, which was to become an iconic graph showing the need for fallow after four years of intensive maize cropping with stover returned. It should be realised, however, that the reported yields far exceed anything seen in farmers' fields in the area. No further yield data were found for this trial.

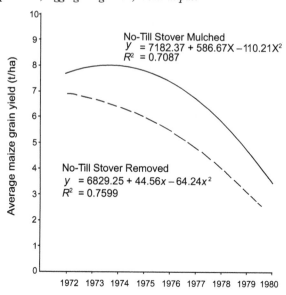

Figure 2.9 Fitted yield curves of maize under no-tillage maize with and without stover returned; kaolinitic Alfisol, Ibadan (adapted from IITA, 1983)

The soil analyses after eight cropping years are worth studying (Table 2.8). Continuous maize with stover returned maintained soil total N at a level comparable with the fallow treatments but organic C and pH had declined strongly after eight years. The plots where stover was removed had become so acidic and low in organic matter and cation exchange capacity (CEC) that manganese toxicity could be expected in sensitive crops. Maize suffered from acute Mg-deficiency. Guinea grass (*Panicum maximum*) cover, cut back three times a year, showed the best results for most soil parameters.

The soil degradation in the intensively cropped maize plots in this trial was symptomatic of the condition of many experimental fields at the Institute in the early 1980s and a general soil fertility survey was therefore carried out in 1982 (IITA, 1983). The properties that had changed most were pH, organic matter, exchangeable Mg and bulk density, especially in fields under continuous maize. The decline of Mg had been more rapid than that of Ca and K, due the decline of pH and SOM content, resulting in a lower effective cation exchange capacity (ECEC) and increased leaching of cations. The ion exchange selectivity of the kaolinitic clay thereby caused preferential adsorption of K and Ca and leaching of Mg (Pleysier and Juo, 1981). A three-year experiment with high ammonium sulphate application rates had already shown these trends to occur at an accelerated pace due to the acidifying effect of $(NH_4)_2SO_4$.

From these findings, it was concluded that continuous cropping would not be sustainable even with residue mulching and minimum tillage. The decline of pH and SOM in the continuous cropping trial (Table 2.9) showed that even for a relatively fertile Egbeda-type soil a fallow phase would have to be included

Table 2.9 Properties of surface soil a (0–15 cm, Egbeda series) after eight years continuous cultivation and fallow (IITA, 1981, 1983). No statistics are provided

Treatment	pH[a] (H₂O)	Organic C 0–7.5 cm (g/kg)	Organic C 7.5–15 cm (g/kg)	Total N (g/kg)	ECEC[b] (cmol_c/kg)	Exchangeable cation Ca (cmol_c/kg)	Exchangeable cation Mg (cmol_c/kg)	Exchangeable cation K (cmol_c/kg)	Bulk density 0–5 cm (g/cm³)
Continuous cropping with minimum tillage									
Maize+residue	5.0	15.8	10.1	1.8	3.23	2.19	0.41	0.35	1.20
Maize–residue	4.7	8.0	6.5	1.1	1.81	1.13	0.24	0.11	1.31
Maize+cassava	5.6	12.8	8.0	1.5	3.40	2.04	0.42	0.32	1.25
Soybean+residue	5.0	8.1	4.1	1.1	3.05	1.65	0.42	0.28	1.23
Natural and planted fallow									
Natural regrowth	6.5	28.1	10.5	1.9	5.14	3.53	0.91	0.41	0.88
Guinea grass	6.7	25.6	12.6	2.6	7.69	4.75	1.28	0.91	1.01
Pigeon pea[c]	6.0	17.9	804	2.3	3.42	2.18	0.64	0.32	1.10
Leucaena	6.1	26.8	11.2	'comparable to natural regrowth'; no data found					

Notes:
a The pH and organic C data are for 1982, the other data for 1980.
b ECEC stands for 'Effective cation exchange capacity'.
c Indeterminate pigeon pea, not really a perennial, declined after one year due to die-back and pests, especially when pruned, and had to be replanted repeatedly.

in the cropping cycle. Furthermore, proper crop rotation and the inclusion of economic tree crops were strongly recommended and a low lime rate would be needed where pH has fallen below 5. The results also strengthened the case for alley cropping, which was expected to perform the same role as fallowing, now simultaneously with crop production.

In respect of the soil degeneration at the Institute's own farm, it was concluded that half the experimental area, after having been under crops for ten years, would have to be put under a planted fallow or mulch farming system to restore its fertility, while 10–15% was so degraded that a deep-rooted shrub or tree fallow was needed to recycle subsoil Ca (Juo and van der Meersch, 1983).

Changes in soil properties under continuous cropping were also being monitored in selected experimental plots at the Onne station. There was increased compaction of surface and sub-surface soil after seven years of tractor cultivation and a decline in organic matter which was similar in ploughed and zero-tilled land. The pH in soil under secondary forest was 3.7, a common phenomenon in highly weathered acidic soil in the high rainfall zone, which explains the beneficial effect of traditional residue burning by raising the pH and adding plant nutrients. In cropped land which had received 1 t/ha of lime seven years earlier the pH was about 4 (Juo and van der Meersch, 1983).

MIXED CROPPING AND ROTATIONS

The work with complex crop mixtures started in the previous period continued for another three years. Principal test crops were maize, cassava and yams (*Dioscorea* spp.), associated varyingly with a range of intercrops such as melon (*Citrullus lanatus*), climbing or erect cowpeas, lima beans (*Phaseolus lunatus*), soybeans (*Glycine max*), determinate (dwarf) and indeterminate pigeon peas (*Cajanus cajan*). It is difficult to derive firm conclusions from this work because of the large number of variables studied simultaneously: crop species, varieties, densities, lay-out, relative planting dates etc. This line of work was largely abandoned in 1978 and henceforth the cropping systems agronomists intended to focus on '... maize and cassava-based cropping systems for upland conditions in the subhumid tropics, cassava- and plantain-based cropping systems for the subhumid tropics, and rice-based cropping systems for hydromorphic areas' (IITA, 1979).

Two more focused tests with grain legumes were started at the Ibadan station in 1975 on Apomu soil, one with the legumes grown in *association*, the other in *rotation* with maize. In the mixed cropping trial, only cowpea was used; in the rotation trial cowpeas, soybeans and pigeon peas. After two years, it was concluded that there was no immediate N-contribution from the cowpeas to the intercropped maize and the mixed cropping test was discontinued. The rotation trial continued for two more years with four simple within-year (main season-short season) rotation patterns, viz. maize-maize, maize-cowpeas, maize-soybeans and maize-pigeon peas, at four levels of applied N (0, 50, 90 and 135 kg N/ha) on an Iregun soil (Oxic Haplustalf) at the Ibadan station. There were no differences between rotations in the yield of first season maize, contrary to the unfertilised treatments, where in the third year maize yields in the maize-grain legume rotations were considerably higher than in the continuous maize treatment (about 5.5 against 3.2 t/ha) (IITA, 1978). The yield was highest after pigeon peas, which may have been due to much greater biomass production, but no indication of significance was given.

The absence of an immediate effect of associated cowpea on maize does not mean that there is no advantage of growing maize and cowpeas in mixture. In a 1981 experiment with mixtures of maize and cowpeas in Ibadan, using cowpea varieties of early (60 days) and medium maturity, there was only a minor depressing effect of cowpeas on maize, while cowpea yield in mixture was more than half that of sole cowpeas. Hence, the 'cowpea-bonus' resulted in a relative yield total[16] for the mixture exceeding 1.5 (Ezumah et al., 1983). That kind of effect of cereal-grain legume intercropping had frequently been found all over the world since the early 1970s (e.g. Papendick et al., 1976).

Integration of herbaceous legumes

The use of herbaceous legumes was going to be one of the mainstays of cropping systems research aiming at soil improvement in the years to come. No explicit reason was given for this choice, but the work resumed a trend which dated back to the pre-independence years in Africa (e.g. Vine, 1953), without actually building on the results from those days. 'Production legumes', such as groundnuts (*Arachis hypogaea*) or soybeans are not meant here; they are dealt with in other sections of this review. The idea of using such species as dual purpose legumes (both for grain and biomass) was not considered at the time, perhaps partly because breeding had gone for maximum investment by the plants in the grain, with too little biomass for significant soil improvement.

There were going to be three approaches to the introduction of herbaceous legumes into cropping systems. The simplest one was to treat the legume as a crop in its own right, planted sole in rotation with other crops, as *rotational legumes*. The second approach was intercropping or relaying the legumes into another crop and letting them take over after harvesting the latter: *relay-cropped legumes*[17]. In both cases, the aim was to establish a protective, soil improving and weed suppressing *cover crop* after the harvest of the main crop, as a short fallow, which would be slashed before planting the next crop and worked into the soil as green manure or left on the surface as *in-situ mulch*. The third method would be to let the legume, once established using either of the above methods, remain in the field permanently, with the following crops planted through the legume leaf carpet as *live mulch*.

COVER CROPS AND IN-SITU MULCHES

Various tests were conducted over the years with a range of herbaceous legumes to examine their potential as cover crops and mulch producers. Tests with *Stylosanthes guianensis, Pueraria phaseoloides, Psophocarpus palustris* (winged bean),

Centrosema pubescens showed that *Stylosanthes* was most effective in suppressing speargrass, followed by *Pueraria, Psophocarpus* and *Centrosema* in that order. The establishment of *Psophocarpus* and *Centrosema* was poor, which was presumed to be due to inappropriate strains of rhizobium. The yield of maize planted in the next year, after killing the legumes with herbicide, was not significantly different for the different legumes, all of them yielding between 4 and 5 t/ha, except in the *Stylosanthes* plots where the legume had established poorly. No yield for a control plot without legumes was reported (IITA, 1980).

Establishing a legume cover crop at minimum cost to the farmer would require interplanting or relaying it into a field crop, rather than planting it as a crop by itself. *Pueraria* drilled into maize rows resulted in an effective cover, although the legume climbed onto the maize where it was planted 7 or 14 days after the maize, but without a negative effect on maize yield. *Stylosanthes* when planted in association with a maize-cassava crop, attracted rodents which damaged the maize and fed on the young cassava tubers.

Mucuna pruriens had also drawn attention because of its luxuriant growth and its potential for soil conservation and weed control. It died off naturally in the dry season which made it easier to drill maize seeds through the rubble in the next season, contrary to some other species, in particular *Pueraria*, which formed a thick live mulch layer. Several exploratory tests were carried out with *Mucuna* in the early 1980s, which showed a drastic suppression of weeds by a *Mucuna* cover. Volunteer *Mucuna* could become a weed in itself, however, smothering the following maize, unless controlled by uprooting or herbicide. This species would play an important role as a potential mulch crop in coming years.

LIVE MULCH SYSTEMS

In the live mulch system tested since 1977, maize was planted into a live sod of herbaceous legumes or grasses, as a low-cost method for soil conservation, weed control and fertility management, which would not require that farmers give up one cropping season for benefits to be expected in the next. In a first exploratory trial, five cover species were tested: *Arachis sp., Centrosema, Desmodium sp., Indigofera sp.* (species not mentioned) and *Psophocarpus*. In 1978, maize was drilled into the sod in both seasons. *Arachis, Centrosema* and *Psophocarpus* were most effective in suppressing weed growth, but maize yield was affected by a lower plant stand, 'due to poor live mulch management at emergence'. The latter two also had to be sprayed with dwarfing hormones to prevent them from smothering the maize. *Indigofera* was worst and also depressed maize yield more than the other species. In following seasons, *Centrosema* and *Psophocarpus* gave the best results in terms of weed control, earthworm activity[18] and maize yield. Yield in the *Psophocarpus* live mulch treatment compared with conventional tillage and no-till is shown in Table 2.10 for two first season crops. Initial differences were rather small, but yield under live mulch was more stable resulting in a higher yield over the entire period of six crops (IITA, 1982). The response to fertiliser-N was erratic. Apparently, no additional weeding had been needed in the live much plots.

Table 2.10 First season maize yields in live mulch compared to conventional and no-till, three N-levels; fifth crop; recalculated from Figure 49 in IITA (1982). No statistics are provided

Fertiliser application rate (kg N/ha)	Maize grain yield (t/ha)					
	Year 2 (3rd crop)			Year 3 (5th crop)		
	Conventional till	No-till	Live mulch	Conventional till	No-till	Live mulch
0	1.24	1.63	2.47	0.52	0.42	0.98
60	1.49	2.70	2.39	0.83	1.32	1.45
120	1.51	2.51	2.41	1.81	1.94	2.55

Two drawbacks of live mulch were noted in 1982: (i) the live mulch species competed with the maize for nitrogen, because of their poor nodulation under the trial conditions and (ii) they tended to climb on the maize smothering the crop. Growth retardants so far were insufficiently effective.

Integration of perennials in alley cropping systems

In 1978, a trial was set up at the Ibadan station to establish a fallow of woody leguminous species, planted at 2 m spacing, initially intercropped with maize and cassava, as was done in the *taungya* system[19] of forest planting. The legume species were *Leucaena leucocephala, Tephrosia candida*, pigeon pea and *Gliricidia sepium*. After allowing the legumes to grow for one year to get established, they were pruned and inter-planted with maize and the prunings were spread over the maize area. The dry matter weight and nutrient contents of the legume prunings and the yield of unfertilized maize are shown in Table 2.11. The tree species (*Leucaena* and *Gliricidia*) produced less dry matter and nutrients after one year than the semi-perennials, as would be expected. In the next year (1980), the tree species produced most biomass. The yield of maize associated with the legumes was better than in the control plots, except for *Tephrosia*, but the yield effect was smaller than the amount of N in the prunings seemed to promise, which may have been due to volatilisation (IITA, 1980) and perhaps also to lack of synchronisation of N release with N demand of the maize. A problem observed with *Leucaena* was its tendency to become a weed, due to its profuse seeding.

Soon after, the idea of establishing fallow species through a *taungya*-like system was abandoned in favour of a system whereby cropping and fallow were meshed by planting the perennials in permanent hedgerows, with annual crops grown in the alleys. The hedgerows were pruned during the cropping season and let go during the off-season in a system named 'alley cropping'.

A first hedgerow field had already been started in 1976 at the Ibadan station on an Apomu soil with *Leucaena*, initially with a grass fallow in the alleys and as from 1978 the field was converted into an alley cropping trial, with two annual maize crops planted in the alleys (Kang et al. 1981b). The factorial trial consisted of fresh

Table 2.11 Initial dry matter and nutrient content of hedgerow prunings and yield of first intercropped maize crop (IITA, 1980)

Treatment	Prunings				Maize yield (kg/ha)
	Dry matter (kg/ha)	N content (kg/ha)	P content (kg/ha)	K content (kg/ha)	
Cajanus cajan	4,100	151	9	68	3,173
Tephrosia candida	3,067	118	7	49	1,912
Leucaena leucocephala	2,467	105	4	51	2,601
Gliricidia sepium	2,300	84	4	55	2,587
Control	NA[a]	NA	NA	NA	2,030
LSD (5%)	Not provided	9	2	8	Not provided

Note:
a 'NA' = 'Not applicable'; 'LSD' = 'least significant difference'.

Leucaena prunings, added at maize planting at 0, 5 and 10 t/ha, partly carried from outside, combined with three N levels (Table 2.12). All treatments received 20 kg/ha of P and K applied to each maize crop. In 1978, similar maize yields were obtained at an application rate of 10 t/ha of fresh *Leucaena* prunings (partly carried from outside) without fertiliser-N, 5 t/ha of *Leucaena* prunings plus 50 kg N/ha, and 100 kg/ha of applied N only (Table 2.12). With a dry matter content of the prunings of 30% and the nutrient composition of Table 2.13, the total application of 7 and 12 t/ha of fresh prunings (the pre-sowing prunings plus an additional 2 t/ha in-season) corresponded with 67 and 115 kg N/ha. This might suggest that the N from the prunings was as effective as fertiliser-N. However, the maize yield reached a ceiling of about 3.5 t/ha at 100 kg N, irrespective of the source and did not respond to further additions. Furthermore, the N efficiency at the lower application rates was very low, hence another factor must have limited yields making conclusions about the efficiency of pruning-N compared to fertiliser hazardous.

In the next two years, only the prunings produced *in situ* were applied, ranging from 5 to more than 7 tons of dry matter, excluding woody stems, which is equivalent to an N-content of about 160–230 kg N/ha. Maize yields were moderate again at around 3.5 t/ha for the best treatment (prunings plus 80–120 kg/ha fertiliser-N). Since the control yielded around 2 t/ha, the agronomic N use efficiency was again quite low[20]. This pattern persisted through 1982, when yields in the plots with prunings were somewhat lower but the control yield had gone down to insignificance.

A screening test was carried out in an Ibadan screenhouse with an Ultisol from eastern Nigeria for the response to lime and P of four tree species: *Calliandra*

Table 2.12 First season maize yield (kg/ha) in an alley cropping-fertiliser trial on an Apomu soil at Ibadan (Kang et al., 1981b; IITA, 1983)

N rates (kg N/ha)	Maize grain yield (kg/ha) 1978 fresh Leucaena prunings			N-rates (kg N/ha)	Maize grain yield (kg/ha) 1980 prunings		1982 prunings	
	0	5 t/ha	10 t/ha		0[a]	in-situ	0	in-situ
0	2,109	2,732	3,221	0	1,350	2,100	610	2,096
50	2,572	3,166	3,256	40		3,115		2,671
100	3,377	3,450	3,432	80		3,520		2,911
				120		3,660		NA[b]
LSD[b] (5%)	296		440	437				

Notes:
a Prunings removed from the plot.
b 'NA' = 'not applied'; 'LSD' = 'least significant difference'.

Table 2.13 Nutrient content of *Leucaena* prunings at end of 1980 rainy season, Ibadan, Apomu soil (Kang et al., 1981b). No statistics are provided

N content (g/kg dry matter)	P content (g/kg dry matter)	K content (g/kg dry matter)	Ca content (g/kg dry matter)	Mg content (g/kg dry matter)	S content (g/kg dry matter)	Zn content (mg/kg)
32	2.7	29.1	8.4	2.9	2.8	52

calothyrsus, *Leucaena* and *Sesbania grandiflora*. The response to P was in the order *Calliandra* > *Leucaena* > *Sesbania* and to lime *Sesbania* > *Calliandra* > *Leucaena*. Overall performance of *Leucaena* was best, irrespective of treatment (IITA, 1983).

These were still exploratory trials which, for instance, did not yet address the important question of the system's N budget, viz. what happened to the large amounts of N accumulated in the biomass, in view of the low agronomic N use efficiency. The results so far, however, seemed promising: a slower decline of SOM than under conventional cropping, enrichment of the soil with nutrients other than N and a favourable effect on soil erosion. These findings or expectations were the start of a vigorous on-station alley cropping research program in the years to come.

The short-lived MIDAS project (1980–1982) started an alley cropping trial in 1981 in the savannah of central Ghana, with maize and water yams grown between *Leucaena* hedgerows. Both crops gave a 40–50% lower yield in the alleys compared with the control at all fertiliser level in 1982 (IITA, 1983; USAID/IITA, 1983). This was attributed to shading of the yams by the hedges and by drought and competition for moisture in the case of maize. This was an early warning signal, that all might not be well with the alley cropping concept, at least for savannah conditions, which went mostly unnoticed. An intriguing finding was that the N use efficiency of dried *Leucaena* leaves applied to maize was equal to (presumably the same amount of) fertiliser-N, irrespective of the application method. That contrasts with findings mentioned above about the low N efficiency of hedgerow prunings in established alley cropping systems.

Wetlands rice production

An exploratory trial was carried out with wet rice growing in valley bottom land at the Ibadan station. Five rice (*Oryza sativa*) crops were grown in two years, with on average about 10% higher yield for irrigated as compared with bunded-rainfed paddy (5.16 vs 4.57 t/ha per crop) and an increasing N effect in successive crops (IITA, 1980).

Tillage studies for wetland rice at Ibadan showed different responses of rice to tillage in coarse textured than in heavier soils. In the lighter soil, puddling gave best results in terms of establishment and yield, while zero-tillage was better in the heavier soil (IITA, 1984).

Steps were also being taken by the Netherlands-supported Wetlands Utilisation Research Project (WURP) to initiate a technology development program in hydromorphic areas across West Africa.

2.3.3 Technology validation

The technologies developed by the soil scientists and agronomists would eventually have to find their place in integrated cropping systems with relevance for the farmers of humid tropical Africa. In this episode, a start was made to validate some technologies in the context of an actual cropping or farming system. Rather than venturing into the real farm and letting farmers try the technology on their own, however, testing mostly took place within the confines of research farms, or in farmers' fields under strict control by the researchers. An interesting exception was the on-farm testing of alley farming during this episode.

On-station technology validation

THE UNIT FARM APPROACH (MENZ, 1980)

In 1977, an activity was started by the economists which was meant to test research findings under conditions resembling those of the real farm. Five 'Unit Farms' were created within the Ibadan station, each consisting of series of plots along a toposequence, which were to be put to their best possible use. The farms were to 'serve as a filter or bridge between research plot experimentation and on-farm evaluation of the most promising technology' (IITA, 1978; Menz, 1980). A rotation or intercrop of maize, cassava and cowpeas was grown in the upland soil, while yams (probably *Dioscorea rotundata*), plantains (*Musa spp.*) and cocoyams (*Colocasia esculenta*) were grown at the lower slope and rice and off-season vegetables in the valley bottoms. The Institute's recommended varieties and cropping practices, including zero-till were to be applied. The farms were run by 'farmers', each with an assistant and hired labourers as needed. What was hoped to be a realistic model for a future farm turned out to be a small technical and financial disaster. The 1979 Annual Report sums up the causes: excessive labour use, extremely high labour needs for rice including bird scaring, major weed problems controlled by expensive herbicides plus hand weeding, high levels of chemical inputs for zero-tillage, mediocre maize yields and low cowpea yields in the second season, and herbicide damage to cassava and cowpeas. And finally: 'the optimum levels of fertilizer and herbicides [were] not yet well defined'. The project was discontinued after two years.

A LIFE-SIZE LAND MANAGEMENT PROJECT (IITA, 1981)

In 1979, a 0.25 ha field at the station was cleared from secondary forest to monitor N release from organic matter (discussed earlier) and to verify 'whether improved soil and crop management practices at the small farmer's scale could extend the occupational period before the land had to be returned to bush fallow'. The practices were: no burning, no tillage, use of mulch, fertiliser, pre-emergence herbicides, pesticides and an improved crop variety. The crop sequence started with soybeans in the second season of the first year, followed by maize in both seasons of the following year. The yields of the soybean and the first maize crop were good, but the second season maize yield was disappointing because of late drought and streak virus, the usual problem with second season maize. In spite of the odd cropping sequence, this could have been a useful exercise to practise a package of recommended practices at a scale resembling that of a farmer in south-west Nigeria. It is not clear what happened next, because nothing more is heard of this validation test.

Technology validation by farmers

ON-FARM TESTING OF CROP PRODUCTION 'PACKAGES'

Some on-farm trials were conducted near the Ibadan station with maize, cassava and cowpea varieties, managed by researchers, not farmers, with the use of fertiliser and insecticides, not farmers' own practices. The kind of problems encountered were: unexplained poor maize stand, damage to second season maize by animals and insects and low yields, except on farms managed by farmer groups, where recommended crop and land management were effectively applied. What were called on-farm trials were, in fact, researcher-managed varietal tests couched in packages of standard technology (fertiliser, insecticides, spacing) provided by the Institute. In other words, they were researchers' trials carried out in farmers' fields. In hindsight, such trials are a recipe for failure, as many on-farm researchers will confirm. Farmers will try to steer the trials to an unfavourable plot, consider them as the researchers' business and gracefully neglect them[21].

Something similar must have happened in on-farm trials in eastern Cameroon, where maize did not respond at all to N and P, its yield hovering around less than 1 t/ha at 80 kg N/ha, 16.3 kg P/ha and 33.2 kg K/ha, whereas it attained up to 5 t/ha in research plots (IITA, 1983).

LEUCAENA-MAIZE ALLEY CROPPING

Ten long-term alley cropping trials were set up on-farm in 1980–1982, in six locations across the Nigerian 'yam-belt' in the Derived and Guinea savannah where availability of staking material was problematic. They consisted of just two plots each, with and without *Leucaena* hedgerows. The trials were designed and managed by the researchers and maintained by the farmers. The criteria for farmer selection was that they planted at least 1 ha of white yams, which requires staking, and were interested to cooperate. 'Some additional trials' were set up in cooperation with the Ayangba and Ekiti-Akoko ADPs. The *Leucaena* was seeded between first-season maize rows and was expected to be used as live yam stakes in the second year. Interesting lessons came out right from the start from this quite realistic on-farm work, described vividly by Ngambeki (1983, 1984) and Ngambeki and Wilson (1984):

- *Leucaena* established well even under 'extremely adverse conditions';
- The hedge rows were said to provide cheap material for staking, firewood or fencing and help in controlling weeds while the yams 'looked much better than in the control plots';
- The 2 m space between the *Leucaena* rows was found to be too narrow and should be increased to 4 m, which was done in the later test plots;
- Those farmers who pruned the yam stakes regularly (six times per season!) obtained yield increments of 18–20%;
- Maize, planted in regularly pruned alleys, registered a yield increment of 10–100%, but from a low basis of less than 1.5 t/ha, except one trial in the derived forest zone with 4.3 t/ha in the control plot and 4.7 t/ha in the alleys; and
- Since 'considerable interest among farmers' was aroused by the idea of controlling *Imperata* through *Leucaena* shade, the idea was raised to alternate a 1–2 year *Leucaena* fallow with 3–4 years of cropping.

Not everything was well, though:

- Farmers commented that the hedges demanded too much attention and took too much labour, especially for pruning (six annual prunings were needed); some did not even crop their alley field for that reason;
- Some pruned too little or gave up pruning altogether after a while so that the *Leucaena* shoots overgrew the yam vines; late pruning damaged the vines, causing yield depression of 8–60%;
- In one farm, the hedges were said to have been burned down by bush fire (perhaps intentionally, as sometimes happened elsewhere); and
- Some farmers found that even at 4 m spacing the hedges occupied too much land or interfered with mechanisation.

There is no further mention of this interesting set of on-farm tests after 1983 (the responsible scientist left in 1984), but some of them may have been part of the 35 on-farm trials mentioned in the 1985 Annual Report, which are discussed in the following chapter.

2.4 Technology delivery and dissemination

During this episode, research at IITA and elsewhere continued to follow the conventional research-extension-farmer model of technology development and dissemination, whereby international institutes would generate prototype technology for further adaptation by the NARS and dissemination by the extension service. Once a technology was deemed to have reached maturity, comprehensive summaries or guidelines had to be prepared with sufficient details for NARS to carry the work forward and eventually translate the technology into extension messages.

2.4.1 Databases, technology digests and guidelines

Fairly detailed accounts of on-going work were published in the Institute's Annual Reports, but 'digested' results were primarily published in prestigious journals. Only the soil physics group wrote comprehensive research summaries on soil erosion (Lal, 1984) and no-till farming (Lal, 1983, 1985), which, according to the Director General's foreword, avoided the pitfall of very long lead times due to reluctance of scientists 'to go out on a limb, before all the evidence is in'. The no-till documents provided clear guidelines about the choice between zero-tillage and other tillage methods under different ecological conditions.

The characterisation of major soils in the humid and sub-humid zones of Tables 2.1 through 2.3 had been a good start, but it stopped short of its goal, perhaps because of the reservations expressed by the 1977 Quinquennial Review about the Benchmark Soils project (TAC, 1978). A really useful compilation of findings from the Benchmark Soils project would eventually have to include:

1   Maps showing the distribution of the soil groups of Tables 2.1 and 2.2;
2   Typical toposequences showing other soil types (Entisols, Inceptisols) associated with the characteristic soils of Table 2.1;
3   Soil analyses of one or two representative profiles for each of the entries in the table, with utilisation history and relevant FCC condition modifiers;
4   Threshold values for major and minor nutrients identified by soil chemical and fertility research; and
5   Comprehensive, well-structured guidelines for use of the data, similar to the USDA Agricultural Handbooks.

The preliminary maps of S and micronutrient deficiencies do qualify as useful summaries of research findings.

For alley cropping, still considered an emergent technology, in need of much more research, a state-of-the-art paper was produced by the end of this episode, entitled 'Alley cropping, a stable alternative to shifting cultivation' (Kang et al., 1984).

Providing digested accounts of research findings for the benefit of the intended users was an explicit goal of the institution. The 1979 Annual Report, for example, stated that 'data are being compiled and summarised with the main objective of providing agricultural researchers with comprehensive information on the quality and limitations of major soils in tropical regions'.

Only modest efforts were devoted to this goal, which translate the fact that the uptake of technologies by end users was not a major focus of the researchers at this time.

2.4.2 Decision support

No decision support material was generated during this episode, perhaps with the exception of Lal's guidelines for choosing appropriate tillage systems in dependence of climate and soil (Lal, 1983, 1985; Section 2.4.1 above).

2.4.3 Dissemination and M&E

The linear model for technology uptake remained dominant during this period: it was assumed that once developed, the technologies would find their way to the farm through extension and development organisations. Besides, some technologies did not target the large majority of current farmers. Research on land management and erosion, for example, was of limited relevance for the current smallholder in the humid zone of West Africa, since 'erosion is generally of no consequence on forested lands and [on] farmland that is cultivated manually' (Lal, 1984). Hence, research was actually targeting a new, so far hypothetical, medium- to large-scale mechanised farm. This farm type, however, was quite common in East and Southern Africa, and to some extent in the West African 'cotton belt', where erosion of agricultural land was much more serious. Hence, erosion and land management research at the Institute was implicitly targeting three categories of users:

1   A future (semi-) mechanised medium- or large-scale agriculture in humid and sub-humid Africa;
2   Development projects like the Nigerian ADPs which intended to promote the emergence of such agriculture; and
3   Medium- and large-scale commercial farms in East and Southern Africa and the West African cotton belt, which was not yet part of the Institute's mandated area at the time.

Collaborative work was being initiated with the Nigerian ADPs, some of which used heavy equipment for land clearing, but so far actual adoption was not taking place.

The only example where the researchers teamed up with extension and farmers to test the adoptability of novel land use technology on the farm in a realistic manner was the alley cropping on-farm project, where lessons were learned quickly about what worked and what did not. This project, however, was not sufficiently anchored within the FSP and disappeared with its lead scientist (an agricultural economist!).

2.5 Outputs and impact

This episode was marked by a considerable amount of research findings on soils, soil fertility management and land use, obtained in surveys and researchers' trials on different soil types, including some attempts at mapping the geographic distribution of erosion hazards and nutrient deficiencies. There were also results on the specific constraints of different soil types and on the barriers to continuous cropping, and increasing attention was devoted to the potential of herbaceous and woody legumes for fertility restoration and maintenance. All these elements were potential components for future recommendations linked to different soil types, but the formulation of such recommendations was only at a rudimentary stage. Some on-farm research was also undertaken, especially with alley cropping in the Nigerian 'yam belt'.

2.5.1 Highlights of research outputs

Important findings from this episode were the following:

- The Alfisols, Ultisols and Oxisols can be meaningfully subdivided on the basis of their geological origin, the mineralogy of the clay and silt fractions and their base-status. Together with the FCC modifier tags, this constitutes an effective summary of soil conditions in the humid and sub-humid zones;
- Some threshold values for available nutrients were determined; they can be incorporated as modifiers into the FCC system as warning signals for possible deficiencies;
- 80% of the wetlands soils in the forest zones are physically and chemically very poor; in the forest-savannah transition zone small inland valleys with poor to reasonable chemical fertility and low P-status occupying 15–20% of the total land surface were little used for cropping. In the savannah zone extensive wetlands are found, in some areas with high activity clays. Nutrient status was generally low;
- Tentative maps were published showing locations where S and micronutrient deficiencies had been found;
- A project to map erosion hazard, based on the erosivity of rainfall and the erodibility of the soil, remained uncompleted;
- The advantage of zero-tillage remained undecided, with several inconveniences making it hard to adopt, especially when combined with a mulch layer;
- High maize and cowpea yields could be maintained for many years in Egbeda soil with modest fertiliser and return of stover, while Apomu soil required much higher rates after only a few years;
- Experimental results and soil degradation of experimental areas at the Ibadan station showed that continuous cropping would not be sustainable even with residue mulching and minimum tillage; a fallow phase would have to be included in the cropping cycle;
- Exchangeable Mg declined more rapidly in degrading soil than Ca and K, because the ion exchange selectivity of the kaolinitic clay caused preferential adsorption of K and Ca and leaching of Mg;
- High lime rates were not effective in acidic Ultisol and an initial lime application of about 1 t/ha was recommended, followed by annual maintenance dressings of 100–300 kg/ha; recycling leached nutrients from the subsoil would require deep-rooting perennials tolerant to high Al levels;
- A Paleustult with low P status only responded to liming if combined with P; Zn deficiency resulted from immobilisation of Zn at pH above 5, unless Zn was applied as well;

- There was no significant difference between burning or mulching crop residue in contrasting soils except in the zero-fertiliser and the zero-P treatments when the mulched plots did better;
- Second season maize needs a higher N rate because of the absence of an early-season N flux;
- Legumes do not contribute N directly to the associated maize, but may increase N availability of the system, when grown either in association or rotation with non-leguminous crops;
- *Pueraria, Stylosanthes, Psophocarpus* and *Mucuna* were most effective in suppressing *Imperata*, while *Centrosema* and *Psophocarpus* performed best in a live mulch system;
- In an established alley cropping system, *Leucaena* contributed 160–230 kg N/ha but the agronomic efficiency for maize yield was low;
- The publication record of 'digested' research results remained below standard, except for the work on erosion and tillage methods; and
- On-farm study on alley cropping in the Nigerian yam belt noted serious complications militating against adoption.

2.5.2 Uptake and impact

No direct uptake or impact was reported during this episode, nor was much energy devoted to support for dissemination by extension and development organisations that could have led to impact. Technology was simply not yet considered ready, nor was there much pressure from development organisations or NARS to provide concrete recommendations. Only the on-farm work on alley cropping in collaboration with ADP personnel came close to dissemination. Only in the northern states of Nigeria annual consultation meetings were held between research and ADP extension personnel, but IITA was not yet involved in work for the Northern Guinea savannah zone.

2.6 Emerging trends

The Benchmark Soils project ended somewhere around 1980. The last reference to the project is in the 1980 Annual Report; after that it was never mentioned again. The failure to bring the Benchmark Soils project to its intended conclusion was highly regrettable. The enormity of the task of describing and analysing a set of soil profiles representative of the major soil taxa of humid and sub-humid Africa and carrying out correlation trials in benchmark soils was probably underestimated and the number of scientists assigned to the task completely inadequate. The project essentially ended after its two main protagonists, Drs Greenland and Moormann, had left the Institute. The legacy of the project consisted of a book of papers by several renowned soil scientists and the complete set of profiles of Nigerian soils from the Benchmark Soils Project (Greenland, 1981). Valuable as these documents are, they fall far short of being usable by practitioners, such as national researchers and extensionists.

There was also regular mention of an 'IITA soil data bank' which should contain many more data sets in addition to those of the Benchmark and the Wetlands projects. It would be worth trying to locate that data bank and make a catalogue of its content for further use.

It is useful to look at what happened to the other 'Benchmark Soils' project, based at the University of Hawaii and working primarily in Asia. A retrospective paper by Beinroth (1984) concluded that:

> yields can be predicted with considerable accuracy on the basis of Soil Taxonomy families, if additional site factors are taken into account', and 'Transfer of agrotechnology to specific farm situations [...] should be based on systems analysis and employ computer simulation techniques to model the soil-weather-crop-management continuum.'

That is precisely the way forward taken by the project's successor: the International Benchmark Sites Network for Agrotechnology Transfer (IBSNAT), a multi-partner undertaking, with participation by other centres (Swindale, 1984). It was not until many years later that IITA embarked in a similar direction (Chapter 5).

Although research on yield responses to soil amendments was only conducted under the controlled conditions of the research farms, the soil scientists were well aware of the strong variation of soil conditions within small distances and their implications for the 'formulation of practical recommendations' (Kosaki and Juo, 1989a). Medium-scale variation was accounted for by replicating experiments on the major soil types in and area, while some exploratory work was carried out at the Ibadan station to characterise the pattern of soil conditions within small fields, using multivariate techniques (Kosaki and Juo, 1989a,b). Attention for within-field soil heterogeneity would emerge again several years later, after research had moved into farmers' fields.

By the early 1980s, it had become clear that the conventional technology transfer model, so successful in the Green Revolution countries, could not be relied upon under the much more varied conditions of African rain-fed agriculture. It had worked well in Asia's irrigated rice areas, where the conditions at the research station and at the farm were quite similar, farmers were more organised, agricultural services were more developed and where governments were actively supporting agricultural development. In Africa, the relevance of findings at the research farms for Africa's smallholders was in doubt. Really good crop varieties can fairly easily find their way to the farm, as they did for maize and cassava (e.g. through the ADPs), but the research-extension model was unsuitable for the development and dissemination of appropriate soil and land use technology and its dissemination. The Institute's social scientists did gather information about the challenges facing smallholder farmers, but they had worked largely in isolation and the physical scientists and their technologies had mostly remained inside the fence.

By 1980, the pressure to move technology out to the farm mounted and the Annual Report for 1981 announced that the FSP was going to put emphasis on:

Developing an integrated approach to on-farm research, [for which] several pilot projects were initiated [...] with national research institutions and World Bank funded ADPs' and 'Developing a package for integrated land and biomass management for sustained food production for farmers in the humid and sub-humid tropics.'

These tasks were going to be assigned, not to the disciplinary scientists who developed the technologies, but to a new group of workers whose task would be On-Farm Research (OFR). They would have to deal with transfer of technology into the real farmer's system.

Meanwhile new research approaches were being advocated in several places around the globe for bringing research closer to the real farm: the 'Farming Systems Approach to Research' known for short as Farming Systems Research (FSR) (Figure 1 in the Introduction). In a nutshell, its key concept was that research should design innovations which are relevant for the conditions of the real farm and test them at that farm, rather than under the controlled conditions of the research plots. Or better still, let farmers do the testing themselves, assisted by the researchers if, and only if, necessary. At IITA, the approach was introduced in the early 1980s as a somewhat alien body, through a Ford Foundation sponsored special program. Its primary task was to promote the adoption of FSR concepts and methods at the Institute and, more importantly yet in the eyes of the funding agency, by NARS in West Africa: the crippled leading the blind so to speak.

Notes

1 By David Norman at IAR, Samaru, Nigeria, Michael Collinson at the Ukiruguru Station, Tanzania and Peter Hildebrand who worked at several research stations in Latin America.
2 For correlation with the soil categories in the World Reference Base, see Annexe II.
3 Louvain and Reading Universities, Rothamsted Experimental Station and the International Soil Museum, Wageningen, for the Benchmark Soils project; the Institute for Soil Fertility, The Netherlands, and the International Fertilizer Development Center, USA, for nitrogen and phosphorus utilisation. Dutch Institutes (STIBOKA and KIT) for the Hydromorphic and Wetlands Utilisation Project (WURP); and the International Livestock Center for Africa (ILCA) on soil restorative processes under pasture.
4 Now consisting of IITA, Leuven and Reading Universities, Rothamsted Experimental Station, and the International Soil Museum.
5 '*Fadamas*' are lowland areas in the Southern and Northern Guinea savannahs that are often used to produce high value crops (e.g. tomatoes, chilli peppers) during the dry season, often with supplementary irrigation.
6 For a more recent review see Sanchez et al. (2003).
7 The MIDAS (Managed Inputs and Delivery of Agricultural Services) project applied this methodology to the substantial amount of soil data collected in the savannah area of central Ghana.
8 For a precise delineation of these zones, see Chapter 5.
9 Called 'parametric assessment of suitability for zero-tillage'.
10 Oxic Paleustalf and Psammentic Usthorthent respectively.
11 It is not clear what caused the low initial yields.

12 The trial design was, in fact, not efficient enough for the simultaneous analysis for the three nutrients.
13 See soil analysis data in Annexe I.
14 There was in fact no liming effect in the first trial year.
15 The seeming decline above pH 5.5 is only based on a single data point.
16 The relative yield total is defined as the sum of the yields in mixture as fraction of the sole crop yields.
17 As was actually common for indeterminate cowpeas in the West African Northern Guinea savannah.
18 Earthworm activity was often mentioned as an indication for healthy soil conditions; research to substantiate this claim was only conducted in later years (Chapter 5).
19 In a '*taungya*' system farmers are allowed to use the space between young saplings in a commercial tree plantation to grow food crops for a few years and take care of the trees along with their own crops.
20 It is not always clear whether the maize yields in alley cropping were calculated for total area, including the space occupied by the hedges, but knowledgeable people assured us that that was probably always the case.
21 For other examples, see Mutsaers, 2007.

References

Atayi, E.A. and H.C. Knipscheer, 1980. *Survey of Food Crop Farming Systems in the "Zapi-Est"*, East Cameroon. IITA/ONAREST, Ibadan.

Balasubramanian, V., J.O. Braide, and E.A. Atayi, 1982. *An Appraisal of the Present Farming Systems of the Atebubu District of Ghana*. 88 pp. IITA, Ibadan.

Beinroth, F.H., 1984. A synopsis of the Benchmark Soils Project. In: ICRISAT (International Crops Research Institute for the Semi-Arid Tropics). *Proceedings of the International Symposium on Minimum Data Sets for Agrotechnology Transfer*, 21–26 March 1983, ICRISAT Center, India.

BNMS-II, 2007. Achieving development impact and environmental enhancement through adoption of balanced nutrient management systems by farmers in the West African Savanna. *Annual Report 2006*. IITA, KULeuven, DGDC.

Buol, S.W. and W. Couto, 1981. Soil fertility-capability assessment for use in the humid tropics. In: D.J. Greenland (Ed.). *Characterization of Soils in Relation to their Classification and Management for Crop Production: Examples from Some Areas of the Humid Tropics*. Clarendon Press, Oxford.

Diehl, L., 1982. *Smallholder Farming Systems in the Southern Guinea Savanna of Nigeria*. GTZ, Eschborn, Germany.

Ezumah, H.C., B.B. Singh and S.R. Singh, 1983. Maize/cowpea intercropping. *IITA Annual Report 1982*, 143–144.

Friesen, D., A.S.R. Juo and M.H. Miller, 1980a. Liming and lime-phosphorus-zinc interactions in two Nigerian Ultisols. I. Interactions in the soil. *Soil Science Society of American Journal*, 44, 1221–1226.

Friesen, D., M.H. Miller and A.S.R. Juo, 1980b. Liming and lime-phosphorus-zinc interactions in two Nigerian Ultisols. II. Effects on maize root and shoot growth. *Soil Science Society of American Journal*, 44, 1227–1232.

Friesen, D. K., A. S. R. Juo, and M. H. Miller, 1982. Residual effect of lime and leaching of calcium in a kaolinitic utisol in the high rainfall tropics. *Soil Science Society of America Journal*, 46, 1184–1189.

Greenland, D.J. (Ed.), 1981. *Characterization of Soils in Relation to their Classification and Management for Crop Production: Examples from Some Areas of the Humid Tropics.* Clarendon Press, Oxford.

Howeler, R.H., 2002. Cassava mineral nutrition and fertilization. In: R.J. Hillocks,, J.M. Thresh and A.C. Bellotti (Eds). *Cassava: Biology, Production and Utilization.* CAB International, Wallingford.

IITA, 1978–1984. *Annual Reports 1977–1983.* IITA, Ibadan.

Juo, A.S.R. and J.A. Lowe (Eds.), 1986. The wetlands and rice in Subsaharan Africa. *Proceedings of an International Conference on Wetlands Utilization for Rice Production in sub-Saharan Africa,* 4–8 November 1985. IITA, Ibadan.

Juo, A.S.R. and M.K. van der Meersch, 1983. Soil degradation. *IITA Annual Report 1982,* 122–124.

Kang, B.T., 1980. *Soil Fertility Management of Tropical Soils.* University of Nairobi/IFDC/IITA Training course on Fertilizer Efficiency Research in the Tropics, 38 pp.

Kang, B.T., 1983. Effect of plant residue mulching and burning. *IITA Annual Report 1982,* 125–126.

Kang, B.T., 1984. Potassium and magnesium responses of cassava grown in Ultisol in southern Nigeria. *Fertilizer Research,* 5, 403–410.

Kang, B.T. and O.A. Osiname, 1985. Micronutrient problems in tropical Africa. *Fertilizer Research,* 7, 131–150.

Kang, B.T., E. Okoro, D. Acquaye and O.A. Osiname, 1981a. Sulphur status of some Nigerian soils from the savannah and forest zones. *Soil Science* 132, 220–227.

Kang, B.T., G.F. Wilson and L. Sipkens, 1981b. Alley cropping maize (*Zea mays* L.) and *Leucaena* (*Leucaena leucocephala* Lam) in southern Nigeria. *Plant and Soil,* 63, 165–179.

Kang, B.T., A.B.M. van der Kruijs and J. van den Hengel, 1984. Comparison of mulching with burning of plant residues. *IITA Annual Report 1983,* 185.

Kang, B.T., G.F. Wilson and T.L. Lawson, 1984. *Alley Cropping: A Stable Alternative to Shifting Cultivation.* IITA, Ibadan.

Kosaki, Takashi and A.S.R. Juo, 1989a. Multivariate approach to grouping soils in small fields. I. Extraction of factors causing soil variation by Principal Component Analysis. *Soil Science and Plant Nutrition,* 35, 469–477.

Kosaki, Takashi and A.S.R. Juo, 1989b. Multivariate approach to grouping soils in small fields. II. Soil grouping technique by cluster analysis. *Soil Science and Plant Nutrition,* 35, 517–525.

Lagemann, J., J.C. Flinn, B.N. Okigbo, and F.R. Moormann, 1975. A case study from Eastern Nigeria. Internal paper, IITA, Ibadan.

Lal, R., 1983. *No-till Farming: Soil and Water Conservation and Management in the Humid and Subhumid Tropics.* Monograph no. 2, IITA, Ibadan.

Lal, R., 1984. Soil erosion from tropical arable lands and its control. *Advances in Agronomy,* 37, 183–248.

Lal, R., 1985. A soil suitability guide for different tillage systems in the tropics. *Soil & Tillage Research,* 5, 179–196.

Menz, K.M., 1980. Unit Farms and farming systems research: the IITA experience. *Agricultural Systems,* 6, 45–51.

Mutsaers, H.J.W., 2007. *Peasants, Farmers and Scientists. A Chronicle of Tropical Agricultural Science in the Twentieth Century.* Springer, Dordrecht, The Netherlands.

Mutsaers, H.J.W. and P. Walker, 1990. Farmers' maize yields in S.W. Nigeria and the effect of variety and fertilizer: an analysis of variability in on-farm trials. *Field Crops Research,* 23, 265–278.

Ngambeki, D.S., 1983. *Leucaena*/maize/yam on-farm trials. *IITA Annual Report for 1982*, 57–59.

Ngambeki, D.S., 1984. On-farm trials of *Leucaena*/maize/yams. *IITA Annual Report for 1983*, 183–184.

Ngambeki, D.S. and G.F. Wilson, 1984. *Economic and On-farm Evaluation of Alley Farming with Leucaena leucocephala, 1980–1983*. Activity consolidated report, IITA, Ibadan.

Papendick, R.I., P.A. Sanchez and G.B. Triplett (Eds), 1976. *Multiple Cropping*. ASA Special Publication 27. American Society of Agronomy, Crop Science Society of America, and Soil Science Society of America, Madison, WI.

Pleysier, J.L. and A.S.R Juo, 1981. Leaching of fertilizer ions in a kaolinitic utisol in the high rainfall tropics: leaching through undisturbed soil columns. *Soil Science Society of America Journal*, 45, 754–759.

RCMD, 1992. *Cropping Systems and Agroeconomic Performance of an Improved Cassava in a Humid Forest Ecosystem*. Resource and Crop Management Research Monograph No. 12. IITA, Ibadan, Nigeria.

Sanchez, P.A., C.A. Palm and S.W. Buol, 2003. Fertility capability soil classification: a tool to help assess soil quality in the tropics. *Geoderma* 114, 157–185.

Swindale, L.D., 1984. Keynote Address. In: ICRISAT (International Crops Research Institute for the Semi-Arid Tropics), *Proceedings of the International Symposium on Minimum Data Sets for Agrotechnology Transfer*, 21–26 March 1983, ICRISAT Center, India.

TAC, 1978. Report of the TAC Quinquennial Review Mission to the International Institute of Tropical Agriculture (IITA). CGIAR TAC Secretariat, FAO, Rome.

USAID/IITA, 1983. *MIDAS Project in Ghana: Small Farms Systems Research*. Terminal Report (1980–1982). IITA, Ibadan.

USDA, 1997. *Predicting Soil Erosion by Water: A Guide to Conservation Planning with the Revised Universal Soil Loss equation (RUSLE)*. Agricultural Handbook Nr. 703. Agricultural Research Service, United States Department of Agriculture.

Vine, H., 1953. Experiments on the maintenance of soil fertility at Ibadan, Nigeria, Part I. *Empire Journal of Experimental Agriculture,* 21, 65–85.

Windmeijer, P.N. and W. Andriesse (Eds.), 1993. *Inland Valleys in West Africa: An Agro-ecological Characterisation of Rice-growing Environments*. Publication 52, ILRI, Wageningen, The Netherlands.

3 New trends, old habits
1983–1988

3.1 Scope, approaches and partnerships

The 1980s were a time of turmoil in tropical agricultural research. Conventional researchers were blamed for working in the isolation of their research plots, where the conditions only remotely resembled those in the farmers' fields, especially in Africa. And when researchers did venture into the real world, they would try to adjust the conditions there to fit their technologies, instead of the other way around, for instance by planting at the agronomically optimal time or by chemically controlling pests. The Farming Systems Research (FSR) approach wanted to reform agricultural research in such a way that its results would directly help farmers solve their constraints and exploit their potential by bringing research itself into farmers' fields. In most international institutes, separate units were therefore set up to bring the necessary change in attitude about, at IITA called On-Farm Research (OFR), because the term 'Farming Systems' was already in use. Perhaps the creation of a separate FSR/OFR unit was inevitable in the beginning, but in the longer term OFR would have to become everyone's affair.

In 1984, the Institute's Farming Systems Program (FSP) was restructured into four research areas: Upland production systems, Wetland production systems, Agroforestry and Plantain research and On-Farm Research, plus a Socio-economic Unit. This arrangement was not free of contradiction: the first two research areas suggest more integration of research efforts around specific systems, whereas setting up OFR and socio-economics as separate units seemed to deny their intended role as integrative approaches. OFR thus started as a more or less parallel donor-driven operation, set up on the initiative of and funded by the Ford Foundation, initially with little input from or impact on the disciplinary research units.

In 1986, in a move meant to steer the FSP into a more genuinely farmer- and farming systems-oriented direction, the new Director General appointed an agricultural economist as program director. The program was renamed Resource and Crop Management Program (RCMP) and, in another reshuffle, five research teams were formed, for:

- Resource management;
- Cassava-based systems;

- Maize-based systems;
- Rice-based systems; and
- Special studies.[1]

There was no longer a separate OFR Unit and this new arrangement was to bring the program's research a step closer to the ideal of integration of research efforts around farmers' existing production systems. However, the vision of the resource management team, working mainly on soil and land use research, was barely touched at this time by the FSR/OFR approach. Soil research did tackle new technical topics, such as land clearing methods and alternative N and P sources, and some of its activities were transferred outside the IITA station, but it continued essentially along the old methodological lines, using the same research methods and making the same assumptions about dissemination of its findings as before. As a result, two research-development models operated side-by-side, as show in Figure 3.1.

In 1986, the new Director General started a *Strategic Planning Study* to redefine the Institute's priorities and its strategic direction for the future (FSP, 1986). The study was conditioned by two points of departure, approved by the Board of Trustees: it 'should be concerned primarily with research relevant to small family farms in Africa, [and] the primary geographic area of focus for the Study should be the lowland humid and sub-humid regions of Africa' (FSP, 1986). The ecologies targeted by the Institute were broadly defined as the humid and sub-humid zones of West and Central Africa. The sub-humid zone was further

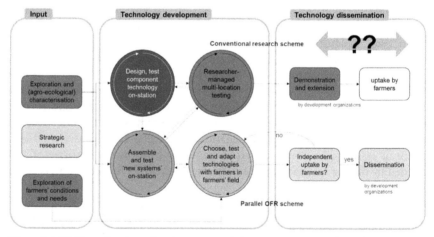

Figure 3.1 IITA's technology development and dissemination model, 1983–1988, with the conventional research-extension scheme remaining dominant. A parallel scheme existed along 'modern' FSR lines

Shading indicates relative emphasis on the various components with darker ones indicating a relatively higher emphasis. The question marks indicate that it was not clear whether and how the transfer process would be facilitated at that time

divided into the forest margins, the Derived savannah and the Guinea savannah. Furthermore, it was thought that the contribution of research would have to come largely from improved varieties, which are productive in the real world of the African farmer, and 'a greatly strengthened farming systems orientation' was to ensure that that would be the case. A lengthy planning process followed, which eventually resulted in a Medium Term Plan 1989–1993, published in 1988. We will discuss elements of this plan in the final sections of this chapter.

In spite of these internal developments, the research agenda remained largely management- and donor-driven. Even before the completion of the Medium Term Plan, major changes were set in motion, with a reduction in the number of soil scientists and an increase in social scientists, reflecting the changing views about the role of the Institute.

During this and the next episode (Chapter 4), IITA's geographical coverage increased significantly, both through increased multi-locational testing of various FSP technologies, and through so-called outreach projects in several countries, contracted with donors (Figure 3.2).

Between 1983 and 1988, there was a steep increase in collaboration with specialised international institutions, through specially funded projects[2]. As before, most of these (the first five of the footnote) resulted from institute-level and individual contacts with advanced institutions, often mediated by Board members, who helped to raise project funds from their home countries. The last-mentioned project was initiated by SAFGRAD (Semi-Arid Food Grain Research and Development) and contracted with the Institute. With the help of these projects, research more and more moved out of the research stations, in particular into the Southern and Northern Guinea savannah of West Africa, but continued in its researcher-controlled experimental mode.

This episode also saw a large increase in specially funded 'outreach' projects[3] in countries across West and Central Africa, initiated by multi- and bilateral donors and contracted for technical assistance. They all had an important applied research component, which in theory would extend research into a wider mandate zone and improve feedback of its results to the centre. In actual fact, they worked fairly independently, with some scientists from the centre providing backstopping and training, for example in OFR methods.

Furthermore, most individual scientists continued to collaborate with scientists in Nigerian institutes in jointly managed trials at research stations and sub-stations attached to those institutes.

3.2 Characterisation of soils, farms and farming systems

Soil characterisation now turned almost exclusively to wetlands, through the Wetlands Utilisation Research Project (WURP). This characterisation formed the basis for the on-farm agronomy work that was started in Nigeria and Sierra Leone, exploring the effects of variation in biophysical conditions and management on inland valley productivity (see Section 3.3).

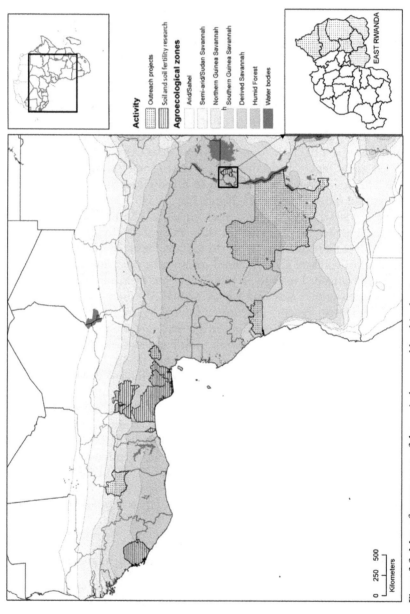

Activity

- Outreach projects
- Soil and soil fertility research

Agroecological zones

- Arid/Sahel
- Semi-arid/Sudan Savannah
- Northern Guinea Savannah
- Southern Guinea Savannah
- Derived Savannah
- Humid Forest
- Water bodies

EAST RWANDA

0   250   500
Kilometers

Figure 3.2 Map of target areas of the period covered by this chapter

3.2.1 Wetlands soils

A systematic inventory across West Africa identified four types of wetlands (Hekstra et al., 1983; Andriesse, 1986):

1   Deltas, tidal and large inland swamps;
2   Large river flood plains;
3   River overflow valleys (small floodplains); and
4   Small inland valleys (stream flow valleys).

The last two categories are part of what is usually referred to as 'inland valleys', each estimated at 4–10% of the 'total inventory area' in humid and sub-humid West Africa. In Sierra Leone, representative sites were identified in each of the wetland types and detailed pedological (Soil Taxonomy) and chemical soil properties were determined for each of these (IITA, 1984; for some examples, see Annexe I). It was further estimated that one-third of the West African wetlands consists of small inland valleys[4], with a total area of between 100,000 and 220,000 km². The project gave priority to this environment with the aim to 'develop low input soil and water management technologies that will enable small farmers to cultivate these valleys more intensively' (IITA, 1984). Benchmark sites were therefore set up in Makeni (Sierra Leone) and Bida (Nigeria), with two valleys each, and detailed land and soil surveys were carried out in these valleys (FSP, 1986).

The 'catchment factor' of the inland valleys, i.e. the ratio of total catchment area and inland valley area was found to vary enormously in West Africa, from 100:1 to 5:1. The catchment factor has a strong influence on the duration of wet conditions in the valleys. It is higher in the drier regions, which is why rice growing is possible in some inland valleys in the Sudan savannah. In the Bida site, the factor is at the high end, and in the Makeni site, at the low end of the range. The amount of erosion also varied greatly, which had resulted in mostly sandy textures in most West African inland valleys (RCMP, 1987).

In order to assess the potential and limitations for crop production, water balance models were developed for the benchmark valleys in Bida and Makeni to simulate their seasonal water conditions and assess their suitability for lowland paddy production. Conclusions were that the cultivation of paddy rice in small valleys near Bida would be risky, even with bunds, while in Makeni, with bunds, two annual paddy crops would be possible (FSP, 1986; Gunneweg et al., 1986).

In 1987, a further detailed transect-based characterisation study was started in the Bida area with the aim of developing a method to relate water balance of the inland valleys to the physical soil conditions and the rainfall pattern (RCMP, 1987).

Iron toxicity is widespread in West African wetlands and a detailed study was therefore conducted in an inland swamp near Bende, south-east Nigeria, about its causes and distribution pattern. It confirmed earlier findings that toxicity is severest close to the seepage source (the surrounding uplands) and lowest where drainage is better and groundwater fluctuation largest, leading to evacuation and oxidation of ferrous ions. For productive development, seepage flow should therefore be intercepted in the upper slope and draining the field at the end of the cropping season to allow oxidation to take place (IITA, 1984; Kosaki and Juo, 1986).

3.3 Technology development

In spite of repeated program restructuring, soil and land use research continued to follow the same old habits and to address the same technical concerns, with little concern about the adoptability of their technologies, which were typically far-off from farmers' current practices. Much of the research was conducted under strictly controlled conditions through experiments on and off-station, increasingly in collaboration with other international organisations and through specially funded outreach projects, but with little involvement other than by the immediate research partners.

The overall picture of the work remained rather fragmented, with research projects carried out on specific technologies by individuals or small groups of scientists with little integration, and reluctance to expose technology to the real farm. Trials were conducted in a small number of locations, mostly under researcher control, thereby avoiding, rather than addressing the variation occurring within and between farms.

3.3.1 Soil and soil fertility management

The soil physics group, while continuing to work on basic erosion questions, increasingly turned their attention to land clearing and subsequent land use, including mechanised farming, and their effect on soil parameters, compaction and productivity. Some large trials combined a number of these aspects in integrated systems, with some farmer involvement in their implementation. Soil and water management research was extended into the Northern Guinea savannah through a SAFGRAD-sponsored program in Burkina Faso. The large amount of work undertaken during this episode was a final upsurge, which ended with the departure of the lead scientist in 1987, after which only limited soil physics work continued in later years[5].

Soil fertility management remained an important line of research. The original concept had been to choose suitable benchmark soils in different ecologies to capture the interaction between environment and technology (Chapters 1 and 2). Three broad agro-ecological zones were distinguished: (i) the forest-savannah transition and Derived savannah zone with Alfisols and associated soil types, (ii) the humid zone with acidic Ultisols and (iii) the Guinea savannah with a wide variety of soils ranging from Alfisols to Ultisols and associated soil types. Targeting soil and soil fertility management research to these zones happened in a rather haphazard manner during this episode, whereby it is often not clear why certain locations were chosen. It is therefore often difficult to draw conclusions with validity outside the immediate trial locations. Only the Inland Valley project followed a systematic approach, by choosing benchmark toposequences representing wetlands conditions in different ecologies and carrying out field research in a few of these benchmarks. In that sense, the wetland management work was exemplary, and foreshadowed the principles which were later advocated by the agro-ecological and benchmark approach (Chapter 4). Researchers also started to carry out more of their work on-farm and explored the variation in effects of technologies under at least partial farmer management.

The work on herbaceous and perennial legumes as potential 'auxiliary species' for farmers' cropping systems gained prominence during this episode, still largely confined to the safety of the scientists' experimental fields. The auxiliary species were expected to strengthen the stability of permanent or semi-permanent systems which were found to crumble in case of the sole use of fertiliser, even in combination with the retention of crop residue (Chapter 2).

Soil erosion control

The investigations on quantification of soil erosion and structural decline continued during this episode, with runoff plots in five locations (Ibadan station, Ijaye village near the Institute, Ikom in Cross River, Heipang on the Jos Plateau and Onne), laboratory studies with a rainfall simulator with soils from 27 locations in Nigeria and calculations with the Universal Soil Loss Equation (USLE). Runoff plots at the Ibadan station studied the effect of slope length and tillage (no-till vs ploughing) and alley cropping on runoff and erosion. In 1986, the test confirmed earlier findings that no-till strongly reduced runoff and soil erosion compared to ploughing, especially in first season maize (*Zea mays*), and that runoff was greater as the slope was shorter, 'because the water on short slopes has less time to infiltrate'. *Leucaena leucocephala* and *Gliricidia sepium* hedgerows planted at 2 m or 4 m spacing were found to reduce erosion in the order

Leucaena 2 m < *Gliricidia* 2 m < *Gliricidia* 4 m < *Leucaena* 4 m << ploughing

Several more sites for erosion studies were set up in 1986 in the forest, transition and highland zones of Cameroon.

The vast amount of data on erosion collected over the years in all these studies were intended to serve the laudable objective of generating predictive tools for soil erosion hazards based on the USLE. To what extent this succeeded is not clear. The earlier attempts mentioned in Chapter 2 were promising, but to our knowledge there have been no further publications bringing together all the available data.

Physical soil management

Seedling emergence, crop establishment and root growth are all affected by soil compaction, as is water infiltration and therefore sensitivity of the soils to water erosion (Lal, 1985). Results up to 1982 had shown the advantages of no-tillage with mulch and light agricultural traffic under the conditions of south-west Nigeria, but there was concern about compaction if heavy equipments were used for clearing. Subsequent appropriate tillage and land use practices, however, might alleviate this effect and several new studies were started in the early 1980s, presumably having medium- and large-scale farming with heavy machinery in mind. Apart from the above large land clearing experiment, which also studied tillage aspects, trials were set up at the Ibadan, Mokwa and Onne stations and in farmers' fields in Ijaye near Ibadan to study the effects of tillage methods on physical parameters, especially compaction, and plant growth. Also, methods were studied to improve already compacted and otherwise degenerated soil, e.g. by the use of the paraplough (widely spaced slanted chisels for sub-soil tillage) and by restorative crops and covers.

TILLAGE AND CROP RESIDUE MANAGEMENT

The use of tractor-mounted or -drawn equipment was expected to lead to soil compaction, reducing water infiltration and therefore crop growth and yield. These effects were simulated on a coarse-textured Alfisol at the Ibadan station by zero, two and four passes of a tractor-mounted 2-ton roller prior to planting. Bulk density was increased and the yields of maize and cowpeas (*Vigna unguiculata*) were depressed by both compaction treatments, while cassava (*Manihot esculenta*) yield was only affected by four passes (Table 3.1). No significance was given for the effects on yield, but yield reportedly correlated significantly with changes in bulk density and infiltration rate. Cassava was less sensitive to compaction by vehicular traffic than the grain crops. It would appear that there was no significant effect of tillage method.

The test showed the sensitivity of a coarse-textured Alfisol to compaction resulting in lower crop yield, due to reduced porosity and infiltration (IITA, 1984, 1985). A trial on a farmer's Entisol (Psammentic Ustortent) near the Institute (Ijaye) comparing the effect of different tillage methods on soil compaction, showed that compaction increased in the order: traditional farming

Table 3.1 Response of crop yield to soil compaction of an Alfisol at the Ibadan station (IITA, 1985). No statistical information is reported

Treatment	Crop yield (t/ha)		
	Maize	Cowpeas	Cassava
No-till			
0 passes	4.7	1.3	18.2
2 passes	3.0	1.0	16.1
4 passes	2.3	0.8	9.5
Ploughed			
0 passes	5.4	1.3	20.6
2 passes	1.6	0.9	24.2
4 passes	1.4	0.9	12.5

Table 3.2 Maize yield in a tillage trial in a compaction-prone Alfisol (IITA, 1983; FSP, 1986)

Treatment	Maize yield (t/ha)	
	1982	*1985*
No-till, light tractor-mounted operations	–	3.64
No-till, hand operations	2.00	4.00
Paraplough once a year followed by no-till	2.73	5.09
Paraplough every two years followed by no-till	2.66	4.24
Chisel planter	–	4.34
Strip tillage	–	4.36
SED[b]	Not given[a]	0.17

Notes:
a Yield for paraplough 'significantly higher at 5%'.
b 'SED' = 'standard error of difference'.

< no-till < tractor ploughed < bare fallow. Erosion was also less under no-till farming and under farmer practices (IITA, 1985).

A test was started in 1982 on tillage methods to alleviate compaction of an Alfisol at the Ibadan station, comparing six modified no-till treatments (Table 3.2). In the first year, the paraplough treatments gave the highest maize yield, and manual no-till the lowest. In 1985, the highest yield was obtained for annual paraploughing and the lowest for tractor-based operations. In the intervening years, the trends were similar but the differences were not significant. It was concluded that paraploughing to loosen the sub-soil combined with narrow strip tillage as part of the planting operation was best in compaction-prone soils (e.g. most Alfisols, and arid-zone soils with a high sand and silt fraction).

Little work had been done so far on the effect of tillage and residue management in cassava. In 1981, a test was therefore set up in Warri (in collaboration with Shell) on a light-textured Ultisol and at the Ibadan station on a gravelly Alfisol with three tillage systems (conventional tillage, strip tillage and no-till), two fertiliser levels (0 and 300 kg/ha NPK 15-15-15)[6] and with or without mulch (6 t/ha field dry weight) carried from outside. The trial lasted for five seasons with no significant tillage effect in any year on either site. Cassava yields in the first and fifth year averaged over the three tillage systems are shown in Table 3.3. In both locations, the yields fluctuated strongly over the years, but declined steadily without fertiliser and mulching. Fertiliser alone prevented a strong yield decline in Ibadan, but less so in Warri. It was concluded that, contrary to Ibadan, continuous cassava production is not possible in Warri without mulching.

Another tillage and residue management experiment was set up at Mokwa in the Southern Guinea savannah in 1981 on a 'naturally compacted' Oxic Paleustalf[7], comparing manual tillage, no-tillage, strip tillage and conventional tillage (disc ploughing) with maize followed by sorghum (*Sorghum bicolor*). In the

Table 3.3 Cassava root yield averaged over 3 tillage methods in the first and fifth season in two locations (RCMP, 1987)

Treatment		Cassava root yield (t/ha)			
		Ibadan		Warri	
Mulch application	*Fertiliser NPK 15:15:15 (kg/ha)*	*1982*	*1986*	*1982*	*1986*
No mulch	0	20.2	9.3	9.1	4.1
	300	23.1	19.8	10.9	8.3
With mulch	0	22.2	18.1	10.7	22.2
	300	26.2	19.1	10.2	21.5
LSD (5%)	5.0	5.7	3.6	4.8	

Note:
'LSD' = 'least significant difference'.

first year the differences in maize yield were not significant, but in the following three years, (1982–1984) the manually tilled and disc-ploughed plots gave the best maize yield. Differences in sorghum yield were inconsistent but appeared to decline with time in all treatments, contrary to maize yield. This may have been due to a build-up of *Striga hermonthica*, an obnoxious, parasitic weed. The effects of crop residue management, comparing mulching, removal of crop residues and residue burning, varied. Both mulching and residue burning gave higher maize yield than removal of crop residues, while burning gave consistently better sorghum yields, probably by preventing transmission of diseases, such as *cercospora* leaf spot, which was observed in the mulched plots. This showed that results on the effect of tillage methods and residue management obtained in the sub-humid zone could not simply be transferred to the savannah and that 'more studies [were] needed' (IITA, 1984, 1985).

A test in 1983 and 1984 at Onne compared the effect of ploughing and no-till with or without mulch on soil physical parameters and on the yield of root and tuber crops. The results are shown in Table 3.4. Bulk density was lowest in the ploughed and the mulched plots, although at the end of the second year, differences were small. Cassava and cocoyam (*Colocasia esculenta*) yields (extremely low for cocoyam) were similar in ploughed and no-till plots, but yam (*Dioscorea rotundata*) yield was lower in no-till plots, as would be expected from the growth habit of this crop. Yam and cocoyam responded positively to mulch (12 t/ha of semi-dry *Chromolaena odorata* carried from outside!), but the effect was doubtful in the case of cassava (positive when no-tilled, negative when ploughed). In conclusion: mulching was beneficial for yams and cocoyams and doubtful for cassava. Tillage method made little difference for cocoyam and cassava, but ploughing was better for yams. In the trial in Warri, discussed above, cassava also did not respond to mulching in the first year, but it did later on.

No-tillage with crop residue left on the surface requires special equipment for planting. A six-row rolling injection planter developed at the Institute was

Table 3.4 Effect of tillage and mulch on bulk density and yields of root and tuber crops, Onne (IITA, 1985)

Tillage	Mulch	Yam				Cassava[a]	
		Bulk density (g/cm3)		Yield (t/ha)		Bulk density (g/cm³)	
1983			Mean		Mean		Mean
No-till	With	1.38		11.1		1.40	
	Without	1.42	1.40[b]	9.8	10.5[b]	1.40	1.40
Ploughed	With	1.19		13.9		1.24	
	Without	1.31	1.25[b]	10.9	12.4[b]	1.34	1.29
Mean	With		1.29		12.5[b]		1.32
	Without		1.37		10.4[b]		1.37

Tillage	Mulch	Cassava				Cocoyam			
		Bulk density (g/cm³)		Yield (t/ha)		Bulk density (g/cm³)		Yield (t/ha)	
1984			Mean		Mean		Mean		Mean
No-till	With	1.37		16.8		1.33		1.1	
	Without	1.41	1.39	12.7	14.8	1.33	1.33	0.3	0.72
Ploughed	With	1.26		13.1		1.24		1.9	
	Without	1.36	1.31	14.5	13.8	1.27	1.26	0.4	1.22
Mean	With		1.32		15.0		1.29		1.5[b]
	Without		1.39		13.6		1.30		0.4[b]

Notes:
a No cassava yields are given for 1983.
b Significant differences at 0.05 level.

compared with a standard no-till planter, both under difficult (soil hard to penetrate) and under favourable conditions. Under both conditions, there were no significant differences in establishment between the two types of equipment. Contrary to the rolling injection planter, other no-till planters need to be equipped with coulters to cut through the rubble (IITA, 1983).

Observations on tied ridging started at the Kamboinse research station in Burkina Faso in the early 1980s, showing striking crop yield increases due to better utilisation of rain water. Further studies of this technique were undertaken from 1985 under the new SAFGRAD-funded project. In an experiment with cowpeas, tied ridges resulted in a (very high) grain yield of 2.3 t/ha, compared with 1.5 t/ha for simple ridges ($LSD_{0.05}$ was 0.65 t/ha!). Only 50% of the additional water stored in the soil was available to the crop, however, due to inability of the cowpea roots to penetrate the compacted subsoil (bulk density 1.6 g/cm^3), 'a characteristic feature of West African savannah soils'. In 1986, the (smaller) yield difference was not significant. Roots of millet (*Pennisetum glaucum*) and Bambara groundnut (*Vigna subterranea*) to the contrary were not restricted by the high density subsoil and Bambara yield was not affected by ridge tying[8] (FSP, 1986; RCMP, 1987).

Top-soil (0–5 cm) compaction was found to be less after two years of maize on tied ridges than under flat planting. The effect did not occur under 'low management' (low density, no fertiliser), however (Table 3.5). Another finding was that tied ridges may cause water logging in soils with a high water table and compacted subsoil (*sols ferrugineux tropicaux[9] lessivés à pseudogley de profondeur*). In a maize trial with or without *Stylosanthes guianensis* as an intercrop, maize yield was 4.70 t/ha with tied ridges, while simple ridges produced 3.25 t/ha, with no effect of *Stylosanthes* on the intercropped maize. Roots of the mixture penetrated deeper than did the sole crop or cowpeas, possibly aiding the associated maize to access sub-soil moisture.

LAND CLEARING AND LAND MANAGEMENT

The work on land clearing and land management illustrates the thinking at the time that intensification of agriculture would require replacement of traditional land use systems with medium- or even large-scale, mechanised methods. Large trials were therefore conducted under fully controlled conditions aiming at the optimisation of mechanised land clearing and subsequent land management while

Table 3.5 Bulk density (g/cm^3) of top soil (0–5 cm) under flat seed bed and tied ridges[a] (FSP, 1986)

Land management	'Low management'[b]	'High management'[b]
Flat seed bed	1.37	1.27
Tied ridges	1.33	1.15

Notes:
a   Effects of tied ridges and management reported significant at 1%.
b   High management: fertiliser and high density; low management: no fertiliser, low density.

alleviating known soil degradation effects. There was no explicit consideration of the feasibility or socio-economic fit for smallholders of the technologies developed.

A large experiment started in 1979 (Chapter 2) comparing four land clearing methods followed by either no-till or disc ploughing, initially with maize+cassava as test crops. After a few years, some of the treatments showed severe soil compaction and erosion and various alternative land use systems were tried as a remedy (Table 3.6), in some of the worst cases after a short *Mucuna pruriens* fallow. Erosion was lowest after clearing by shear blade and highest after tree pusher use and after manual clearing. In 1986, crop yields were lower in the treatment combinations with high erosion, with the exception of alley cropping, which combined low yield with low erosion (RCMP, 1987).

After completion of the trial, detailed measurements were taken in three successive years (1989–1991) of a range of physical and chemical soil properties. In combination with the properties of the original soil, these showed that the mechanical clearing and tillage had resulted in the loss of topsoil and exposure of the subsoil (Hulugalle, 1994). It was concluded that mechanised clearance had resulted in long-term physical degradation, aggravated by mechanical tillage. The initial advantage of shear blade over tree pusher clearing was undone by mechanical operations in following years.

A somewhat less complex but still large trial was set up with the Ilorin Agricultural Development Project (ADP) in the Guinea savannah in 1983. It consisted of eight runoff plots of 1000 m^2 each, one of which was left fallow, while the other seven were cleared by a shear blade/tree pusher combination, followed by the land use treatments of Table 3.7. The land clearing method was presumably chosen in conformity with the ADP's land development strategy based on mechanised farming, while the tillage and land use practices were probably considered best-bet practices for this environment. Root cutting or ripping, for example, would be needed to enable conventional tillage after mechanical clearing. The trial was rather disastrous in the first year with failed hedgerow establishment and low maize yield. The only treatment doing well was number 7, with a (barely credible) yam yield of more than 40 t/ha. No more mention is made of this trial in following Annual Reports.

Another large, 24 ha, 'real-life' replicated trial was set up in Okomu, near Benin City in 1985, supported by the United Nations University, which compared a number of current and potential land use systems following mechanical forest clearing. The purpose was again to study the effect on soil and environmental parameters and productivity. The justification for this work was the rampant clearing of forest land for food production all over the humid tropics and 'the urgent need for developing sustainable soil and crop management systems for an intensive land use' (Lal, 1993).

All the land for the trial was mechanically cleared by shear blade, except in a 'traditional food crop system' treatment which was manually cleared. The vegetation in the other plots was burned in windrows, as is customary in oil palm (*Elaeis guineensis*) plantations in the area. The land use treatments following land clearing are shown in Table 3.8.

Table 3.6 Land clearing, tillage methods and subsequent cropping sequences and crop yield[a] in 1983 and 1986, Ibadan station (IITA, 1984; RCMP, 1987, Hulugalle, 1994)

Land clearing (1979)	Tillage through 1982	Cropping system from 1983[a]	Yield (t/ha)			
			1983		1986[b]	
			Maize	Cowpea	Maize	Cowpea
1 Traditional methods	Traditional	Pasture	No crop	No crop	No crop	No crop
2		Maize-cowpeas	4.44	0.39	2.00	0.74
3 Manual clearing	No-till	Pasture	No crop	No crop	No crop	No crop
4	Disc ploughing	Pasture	No crop	No crop	No crop	No crop
5		Alley cropping, maize-cowpeas	No crop	No crop	0.91	0.27
6 Shear blade	No-till	Pasture	No crop	No crop	4.10[c]	0.68[c]
7		Maize-cowpeas	4.17	0.43	2.10	0.67
8 Tree pusher/root rake	No-till	Pasture	No crop	No crop	1.80[c]	0.59[c]
9		Maize-cowpeas	4.05	0.41	2.00	0.15
10	Disc ploughing	Alley cropping, maize-cowpeas	No crop	No crop	0.73	0.23
LSD[d] (5%)			0.22	0.05	NA[d]	NA

Notes:
a The maize-cowpea rotation was not applied in all years; short Mucuna fallow was planted in treatments 7 and 10 in 1982, in treatments 4 and 9 in 1983, and again in treatment 7 in 1986; sweet potato-maize was planted in treatments 2 and 7 in 1983.
b Tentative; the exact combination of treatment and crop yield was not always clear from the documentation; no statistical differences given.
c In 1986 the pasture land was cropped again.
d 'LSD' = 'least significant difference'; 'NA' = 'Not available' (in the source document).

Table 3.7 Land clearing, tillage and land use practices, Ilorin Agricultural Development Program collaborative trial

Treatment	Tillage	Land use
1	Hand hoe tillage	Maize cropping
2	Root ripper-disc ploughing	*Gliricidia* alleys+maize
3	No tillage	*Leucaena* alleys+maize
4	No tillage	Maize with *Mucuna* in alternate years
5	Root cutting-disc ploughing	Maize with relayed *Mucuna*
6	Manual clearing-stumping-disc ploughing	Maize cropping
7	Traditional ridge farming	Yams
8	–	Bush fallow

Table 3.8 Forest clearing, tillage and land use practices, Okomu trial (RCMP, 1986)

Treatment	Land use system	Crop rotation
1	Forested control	–
2	Traditional food crop system	yr 1: yam+maize+melon yr 2-etc.: cassava + associated crops (farmer practices)
3	Alley cropping with *Gliricidia*, *Acioa* and *Alchornea*	yr 1: rice yr 2–3: maize/cassava-cowpea yr 4: rice-soybean or castor (!)
4	Improved forestry	*Senna siamea, Khaya senegalensis, K. grandifolia, Casuarina, Gmelina arborea*
5	Oil palm	yr 1: maize through oil plam yr 2–3: cassava+melon-cowpea yr 4: cocoyam
6	Grazed pasture with coconut	*Cynodon, Nchrisi, Panicum maximum*
7	Plantain	continuous plantain
8	Cassava-based system	yr 1: cassava yr 2: maize-cowpea yr 3: yam+melon+maize-cowpea yr 4: cassava
9	Improved forestry with 'mixed species'	*Gliricidia, Dacryodes edulis, Treculia africana, Pterocarpus* sp., *Dialium guineense*

The treatments are interesting as they show the thinking at the time about best-bet technologies (as well as the continued avoidance of real farmer testing of those technologies). In the first year runoff and erosion were negligible in all treatments. In 1986, the conclusion from physical measurements was, unsurprisingly, that 'the least degradation was observed in traditionally farmed and forested plots'. The effect on bulk density of mechanical clearing compared with manual clearing and (undisturbed) forested control (Table 3.9) was strongest at 15 cm depth but after three years the differences had all but disappeared[10]. Below 30 cm, bulk density remained at 1,35 g/cm³ in all plots. Differences between cropping treatments were small in 1987. Infiltration rates were substantially lower in mechanically than in manually cleared plots. In 1987, they were 49 and 156 cm/h respectively, compared with 220 cm/h in the forested control.

Another rather obvious result was that throughout the trials, chemical soil properties were more favourable in the windrow than in the non-windrow areas of the plots, where all vegetation had been removed. After two years, there was no significant effect of cropping system on any of the chemical soil analyses, averaged over windrow and non-windrow parts, After four years, there were significant differences only in pH, which was highest under continuous (non-grazed and uncut) pasture, and lowest under oil palm and 'improved forestry'. The exchangeable cation status was always higher in the topsoil of the cropped plots than in the intact forest, because of enrichment by the ashes of the former.

Crop yields were given for the different cropping systems (Lal, 1993), but it is not clear how they should be compared, given such disparate land use systems and crops, plus the usual pest problems with cowpeas in this environment. An interesting result was the good performance of plantains (*Musa* sp.) in the windrow areas and the practically zero yield in the non-windrow areas, which is consistent with frequent observations of the crop's sensitivity to soil fertility and its disappearance from intensified fallow-based systems without external inputs. All crop yields declined steeply with cropping duration, particularly in the non-windrow locations.

Table 3.9. Bulk density and infiltration rates in deforested plots averaged over cropping treatments and in forested control, Okomu, Nigeria (IITA, 1988). No statistical information is reported

Depth	Year	Mechanically cleared		Manually cleared		Forested control	
		Bulk density (g/cm³)	Infiltration rate (cm/h)	Bulk density (g/cm³)	Infiltration rate (cm/h)	Bulk density (g/cm³)	Infiltration rate (cm/h)
5 cm	1985	1.29	–	1.28	–	1.09	–
	1986	1.19	–	1.22	–	1.11	–
	1987	1.22	–	1.15	–	1.00	–
15 cm	1985	1.51	–	1.38	–	1.18	–
	1986	1.44	–	1.43	–	1.36	–
	1987	1.39	49	1.36	156	1.32	220

In a separate test comparing the development of *Senna siamea* and *Gliricidia*, the former produced considerably more biomass, confirming its vigour as observed in several other studies as well.

Soil fertility management

In the early 1980s, N and P response trials were conducted in various locations. There was a growing interest in alternative sources of the major nutrients N and P, and tests with different sources were conducted across Nigeria. An increasing number of these trials were carried out in the Southern and Northern Guinea savannah zones of West Africa, signalling an increasing interest in these zones as potential bread baskets. Trial locations were mostly laid out on representative soil types.

Improving the efficiency of applied fertiliser, especially N, was becoming a major concern in the 1980s, especially for the humid zone, because of the sensitivity to leaching of the acidic soils, while studies on soil acidity and Al-toxicity continued. With the increasing number of nutrient response trials, large variation was observed in control yields, nutrient response and nutrient-use efficiency. This, however, received surprisingly small attention, let alone scrutiny to investigate underlying causes, and did not influence subsequent research. This is the more surprising as the expanded program of nutrient trials was supported by two specially funded nutrient efficiency projects.

NUTRIENT RESPONSE OF MAIZE

At Ikenne, on an Oxic Paleustalf (Alagba Series), there was a linear response of maize to N up to 120 kg/ha in 1982 and up to 80 kg/ha in 1983. The magnitude of the response was quite low, however, with maize yielding 2.6 t/ha without N and 3.3 t/ha at 80 kg N/ha. In Okolu (probably Ultisol), maize responded up to 160 kg N/ha, going from 2.4 t/ha at 20 kg N to 3.5 t/ha at 80 kg N and 4.4 t/ha at 160 kg N/ha (FSP, 1986).

At Mokwa on an Oxic Paleustalf, maize yield went from 0.2 to 2.2 t/ha and in Zaria from about 1.2 to 2.5 t/ha at N-rates of 0 to 80 kg, i.e. with an N-use efficiency of respectively 25 and 16 grain/kg applied N (IITA, 1982–1984). Maize responded up to 120 kg N/ha in Jos, Guga and Saminaka with an N-use efficiency in the linear part of the response curves varying between 20 and 30 kg grain/kg applied N for the best-yielding varieties. Yield levels without N were very different at these sites, however: in Jos, at mid-altitude, it was a high 2.8 to 3.6 t/ha, while in the low altitude locations, it was between 1 and 1.6 t/ha (FSP, 1986). In another trial at Mokwa, maize responded up to 320 (!) kg N/ha, with yield going from 2.7 kg/ha at 20 kg N to 4.0 kg at 80 kg N and 5.0 t/ha at 320 kg N/ha (FSP, 1986). Thus, N-use efficiency was very different across sites, a phenomenon that went uncommented and unexplained.

In 1983 and 1984, maize responded significantly to P up to 60 kg P_2O_5 (26.2 kg P/ha) applied as TSP in Ikenne. At two sites in Ondo State (with the Ekiti-

Akoko ADP), there was only a small effect of P, but at the third site (Ikole-Ekiti), the response was strong, especially where the S-containing phosphorus sources PAPR50 (partially acidulated rock phosphate) with 50% of the acid, needed to fully unlock the phosphate in RP, been used in the preparation of this compound and SSP were applied. S-deficiency was observed where NPK 15:15:15 or DAP was applied (IITA, 1984). No information was given on the soil types, which are broadly in the Alfisol zone. In 1983 and 1984, maize responded significantly to P applied as SSP up to 90 kg P_2O_5/ha (39.3 kg P/ha) at Mokwa and Samaru, but in Mokwa, yield was unexplainably depressed at 30 kg/ha (significance not indicated). In other exploratory trials in central and northern Nigeria (Jos and Guga), maize responded to P up to about 60 kg P_2O_5/ha (26.2 kg P/ha) in (FSP, 1986).

In acidic soils in the humid zone of southern Nigeria (Nteje and Owerri, both Ultisols), maize yields were low. In Owerri, N-response was more or less linear up to about 80 kg/ha, with maize yield increasing from 200 to 1200 kg/ha. In both sites, maize responded significantly to P up to 60–90 kg P_2O_5/ha (26.2–39.3 kg P/ha) (IITA, 1984), but yield was depressed at the higher P rates in Owerri, perhaps due to induced Zn-deficiency. In a trial at Warri, maize responded to N up to 160 kg/ha with grain yield going from 0.69 t/ha at 20 kg N to 1.63 t at 80 kg N (FSP, 1986). In both these trials, agronomic N-use efficiency was quite low: 12.5 and 15.8 kg grain/kg N respectively. Again, no explanation was given. From the comparison of P-sources with and without S in four locations in the Derived and Southern Guinea savannah (SGS), it was concluded that sulphur is needed in many savannah locations (see below).

Trials with different types of nitrogen fertiliser were conducted between 1982 and 1984, using CAN, calcium cyanamide and different forms of urea. At Onne, Ibadan and Owerri, they all showed about the same effectiveness on maize. In Ikenne in the forest zone, prilled urea was somewhat better than the others at 80 kg/ha, CAN and straight urea were best at Mokwa and Zaria (IITA, 1983–1984). Calcium cyanamide showed promise because of lower leaching losses and the additional benefit of its weed control properties (Pleysier et al., 1983). It has to be applied at least two weeks before sowing.

The effect of different P-sources on maize yield was tested in Ikenne (humid forest zone), Nteje and Owerri (acidic soils, humid forest ecology), Mokwa (Southern Guinea savannah) and Samaru (Northern Guinea savannah). The P-formulations used were SSP, Togo phosphate rock (PR), and partially acidulated rock phosphate (PAPR25 – partially acidulated phosphate rock with 25% of the acid, needed to fully unlock the phosphate in RP, been used in the preparation of this compound – and PAPR50). The responses in two years were very variable, but generally the higher the P-solubility the better the response. SSP mostly came out on top, except in Owerri where PAPR50 came out highest. In the latter case, the highest application rates depressed maize yield, possibly because of P-induced Zn deficiency (IITA, 1985).

An interesting test was carried out in four savannah locations comparing S-containing P sources (SSP and PAPR50) with formulations without S (15-15-15 compound fertiliser and di-ammonium phosphate, DAP). Yields of maize for the different P forms were converted into a Relative Agronomic Effectiveness (RAE) parameter (effectiveness relative to SSP):

$$RAE = \frac{\text{Yield with P formulation} - \text{Yield without P}}{\text{Yield with SSP} - \text{Yield without P}}$$

Figures below 100 mean lower effectiveness than SSP, which contains S in the form of $CaSO_4$. Table 3.10 shows the RAE values of trials carried out in two Derived savannah locations (Ikare, Ikole) and two Southern Guinea savannah locations (Ilorin and Bida). Low RAE values for DAP and 15-15-15 imply a higher response to S-containing P-sources, which is taken as proof for S-deficiency in all these savannah locations.

In an N × K factorial on a K-deficient soil at Onne (exchangeable K of 0.08 $cmol_c/100g$, pH 4.4) in 1985, sole cassava produced the highest root yield (25 t/ha) at either 100 kg K_2O (83 kg K) or 50 kg K_2O (41.5 kg K) plus 100 kg N per ha (FSP, 1986). Surprisingly, in another trial at Onne in 1982, cassava did not respond at all to N, and K only had a small effect on dry matter percentage (Kang and Juo, 1983).

Soybean (*Glycine max*) responded significantly to P up to 45–60 kg $P_2O_5/$ha (19.6–26.2 kg P/ha) at Samaru, Guga and Jos, with yield increasing from 500 to 1500–2000 kg/ha at Samaru. Yield declined at higher rates, probably by causing nutrient imbalances. In Sika, an early maturing (60-day) cowpea variety responded up to 20 kg/ha only with declining yield at higher rates, while a medium-maturity variety responded up to 60 kg/ha. No information was provided on the nature of the soil in these trials. In another test carried out in Mokwa, (Southern Guinea savannah, Oxic Paleustalf) there was no significant effect on soybean yield of either P or Zn[11] (RCMP, 1987).

In a factorial test on a 'sandy clay loam' with pH 5.3 in Gandajika, Zaire (now Democratic Republic of Congo), both groundnuts (*Arachis hypogaea*) and soybean responded to P, but in two out of three trials only up to a maximum of about 30 kg P_2O_5 (RCMP, 1987).

A long-term trial (see also Chapter 2) studied the effect of liming on maize and cowpeas in a highly acidic Ultisol at Onne. In 1986, the full range of results

Table 3.10 Relative agronomic effectiveness (RAE) of P-sources on maize in four savannah locations in Nigeria (IITA, 1985). No statistical information is reported

Location	Relative agronomic efficiency (%)		
	PAPR50	DAP	NPK 15-15-15
Ikare	120	76	71
Ikole	120	75	71
Ilorin	70	55	45
Bida	99	17	10

Note:
'PAPR50' = 'Partially acidulated phosphate rock;' 'DAP' = 'Di-ammonium phosphate'.

was analysed for the relationship between maize (TZPB) yield and pH, Al-saturation and Ca-saturation. Curvilinear regression analysis was carried out for each of these factors individually. An example is shown for Al in Figure 3.3, suggesting a threshold value for 85% of maximum yield at an Al-saturation of about 30%. The full analysis showed that 85% of maximum yield (5 t/ha) could be attained with: pH > 4.8, Al-saturation < 38% and Ca-saturation > 45%.

For this soil, a Typic Paleudult with pH 4.3 and effective CEC of 3 cmol/kg in the topsoil, 45% exchangeable Ca meant 1.35 cmol/kg. These conditions could be created by a moderate rate of lime of 0.5–1.0 t/ha of $CaCO_3$ once every three years (RCMP, 1987), essentially confirming the findings of the first five years. It should be noted, however, that the complete 10-year data set came from a single experiment at one location.

In on-station tests on acid soils in Bas-Zaire, similar responses to lime were obtained with considerable yield increases of groundnut at an application rate of 200–500 kg/ha of burnt limestone (52% CaO). Only a small increase in maize yield was obtained, but the overall yield was low at about 1 t/ha, probably due to other deficiencies. In 17 researcher-supervised on-farm trials in Bas-Zaire, groundnuts also responded to liming, the average yield with lime being 0.55 t/ha and 0.38 in the unlimed plots with a high percentage of empty pods. The crop only responded to fertiliser in the presence of lime (RCMP, 1987).

Long-term trials at the Institute had shown serious soil degradation to occur earlier as cropping methods were less conservationist. The use of acidifying N-sources was suspected to accelerate the degradation process and a verification test was set up in 1982. Three types of N-fertiliser (CAN, urea and ammonium sulphate) were applied to maize at 150 kg N/ha in two annual cropping sequences: maize-maize and maize-cowpeas. Soil management was no-till with crop residues returned. First season maize yields, pH and extractable Mn in the fourth cropping year (Table 3.11) showed an acidifying effect of all N-sources

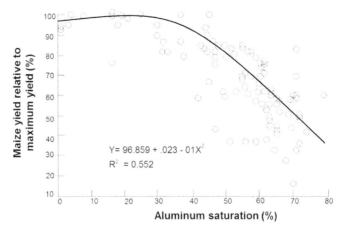

Figure 3.3 Relationship between Al-saturation and % of maximum maize yield; 10-year liming experiment at Onne (adapted from RCMP, 1987)

Table 3.11 Effect of N-sources on first season maize yield, pH and Mn in fourth cropping year (1985), Alfisol, Ibadan (FSP, 1986)

N-fertiliser (150 kg N/ha)	First season maize yield (t/ha)	pH	KCl-extractable Mn (mg/kg)
Continuous maize			
CAN[a]	5.24	5.6	96
Urea	5.07	5.5	99
Ammonium sulphate	3.53	4.9	183
0-N check	2.54	6.1	53
Maize-cowpeas			
CAN	4.78	5.5	81
Urea	5.41	5.6	74
Ammonium sulphate	3.49	5.3	129
Control without N	4.62[b]	6.0	37
LSD[a] (5%)	0.76	0.2	n.a.

Notes:
a 'CAN' = 'calcium ammonium nitrate', 'LSD' = 'least significant difference'.
b This yield looks suspiciously large.

at this application rate, most strongly for ammonium sulphate in continuous maize. The lower maize yield in the ammonium sulphate treatment was possibly mediated through Mn-toxicity.

The critical level for Mn had not yet been established, however. The high yield of the 0-N check in the maize-cowpea sequence looks suspicious even with the N-contribution of the cowpeas to the following maize crop. Cowpea yields were not given. The results in the following year (1986) were similar, at a 10% lower overall yield level, confirming the conclusions.

Lysimeter studies were carried out in Onne to quantify the losses of cations in relation to that of nitrate and the soil's cation content, for cropped and uncropped soil (IITA, 1984; FSP 1986; Wong et al., 1992). All plots received 1 t/ ha of lime (calcium hydroxide), and all cropped and half of the uncropped ones received fertiliser. The analyses showed that 60–70% of fertiliser-N applied to bare soil (135 kg N/ha) had been leached below 120 cm at the end of the second rainy season and almost 50% when a crop was grown. Even larger amounts of N from mineralisation of soil organic matter (SOM) were leached, viz. 156 kg N/ ha from bare soil and 87 kg N/ha from cropped soil.

In the cropped treatments, 312 kg/ha of Ca was lost due to leaching below 1.35 m (the depth of the lysimeters). In all treatments, the charge ratio of leached cations Ca+Mg and nitrate was less than unity in the first rainy season, due to leaching of other anions, but it approached unity during the second season. The Ca/Mg ratio in the leachate was similar to that in the solid phase, showing equal affinity of this soil for the two cations. The Mg/K ratio in the leachate, however, was much higher than that in the soil, because of the stronger affinity of the soil for monovalent cations. This, in turn, was related to the soil's low degree of K-saturation. It was concluded that under the high rainfall conditions and sandy soils in Onne, leaching of divalent cations with nitrate will rapidly decrease base saturation of the soil. An annual lime application of 500 kg/ha would be required to replace lost Ca, similar to the findings from the long-term field trial above (Wong et al., 1992). It should be noted, however, that the soil in both cases was a Typic Paleudult from the same experimental station.

Grimme and Juo (1985) showed, however, that losses of N due to leaching may be less than expected under high rainfall conditions in a vegetation-covered soil, because of a high density of macro-pores, which take the excess moisture down before equilibrium can be established with the N in the soil.

Management of wetlands soils and soil fertility

Although the primary goal of the Wetlands sub-program was more productive use of temporarily inundated valley bottom land, it was clear that wetlands could not be studied in isolation of the adjacent slopes and uplands. One of the premises of the wetlands project was that 'improved sawah systems [were] one of the most sustainable wetland production systems [..] from temperate monsoon to humid tropics in Asia [and that] there are no physical and environmental limitations [for] the sawah system in most of West African inland valleys' (RCMP, 1987).

Apart from a short-lived experiment at the Ibadan station, all the work on wetland utilisation was carried out on-farm, some of it for technology development, others for dissemination purposes or both. In fact, the Bida wetlands activities could be seen as coming close to the new style of research, according to the 'parallel OFR scheme' of Figure 3.1, whereby research formed a continuum from technology development to dissemination. The distinction between on-station and on-farm research thereby gets blurred. Since most activities had an exploratory nature, we will discuss it here rather than under Section 3.4 (technology delivery and dissemination).

In 1983, an experiment was started at the Ibadan station to develop lower slope and valley bottom land into an intensive production environment with irrigation, fed from an upstream water reservoir, drainage canals and small levelled paddies. Crops would be grown according to the position on the toposequence and the season. The project resembled the earlier 'Unit Farm' cropping scheme (Chapter 2), this time with the added sophistication of irrigated paddies. The idea was that the Institute should also experiment with more sophisticated techniques for the exploitation of inland valleys. In following years, there is no further mention of this project, which seems to have been dropped silently, probably because of the scientists' lukewarm reactions, and all further efforts were focused on the on-farm work.

Research in the Bida area was carried out jointly with staff of the Bida Agricultural Development Program (BADP) and the National Cereals Research Institute (NCRI). All field trials were on-farm with varying degrees of farmer management. Based on the dynamics of ground and surface water, it was estimated that, in the Bida area, three types of paddies could be distinguished: (i) paddies with more than 130 days flooding, (ii) paddies with 100–130 days flooding and (iii) the most risky paddies with less than 100 days flooding. In one of the pilot valleys (Gara), 85% of the valley bottom paddies were in the first two categories, while in the other valley (Anfani) it was only 70%. The 'fringe land' was found to be too risky for wet rice in both sites.

Farmer-managed rice variety × 'management' trials were conducted with 12 farmers in 1985. Improved management consisted of fertiliser (NPK 90-60-30) and herbicide application only, all other operations were left to the farmers. The average yields are shown in Table 3.12. Iron toxicity varied strongly among the 12 farmers (the local variety was less sensitive), and so did seedling age, weeding and water regime, but the effects of these differences, although noted, were not used in the analysis of the results. It was also noted that growing an early maturing variety could conflict with sorghum harvest in the uplands. Remedying this by planting earlier would cause other timing conflicts, thus showing a rather profound impact on a farmer's system of only a small change.

In 1986, another management trial was carried out in Gara, now with improved water management, consisting of improved bunds and waterways. The results are shown in Table 3.13. There was a large effect of improved water management. The high yield of the researcher-managed treatment without fertiliser (consistent

Table 3.12 Mean on-farm rice yield for three varieties under farmer's and 'improved' management in Bida fadama land (IITA, 1984)

Variety	Rice yield (t/ha)	
	Farmer's management	*'Improved' management*
Local	2.33	3.76
ITA 306	3.37	3.53
Faro 29	2.59	3.90
LSD (5%)	0.84	

Note:
'LSD' = 'least significant difference'.

Table 3.13 Rice yield in a fertiliser-management trial, averaged over varieties; Gara (Bida area). RCMP, 1987. No statistical information is reported

Application rates of NPK fertiliser[a]	Rice yield (t/ha)		
	Farmer cultivars, farmer management	*Improved cultivars, farmer management*	*Improved cultivars, researcher management*
0-0-0	–	–	4.6
15-15-15	1.9	3.7	–
90-60-60	2.8	5.2	5.6

Note:
a Expressed as rates of N, P_2O_5 and K_2O.

for all four varieties) is peculiar and was not explained, but probably highlighted a large discrepancy between the researchers' best management, and the reality under which smallholders are growing crops.

In the first year of field testing (1985), maize variety × fertiliser trials were conducted in the uplands adjoining the valleys with maize, whereby the maize was intercropped with late maturing sorghum, as farmers usually did. The (farmers') rationale for this crop mixture is similar to that of maize intercropped with cassava in the more southern areas, whereby the long season crop (sorghum or cassava) takes over once the maize has been harvested, in a 'flying start' (as C.T. de Wit called it when visiting the institute). The results from 10 of the original 15 farmers are shown in Table 3.14. The other five were abandoned due to weed, Striga and monkey infestation. Yields were quite poor and in 1986 even lower yields were recorded (less than 1500 kg/ha at the highest fertiliser rates). Hence, the nutrient use efficiency was low, suggesting other limiting factors, which could perhaps have been identified from the variability across farms. A hybrid maize variety yielded more than 2 t/ha in 1986, but also barely responded to N.

A large number of vegetable species and varieties grown during the dry season produced good yields under researcher management but did poorly in farmers' hands due to poor management. More management research was therefore deemed necessary.

The pilot valleys at Makeni were very different from those in Bida, with the water table going below 30 cm only during April and May, resulting in a 10-month growing season which allows two swamp rice crops a year. On the other hand, after only one day 75% of a rain storm would already have left the basin as runoff, which caused regular moisture stress to swamp rice. In a survey, farmer yields were found to be quite low, at an average of 1430 kg/ha at the centre of the swamp, going to 525 kg/ha at the fringes (RCMP, 1987).

Contour bunds were tested in 1986 to better distribute the water over the valley and increase the area with adequate water depth, resulting in an estimated 30% increase in production. Although farmers were said to be interested in this technology, there is no further mention of this work[12].

Table 3.14 On-farm maize yield in Bida inland valley uplands (RCMP, 1986). No statistical information is reported

Variety	Maize yield (t/ha)		
	Fertiliser NPK[a] application (kg/ha)		
	0-0-0	45-45-45	132-45-45
TZESR-W	1.47	1.60	2.34
EV 8430-SR	1.26	2.30	2.24
Farmers'	0.96	1.89	1.94
Mean	1.23	1.93	2.16

Note:
a Expressed as rates of N, P_2O_5 and K_2O.

3.3.2 Cropping systems

A broad range of cropping systems research was carried out in this period, which addressed an equally broad range of fertility and productivity-related questions. The studies continued to be carried out mostly in the isolation of the research plots, using 'model' crop combinations which were chosen mostly for their suitability to study fertility processes. Prominent were the studies on 'auxiliary species' which could be integrated in (semi) permanent cropping systems to mimic several functions of traditional fallow, such as weed suppression, soil protection and recovery of leached nutrients.

Continuous cropping, mixed cropping and crop rotation

Many mixed cropping trials were carried out over the years with cassava as the major component, relayed into maize or into a grain legume, reflecting farmer practices in the (sub) humid zone. These trials were concerned with planting patterns and densities, the inclusion of additional components such as melon (*Citrullus lanatus*) or sweet potatoes (*Ipomoea batatas*), time of planting of the components and fertiliser use. The most significant work from a soil fertility perspective is summarised here.

Mixed cropping experiments during this episode emphasised combinations of cassava with other crops under the Institute's mandate, in particular maize and grain legumes, which are often grown in association with cassava in the transition and Derived savannah zones of West and Central Africa (IITA, 1985). Rather than going into the details of the individual experiments reference is made here to a review of cassava-based intercropping research of wide provenance including IITA (Mutsaers et al., 1993). Figure 3.4 summarises the results of many trials carried out between 1975 and 1992, expressed as the relative yields of cassava and

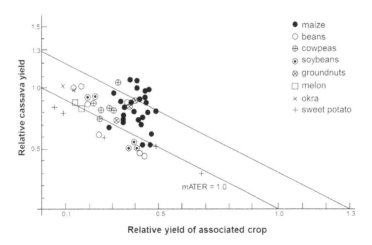

Figure 3.4 Relative yields of cassava and associated crops, adjusted for growth duration, and their sum, the modified Area-Time Equivalent Ratio (mATER); data from various authors (adapted from Mutsaers et al., 1993)

associated crops, and the *modified Area-Time Equivalent Ratio* (mATER), a measure of the productivity of a mixture compared with sole crops if both would occupy the same time span[13]. It exceeds unity for advantageous mixtures. Those mixtures including groundnuts or cowpeas were mostly favourable, while mixtures with soybean often were not, because of strong early season competition. Interestingly, maize showed a varied pattern, from strongly positive to negative. Detailed analysis of the dataset showed that cassava did not fully recover from early growth reduction caused by vigorous maize, the threshold maize yield being estimated at 3.5 t/ha. Densely planted maize and heavy fertiliser application would have that effect[14]. There were indications that P-conditions and P-fertilisation could influence the relative productivity of mixtures of cassava with legumes. In an example by Leihner (1983), the higher ability of cassava to extract P from the soil and the stronger response of cowpeas to applied P resulted in increasing yields but declining ATER with increasing applied P-rates.

A test at the Ibadan station on direct N-contribution by cowpea to intercropped maize confirmed the earlier finding that such contribution was negligible. The often-observed yield advantage of intercropping a cereal with a grain legume compared to the sole crops is a result of differential competition for resources. Direct contribution of nitrogen to maize would only be expected from decomposed residue of a rotational cowpea crop, although that would also be minor for the early maturing cowpea varieties (FSP, 1986).

The response to N of different crop combinations was studied in a trial on a newly cleared field (Typic Paleudult) at Onne in 1982–1984 (van der Heide et al., 1985). The fallow vegetation was manually cleared and incompletely burned. There were two rotations: maize-cowpea and maize-leguminous cover crop, and a mixed crop of maize+cassava (Table 3.15). Five levels of N were applied to all treatments and other nutrients (P, K, Mg and Zn) were applied in presumably adequate quantities. In a fourth treatment with maize+cassava+pigeon peas (*Cajanus cajan*), the pigeon peas were severely suppressed, from which it was concluded that it was not suitable as a supplementary crop for the maize+cassava mixture. In the first year, maize yield was significantly lower in the mixed crop with cassava than in the sole crop at all N-levels. Apparently, cassava dominated maize in the recently cleared field. The difference persisted in following years, but its magnitude declined with increasing N-levels (Table 3.15). Maize yield was significantly higher after the green manure legume than after cowpea in 1983, but in 1984 both were practically the same. *Stylosanthes* was replaced in the second year by *Mucuna*, because of the former's poor early growth. The trend in cassava yields was erratic and non-significant, oscillating around the average at all N-levels, while maize yield increased with N-level. An effect of fertiliser-N on cassava yield was probably undone by the increasingly vigorous maize. The lower maize yield in the intercrop was amply compensated by the cassava yield, but a full comparison of the two systems should include cowpea yield, which was not given. Considering the usually low cowpea yields, it is unlikely that the rotation with cowpea was more favourable than the maize+cassava intercrop. Agronomic N-use efficiency of maize at low N-rates was low again, possibly due to, or aggravated by, nitrate

Table 3.15 Maize and cassava yield in mixed cropping and rotations with herbaceous legumes[a] at 5 N fertiliser levels at Onne (IITA, 1983, 1985; van der Heide et al., 1985)

Treatment	Maize grain or cassava root yield (t/ha)					
	N-rate (kg/ha)					
	0	45	90	135	180	Mean
Maize-cowpea						
1982	2.23	2.97	3.50	3.56	4.06	3.26
1983	0.88	1.61	2.27	2.69	2.75	2.04
1984	0.65	1.66	2.33	3.17	3.53	2.27
Maize-herbaceous legume						
1982	2.23	2.97	3.50	3.56	4.06	3.26
1983	1.06	1.87	2.49	2.73	3.19	2.27
1984	0.69	1.99	2.60	2.85	3.36	2.30
Maize+cassava						
1982	1.32	2.60	2.25	2.77	2.08	2.20
Cassava yield[b]	9.51	7.44	7.04	8.12	5.61	7.83
1983	0.48	1.13	1.47	1.98	2.24	1.30
Cassava yield[b]	6.66	8.77	7.32	6.71	7.46	7.26
1984	0.55	1.55	2.22	2.92	2.98	1.80

LSD[c] (5%) values:

1983	Between N-rates: 0.251; between systems: 0.174 Between systems across N-rates: 0.369
1984	Between N-rates: 0.251; between systems: 0.139 Between systems across N-rates: 0.340

Notes:
a   In year 1 and 2 *Stylosanthes*, in year 3 *Mucuna*.
b   Cassava harvested in the following year.
c   'LSD' = 'least significant difference'.

leaching in this high rainfall area. Also, the apparent yield ceiling at high N-rates fell short of potential yield, the gap between the two increasing with time.

Integration of herbaceous legumes

Initially, herbaceous legumes grown as rotational- or relay-crops had been primarily intended for fertility improvement (Chapter 2). Some work with that objective continued in this episode, but gradually the potential of legumes for weed suppression, especially *Imperata cylindrica*, came to the foreground.

There was continued interest in this period in herbaceous legumes which could be interplanted into first season maize and would develop into soil improving covers after the maize harvest. For instance, a new entry *Canavalia ensiformis* was tested at the Ibadan station which did not affect maize yield significantly at Ibadan even when planted at the same time and had a positive effect on next year's maize yield.

At the Kamboinse station in Burkina Faso, a range of leguminous cover crops were also tested for their effect on soil properties and yield of the following maize crop. The most promising in respect of ground cover, root penetration and effect on maize yield (ranging from 0.8 to 2.4 t/ha) were: *Macroptilium lathyroides* (2.4 t/ha) > *Macroptilium atropurpureum* > *Psophocarpus palustris* > *Lablab purpureus* (0.8 t/ha). The yield increases were explained mainly from improved subsoil physical properties (RCMP, 1988).

A similar test was carried out in Kagasa, Rwanda, with four herbaceous legumes planted as short legume fallow, viz. *Crotalaria juncea, Sesbania sesban,* pigeon pea and *Mucuna.* Sorghum planted in the next season produced higher yields after all these legumes than after natural fallow, in the order: *Mucuna* (1.9 t/ha) > Pigeon pea >*C. juncea* > *Sesbania* > natural fallow (1.2 t/ha).

In 1983, a study was started in *Imperata*-infested fields near the Institute, comparing the effect of several control methods including herbicides and legume cover crops, viz. *Mucuna utilis, Pueraria phaseoloides* and *Psophocarpus.* Among the legumes, only *Mucuna* provided sufficient cover early enough to significantly reduce weeds; the other two did not develop rapidly enough.

In spite of the legumes' assumed potential for fertility improvement, a point to consider was whether it would be wise for a farmer to give up a cropping season for the future benefit expected from the legume. At the Cotonou station, it was shown in 1986 and 1987 that the yield increase of two successive maize crops following *Mucuna* or *Pseudovigna argentea* fallow did not compensate for the maize foregone during the legume fallow. As will be seen in the next chapter, when farmers were introduced to legumes as a cover crop in Benin, they were less impressed by the fertility aspect than by their potential for *Imperata* control.

LIVE MULCH SYSTEMS

Research on live mulch, a genuinely novel concept, continued, but only on-station. In previous years *Psophocarpus and Centrosema pubescens* had emerged as the best species, with the proviso that growth regulators had to be applied to prevent them from smothering the maize (Chapter 2). In a trial started in 1981 (on Alfisol, Iwo series), a single maize crop was grown annually in a live mulch system compared with conventional- and no-tillage at three fertiliser-N levels. Maize yields obtained in 1983 are shown in Table 3.16. In live mulch, yields were maintained at a reasonable level, with little or no need for weeding, but the fertiliser-N use efficiency was quite low, possibly due to competition. In the fifth year, maize was grown without further fertiliser-N and yields, averaged

Table 3.16 Effect of ground cover management and fertiliser-N on maize yield (t/ha) in 1983, and after-effect in 1985 of fertiliser-N after five years of continuous maize (IITA, 1984; FSP, 1986)[a]

Treatment	Maize yield (t/ha)			
	Fertiliser-N (kg/ha)			Mean yield in year 5, no-N
	0	*60*	*120*	
Conventional tillage without maize stover	1.6	2.7	3.1	1.6
Conventional tillage with maize stover	1.9	3.4	3.9	2.5
No-tillage without maize stover	1.1	2.1	2.9	2.1
Live mulch				
Arachis repens	2.4	2.0	3.1	NA[b]
Centrosema	2.4	3.2	3.2	3.2
Psophocarpus palustris	2.8	3.2	3.4	3.0
LSD[b] (5%)	fertiliser means: 0.9 groundcover means: 1.1			

Notes:
a Fertiliser means: 0.9; groundcover means: 1.1.
b 'NA' = 'not available'; 'LSD' = 'least significant difference'.

over the N-treatments applied in previous years (last column of Table 3.16), remained at around 3 t/ha (FSP, 1986). In a final instalment of this trial in 1986, the live mulch was killed and maize was planted without fertiliser in all plots. The results are shown in Table 3.17. All measurements were averages over previous fertiliser levels. The after-effect of live mulch on soil fertility and maize yield was quite spectacular.

Table 3.17 Effects of ground cover management on soil properties and maize yield in 1986, after five years of continuous maize growing (IITA, 1987)

Treatment	Organic C (%)	Cation exchange capacity (cmol/kg)	Weed dry matter, 4 WAP[a] (kg/ha)	Maize yield (t/ha)
Conventional tillage	1.0	4.19	186	1.87
No-tillage	1.4	5.60	265	1.31
Live mulch				
Centrosema	1.9	6.11	11	4.12
Psophocarpus	2.5	6.24	50	3.12
LSD[a] (5%)	0.9	NA[a]	67	0.64

Note:
a 'LSD' = 'least significant difference'; 'NA' = 'not available' (in the source document); WAP = 'weeks after planting'.

With the results so far, the live mulch system would appear to be ready for on-farm testing, except that smothering of the crop by the legume remained a major problem. Growth retardants had to be used, and the 1986 Annual Report stated that 'future research [...] will focus on management practices that will make it easier for smallholder farmers to manage the live mulch without the use of herbicides'. The search for non-climbing species continued. In 1987, *Crotalaria verrucosa,* a proliferous non-climbing species was chosen and interplanted into maize at the Ibadan station. At least one weeding was needed for proper establishment of the legume.

C. verrucosa resurfaced later for different purposes (Chapter 4), but research on live mulch faded away after 1988, without it being formally shelved as unsuitable for dissemination.

Integration of perennials in alley cropping systems

The 1980s and early 1990s could justifiably be called the alley cropping era. There were great expectations about the technology for its perceived, although not unequivocally proven, potential as a genuine alternative to shifting cultivation. Much research was undertaken during this episode in Nigeria, and elsewhere, to further refine the technology, and the list of promising species expanded.

AGRONOMIC PERFORMANCE OF ALLEY CROPPING SYSTEMS

In the early long-term trial with *Leucaena* on Apomu soil at Ibadan, which started in 1976 (Chapter 2), maize yield in the alley plots without fertiliser-N where prunings were removed, was down to 610 kg/ha in 1982 and 260 kg/ha in 1983, while the yield with the prunings retained and no fertiliser-N remained at about 2 t/ha in both years. Since the N-content of the *Leucaena* prunings was about 200 kg/ha, the N-efficiency was quite low. The entire trial field was left to a one-year fallow in 1985 and was cropped again in 1986 and 1987 with the results shown in Table 3.18. Maize in the alleys only did substantially better than in the non-alley plots when no fertiliser-N was applied, with a yield that was similar to the 40 kg/ha treatment without alleys. The relatively low yields and the absence of an N-response at the higher application rates in the non-alley plots suggests other limiting factors.

A trial in 1982 to compare the effect of fertiliser-N with an equivalent amount of N from *Gliricidia* prunings also showed a substantially lower effect of the N from the prunings than from fertiliser-N at an application rate of 40 kg N/ha, but at 80 kg N/ha the yield from the *Gliricidia* prunings was actually higher (about 2.5 t/ha against 2.2 t/ha). Again, in view of the very small yield increment from 40 to 80 kg fertiliser-N, another factor must have been at play here.

Another alley cropping trial was started in 1981 at the Institute on an eroded Egbeda soil, comparing two leguminous (*Gliricidia* and *Leucaena*) with two non-leguminous species (*Dactyladenia barteri* and *Alchornea cordifolia*) at 2 m and 4 m spacing with and without fertiliser, and with maize followed by cowpeas as test

Table 3.18 Maize yields 1986[a] and 1987[b] in the long-term *Leucaena* alley cropping trial after a 1-year fallow (RCMP, 1987, 1988).

Treatment		Maize yield (t/ha)		
Fertiliser N	Pruning	No hedgerows	With hedgerows	
applied (kg/ha)	management	1986	1986	1987
0	Removed	NA[c]	0.64	0.57
0	Retained	1.13	1.98	1.61
40	Retained	2.29	2.39	2.25
80	Retained	2.89	3.01	2.70
120	Retained	2.58	NA	NA
180	Retained	2.97	NA	NA
LSD[c] (5%)		1.06	0.40	SE[d]: 0.09

Notes:
a   Recalculated from Figure 5.1 in RCMP, 1987.
b   In 1987 maize in the check plots without alleys was severely damaged by grasscutters.
c   'LSD' = 'Least significant difference'; 'NA' = 'not available'.
d   Probably standard error of the means.

crops. The leguminous hedgerow species grew more vigorously than the non-leguminous ones, which both originated from the acidic Ultisol zone of south-east Nigeria. Maize yield was depressed by the hedgerows, especially at the 2 m spacing, due to shading and perhaps root competition. The pruning height was therefore reduced in 1982 to 75 cm and the 2 m spacing treatment was dropped. All plots received basal dressings of P and K, and N rates of 0 and 45 kg/ha in the alley plots and 0, 45, 90 and 135 kg in the check plots. Finally, all plots were split, with one being tilled and the other under no-till. The yields from 1985 to 1987 are shown in Table 3.19. Maize responded to N in all alley plots, but in 1986 and 1987 the response was not significant in the leguminous alleys. The rating of the maize yield was the same for fertilised and the unfertilised plots, viz.: *(Leucaena, Gliricidia)* > *(Dactyladenia, Alchornea)* > Control. The yield of the non-legume alley plots fertilised with 45 kg/ha N was about the same as that of the control at 90 and 135 kg N/ha, while the yield of the leguminous plots was even higher.

In 1990, the non-leguminous alleys did not produce better than the control in the absence of applied N, while the leguminous alleys maintained their advantage. The results in 1991 and 1992 for the leguminous alleys were similar to 1986/87, but at a somewhat lower overall level (RCMP, 1992; RCMD, 1993). The tilled plots did considerably better than no-till across the board, which is remarkable in view of other favourable experiences with no-till. This may have been caused by stronger root competition in the untilled alley plots, but it does not explain the same effect in the control plots, unless these were also invaded by hedgerow roots from adjoining plots (e.g. Hauser et al., 2004). An indication that this may have been the case is provided by a (not replicated) comparison

Table 3.19 Maize yields in tilled and no-till alley cropping plots at two levels of fertiliser-N, Ibadan station (RCMP, 1986–1988; RCMP, 1991–1992).

Treatment	N (kg/ha)	Maize grain yield (t/ha)								
		1985	1986		1987		Mean, 1986/87		1990[a]	1991
		tilled	no-till	tilled	no-till	tilled	no-till	tilled	tilled	NA[b]
Acioa	0	1.86	1.94	2.59	1.96	2.09	1.95	2.34	a	NR
	45	3.60	2.78	3.28	2.66	3.33	2.72	3.31	NA	NR
Alchornea	0	1.64	1.46	2.56	1.66	2.80	1.56	2.68	a	NR
	45	3.13	2.50	3.29	2.93	3.36	2.72	3.33	NA	NR
Gliricidia	0	2.32	2.17	3.21	2.75	3.53	2.46	3.37	b	2.57
	45	3.53	3.01	3.59	3.08	3.44	3.04	3.52	c	3.10
Leucaena	0	3.32	2.72	3.35	2.63	3.20	2.68	3.28	b	3.12
	45	4.32	2.91	3.52	3.18	3.54	3.05	3.53	c	3.48
Control	0	1.10	0.93	1.63	0.92	1.54	0.93	1.59	a	NR
	45	1.92	1.78	2.42	1.68	2.48	1.73	2.45	b	1.91
	90	3.26	2.39	3.47	3.01	3.26	2.70	3.37	c	NR
	135	3.20	2.89	3.32	3.06	3.42	2.98	3.37	c	2.59
LSD[b] (5%)		0.96	0.74		SE[b]: 0.226					

Notes:
a Yields not given; the same letter means no significant difference, a 'higher' letter means higher yield (RCMP, 1991); 'NR' means 'not reported', possibly other species were discontinued.
b 'LSD' = 'least significant difference'; 'SE' = 'standard error'; 'NA' = 'not available' (in the source document).

of the yields of maize grown in established alleys and under conventional and zero-tillage on large runoff plots at the Institute. No-till did better here than both conventional tillage and alley cropping (Table 3.20). The large plot size would have eliminated or strongly reduced sub-soil influence from adjoining hedgerows on the non-alley crops in this case.

Another indication of the sometimes elusive effects of alley cropping is provided by a researcher-managed on-farm trial in Alabata, with maize and cowpeas grown with and without hedgerows. The yields of both crops at two fertiliser levels (0 and 300 kg/ha NPK 15-15-15) were depressed by *Leucaena* alleys, while *Gliricidia* alleys had no effect compared with no alleys (RCMP, 1987).

Three potential new hedgerow species, *Senna*, *Flemingia congesta* and *Calliandra calothyrsus*, were screened in the early 1980s for biomass, N production and coppicing behaviour, compared with *Gliricidia*. *Senna*, a non-nodulator, produced most biomass and had a slower decomposition rate than *Gliricidia*, while *Flemingia* had the lowest biomass production and the slowest decomposition rate. *Gliricidia* produced considerably more N than did the other two, but *Senna* caused the greatest increase in maize yield, possibly because of its better distributed N release, moisture conservation and weed suppression (IITA, 1984; RCMP, 1987–1988). *Calliandra*, with foliage edible by ruminants, was found to grow fast and to coppice readily. In 1986, plots receiving *Calliandra* prunings produced 29% more maize than plots without prunings. In 1987, the yield increase due to the prunings was 45% when no fertiliser-N was applied, while removal or retention of prunings apparently did not make a difference when the crop was fertilised (Table 3.21).

Table 3.20 Maize yield in a large scale unreplicated tillage observation (RCMP, 1988). No statistical information is reported

Treatment	Maize grain yield (t/ha)
Conventional tillage	2.62
No-till	3.11
Gliricidia sepium	2.54
Leucaena leucocephala	2.35

Table 3.21 Maize yield in an alley cropping trial with *Calliandra* (IITA, 1988)

Pruning management	Maize grain yield (t/ha)		
	N applied (kg/ha)		
	0	45	90
Prunings removed	2.08	2.93	3.16
Prunings retained	3.02	2.73	3.28
SE[a]	0.24		

Note:
'SE' = 'standard error'.

Cowpea yield did not respond to either pruning removal or to N applied to the maize. The apparent yield ceiling of maize in this trial was rather low at 3000 kg/ha. Together with the absence of an effect of the prunings in fertilised plots, this seems to point to other limiting factors.

Finally, a long-term trial with *Leucaena* and *Senna* was started in 1986 at the Ibadan station, which would run for 20 years and was going to yield some important insights in the long-term dynamics of alley cropping and the role of SOM. The treatment of the design and results of this trial is postponed to Chapter 4.

In the early 1980s, several research activities were started with plantains and cooking bananas at the humid forest station in Onne. Plantains are a very important component of the cropping systems in many (forested) humid areas where soil conditions are favourable, but they decline rapidly after the first bunches, presumably by loss of soil organic matter (IITA, 1983, 1985).

Earlier exploratory work had shown that mulching was probably essential, both for fertility maintenance and erosion control, especially in the early stage of crop development. The effect on plantain yield of three types of mulch brought in from outside increased in the order *Flemingia* < *Chromolaena* < *Pennisetum purpureum*. Two approaches were considered to generate enough mulching material for sustained plantain production: (i) intercropping plantains with cover crops (*Chromolaena, Flemingia, Pennisetum*) and (ii) selective maintenance of vegetation strips left between rows of plantains, as a modified alley cropping system.

In a test of the first option, it was found that it was not possible to produce sufficient mulching biomass even on 40–50% of the area within a plantain plot and that additional material had to be carried from outside (IITA, 1985). A trial to test the second option, maintaining strips of the natural vegetation, started only in 1987 (Chapter 4).

Two more tests with *Flemingia* were initiated in the early 1980s. In the first, a new plantain crop, associated with *Flemingia* as a live cover crop, was compared with plantain mulched with *Pennisetum* carried from outside and with no mulch. In the first year, no treatment effect was found on yield. In the second year, the yields in the three systems were 14.5, 19.9 and 15.5 t/ha respectively, so plantain with *Flemingia* did not do better than the control without mulch and much worse than plantain with *Pennisetum* mulch brought in from outside.

In a second trial started in 1983, *Flemingia* was interplanted into existing plantains, to initiate a short (2-year) fallow. After a year the plantain was removed and the *Flemingia* continued as a fallow for two years, after which it was treated in four different ways:

1　*Flemingia* killed and residue removed; plantain fertilised;
2　*Flemingia* killed and residue retained; plantain fertilised;
3　*Flemingia* hedgerows cut back to 50 cm, residue retained; plantain fertilised; and
4　*Flemingia* hedgerows cut back to 50 cm, residue retained; plantain *not* fertilised.

In 1986, the first crop of the newly planted plantains, there were no differences yet among the first three treatments, while treatment 4 produced less than the others, hence mulching did not affect plantains in the first year. There was no further mention of this work in the following years, possibly because the lead scientists left the Institute in 1987.

ALLEY CROPPING IN OUTREACH PROJECTS

In Gandajika, Zaire, successful establishment of *Leucaena* hedgerows was obtained irrespective of the associated crop (cassava, maize or soybean). The initial depressing effect of the mature cassava on hedgerow growth disappeared later on when all plots were sown to soybean. In a cropping trial with the *Leucaena* alleys, yield of the first maize crop in the established alleys was depressed, probably due to drought and moisture competition. In the following year, there was no effect of the alleys on a total land area basis, the yield depression in the rows close to the hedges being compensated by higher yield of the inner rows. Pruning frequency was increased in following years to once a month, probably to reduce the effect of shading on the crops.

SUPPORTING STUDIES

Further refinements of the alley cropping technology were sought through better timing of pruning of the hedgerows to optimise N-availability to the crops. Preliminary tests in 1986 showed that *Leucaena* and *Gliricidia* prunings were best applied close to planting of the crop, while for *Flemingia*, with slower decomposition rate, the pattern was unclear. The test was repeated in 1987 with similar results. In order to disentangle the effect of decomposition of the prunings on one hand, and growth and competition for N by the hedgerows on the other, an experiment was set up in 1987 whereby prunings of *Leucaena*, *Gliricidia* and *Senna* were applied at 25 t/ha fresh weight in plots without alleys and their N loss was simultaneously monitored. For each species, the N applied in the form of prunings was most efficiently used by the maize when applied at planting, but N-utilisation (the difference between maize-N in plots with and without prunings) was low at less than 10% of the released N. In a similar test in Alabata, even lower efficiencies were measured (RCMP, 1987).

In 1984, the International Livestock Center for Africa (ILCA), in collaboration with the Nigerian Federal Livestock Department, started a project in two villages in the Derived savannah near Oyo town to introduce *Leucaena* and *Gliricidia* for the dual purpose of fertility improvement and the production of fodder for small ruminants. After two years, the trees in some fields had developed poorly and showed signs of nutrient deficiency. Pot trials were conducted in Ibadan with soil from a poorly and from a well-developed alley to study the effect of *Rhizobium* inoculation and N and P fertilisation. Although the test results were erratic (RCMP, 1987), it was concluded that low N- and P-content of the soil were partly responsible and that the indigenous rhizobia were not very effective, especially in *Leucaena*. It was further concluded that fertiliser-P would be needed for good nodulation, in spite of the high degree of mycorrhizal infection of the trees in the poor soil. Since the relative symbiotic efficiency (RSE) of *Gliricidia* was higher than that of *Leucaena,* the former should be recommended to avoid the need for inoculation.

Leucaena alleys in an acidic soil in M'Vuazi, Zaire also developed very poorly and had no effect on crop yield after three years. Inoculation tests with different strains of rhizobium showed an effect on plant height after 90 days for various strains, including an isolate from M'Vuazi. Since the M'Vuazi isolate had no effect on nodulation, however, and the plants remained yellow, it was concluded that it was not effective.

These are striking examples, giving researchers an early warning that the technology developed on-station might not perform under farmers' conditions. Researchers responded by attempting to adapt the farmers' environment to suit the technology, looking into inoculation or fertilization, rather than investigating what environments would fit the technology. Erratic performance under farmer conditions would later result in abandonment of the technology.

During many years of research, no indications were found that any of the auxiliary species used in cropping systems trials caused a serious build-up of plant-parasitic nematodes (Lowe, 1992). *Leucaena* fallow and alley cropping reduced the populations of all nematodes, but spiral nematodes were found to increase slightly under *Mucuna. Pennisetum* maintained populations of root-knot, root lesion and stunt nematodes and *C. juncea* significantly increased root-lesion nematodes, as did maize. For any future promising rotation or cover crop system, the effect on nematode populations would have to be monitored.

Several screening observations and tests were conducted over the years in the Ultisol zone of south-east Nigeria with potential hedgerow species for their adaptation to acidic soil conditions. In 1983, a long-term test was set up at the Onne station with five species, three leguminous (*Senna, Flemingia* and *Acacia mangium* Willd.) and two non-leguminous (*Dactyladenia* and *Gmelina arborea* Roxb.) with cassava as test crop. In 1986, cassava yields with *Dactyladenia* and *Senna* were not significantly better than the control, while *Gmelina* seriously suppressed cassava, and *Acacia* did not survive pruning. No information was given about *Flemingia*. Results for the years 1987–1989 were not found[15].

In collaboration with ILCA's small ruminants project, 13 varieties of *Gliricidia* were compared with three other (potential) hedgerow species in the Derived savannah of Oyo-North, viz., *Leucaena, Senna* and *Senna floribunda*. The latter two are commonly planted in settlements in the area. *Leucaena* and the two *Senna*s were most vigorous and appeared better adapted, but some *Gliricidia* varieties (ILG 55 and ILG 59) showed promise. Four species were tested for suitability as hedgerows in the semi-arid area of Rwanda, viz *Senna spectabilis, L. leucocephala, L. diversifolia* and *Calliandra*, with beans and sorghum as test crops. In the third year, *S. spectabilis* had a significant positive effect on the yields of both crops. In Chapter 4, all available screening results up to 1994 will be summarised.

ISSUES ON TRIAL DESIGN AND YIELD MEASUREMENT IN ALLEY CROPPING

In an article on a *Leucaena* alley cropping trial in the 1987 Annual Report, maize yield was calculated separately for rows bordering the hedges and for middle rows, both on a per hectare basis, to measure the border effect by the hedgerows on the crop. This might suggest that the reported yields were for net area, i.e. excluding the space occupied by the hedges. If that were the case in this and other trials, it would have a strong influence on the reported productivity of alley cropping. According to knowledgeable present and former staff, however, this would have been rarely the case, if at all. There are other pitfalls in experimentation with alley cropping, for example the requirement that the ratio between hedgerow- and cropped area is the same in the trial plots as in a full size alley cropping field. Furthermore, there is the issue of hedgerow roots and canopies extending beyond the plot boundaries, putting non-alley plots at a disadvantage. These are important questions for a fair assessment of the alley cropping technology, which have frequently been unheeded (Hauser et al., 2004). The training document developed by the Alley Farming Network for Tropical Africa (AFNETA) does not explicitly mention such issues (AFNETA, 1992).

3.3.3 Technology validation by farmers

It is useful at this point to make a distinction between technology validation carried out by researchers and by farmers in what used to be called on-farm research (OFR). The former is part of the conventional research-extension model whereby in the final stage of its development a technology is exposed to the biophysical conditions, but not to the management of the real farm. Most of the on-farm tests carried out during the first 15 years of the Institute's existence were of this type, with the exception of some of the alley cropping tests (Ngambeki and Wilson, 1984). On-farm research in the 1980s meant an approach whereby an important part of applied research is carried out at the real farm, technology is chosen or even designed in collaboration with the farmer according to their needs, and tested as much as possible by the farmers themselves. Station research would mainly have back-up support function (Figure 3.1).

In the early 1980s, IITA felt the need to create an in-house on-farm research capability, positioned somewhere between these two approaches, in order to expose its station-developed technologies to the conditions of the real farm, in other words: 'technology-driven OFR'. The Ford Foundation provided for two OFR positions, but only one agronomist was recruited. His (Ford Foundation-defined) mandate was to disseminate FSR/OFR concepts and methodologies in national research institutes in Nigeria and Ivory Coast, rather than test the IITA's technologies in farmers' fields. The latter task continued to be carried out by the scientists who had developed the technologies, in collaboration with FSP's social scientists and with scientists in the outreach programs. As a result, for a while, there were two distinct OFR flavours within the FSP, one dealing with testing of IITA technologies in farmers' fields and analysing the conditions to be satisfied for their adoption, the other promoting the adoption of OFR approaches by national research institutions through training and by working directly with NARS scientists to apply those approaches in the field. In the mid-1980s, the Ford Foundation-supported OFR unit also started their own OFR activities in collaboration with the University of Ibadan in a cluster of villages in south-west Nigeria (the 'Ayepe project'), as a facility for methodology development and training.

In the late 1980s, the two approaches began to converge, the Institute's OFR 'Unit' in Ibadan became integrated in RCMP's new 'systems teams' and AFNETA (Alley Farming Network for Africa) and most 'outreach' projects in West and Central Africa increasingly adopted 'farmer-participatory' approaches. One of them, the RAMR (*Recherche Appliquée en Milieu Réel*) project in Benin, in fact was in the forefront of the application and further development of OFR methods.

Soil fertility management

An obvious target for simple on-farm technology testing were variety-fertiliser combinations, which always appeal to farmers, and the Ayepe project and most 'Outreach Programs' across West and Central Africa embarked on such tests. Since they were usually carried out by many farmers, they opened up opportunities to study the variability of treatment effects across farmers and hopefully explain differences among them. The technique most used was Hildebrand's adaptability analysis (Hildebrand, 1984), whereby the yields for different treatments in each farmer's field are plotted against the average yield over all treatments in the same field, called the 'environmental index', or more neutrally the 'site-mean'. An example from the RAMR[16] project in Benin Republic is given in Figure 3.5. The fertiliser effect (the vertical distance between the solid and the broken lines) increased as the overall yield was higher, with the N-use efficiency increasing correspondingly (from 11.4 to 37.8 kg/kg). This pattern was very generally observed for a range of production factors and stressed the point that mean yields in (farmer-managed) on-farm trials are rather meaningless and need to be disaggregated, because treatment effects often vary strongly across farmers. The technique also allowed graphical estimation of the yield increment needed for a farmer to pay for the cost of the technology, and what percentage of the farmers actually did benefit.

Sometimes on-farm tests were accompanied by soil analyses to try and attribute the large variation in treatment effects to particular soil-related factors, often with little success, the soil effects being overridden by 'management factors' (e.g. Mutsaers and Walker, 1990). Also, the farmers' field choice will be based on their appreciation of a field's *overall* fertility status, which will often result in the absence of significant effects of variables, which have been found to be important in controlled trials. Length of fallow, for instance, will be correlated with crop yield in station trials, but not necessarily in on-farm trials, because farmers will choose fields which are 'ready' for a new cropping cycle, and the fallow needed to reach that point will be shorter in inherently fertile than in less fertile land. Also the negative effect of one fertility-related factor may be

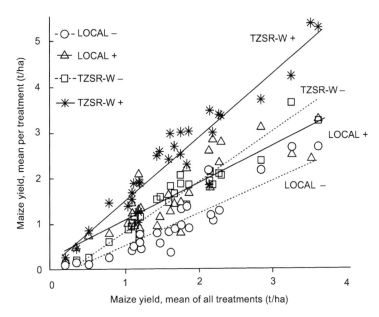

Figure 3.5 Individual treatment mean vs site mean yield, on-farm maize variety ×
fertiliser trial (adapted from Versteeg and Koudokpon, 1993) . '+' refers to 'with fertiliser'
and '-' refers to 'without fertiliser'

compensated by the positive effect of another, thereby reducing the statistically
measured effect of any single factor.

An interesting example of disaggregated test results, where the effect of
overall soil fertility and that of a single factor (soil P-status) stood out clearly
is from the Ndop Plain in north-west Cameroon, where a 'stepwise trial'
with maize variety (always associated with groundnuts), fertiliser and planting
density was carried out with 21 farmers scattered across the plain (RCMP,
1987). On the basis of mean maize yield, farmers were divided in three groups
with low, medium and high yield. The low yielders were all in the south with
'infertile red soils', while six of the seven high yielders were from the north
with 'more fertile brown/black soils'. Only the low yielders showed a P-effect.
The reason why these effects came out clearly was that the major difference
in soil fertility was a regional phenomenon, affecting all farmers rather than
something subject to farmers' choice. This demonstrated the (potential) power
of on-farm research to stratify farmer populations on the basis of their physical
(or economic) resources.

COVER CROPS AND LIVE MULCH

In spite of the large volume of research devoted to cover crops and live mulch,
no on-farm testing took place with these technologies during this episode,

even though smallholder farmers were frequently mentioned as the intended beneficiaries. The influential on-farm work on cover crops carried out by RAMR in Benin, started in 1988 (Versteeg and Koudokpon, 1993) and will be discussed in Chapter 4.

ALLEY CROPPING

In Nigeria, 32 existing alley cropping demonstrations and on-farm tests were evaluated in 1985, probably including some of the on-farm trials mentioned in Chapter 2, but only little concrete information was given on their status.

Three farmer-managed tests were started in 1984 in Ajaawa, near Ogbomosho in the Derived savannah. In 1985, maize in *Leucaena* alleys produced significantly better than in the unfertilised control. Yield in the fertilised control was not significantly higher than in the (non-fertilised) alleys (Table 3.22). In four sites in Zakibiam, maize also produced an additional 500 kg/ha of grain in the (unfertilised) alleys compared with the control, but there was no significant difference between alleys and no alleys when (a high rate of) fertiliser was applied to both. The yield figures point to an effect of hedgerow prunings which was equivalent to less than 75 kg of fertiliser-N/ha. The efficiencies of applied N were again quite low in both villages, especially in Zakibiam. Results from these trials were mostly interpreted through their mean treatment effects across participating trials, without considering the variability observed between farmers.

Many more on-farm alley cropping tests were started in Nigeria and in several outreach projects from 1985 onwards, testifying to the faith put into this novel technology. In 1985 and 1986, 12 alley cropping on-farm tests with *Leucaena* were started in 'IITA's own' on-farm research villages of Alabata/Ijaye, close to the Institute, of which eight were installed successfully. The degree

Table 3.22 Average maize yields in alley cropping on-farm tests in two savannah villages (Ajaawa and Zakibaiam) in Nigeria (RCMP, 1988).

Treatment	Ajaawa (3 sites)		Zakibaiam (4 sites)	
	Fertiliser application (kg/ha)[a]	Maize yield (t/ha)	Fertiliser application (kg/ha)[a]	Maize yield (t/ha)
No alleys	0	2.52	0	0.90
	75-30-30	3.26	90-45-45	2.05
Leucaena alleys	0	3.01	0	1.37
			90-45-45	2.22
SE[b]	0.138	0.185		

Notes:
a Applied as NPK 15:15:15 compound fertilizer plus a top dressing of urea; rates are in kg/ha of N, P2O5 and K2O.
b 'SE' = 'standard error'.

of researcher involvement in these tests could not be verified. On-farm alley cropping trials were also started in Benin Republic. More on these trials is reported in Chapter 4.

3.4 Technology delivery and dissemination

Technology delivery and dissemination is, of course, the ultimate aim of applied research, as international scientists were well aware. The conventional linear research-extension model, (Figure 3.1) however, kept technology development too far removed from the real farm. Nevertheless, there were some serious efforts during this episode to pave the way for technology to reach the farm, but they remained rather inconsequential.

3.4.1 Databases, technology digests, guidelines and decision support

On-farm researchers and extensionists need digested research results with clear indications where and how a technology may fit. It is the task of research to prepare such documentation in order that a technology can be put to the reality test of the farmer's field. In the early 1980s, guidelines were prepared for tillage systems including no-till (Lal, 1983, 1985) and for alley cropping (Kang et al., 1984). A decision matrix for tillage system in dependence of soil texture and climate by Lal (1985) is shown in Figure 3.6. Also a bulletin on land clearing methods for mechanised farms was announced, and a reduced tillage package was proposed for medium- and large-scale food crop farming in the Alfisol zone, consisting of three annual maize-cowpea cycles followed by a year of *Mucuna* fallow. Neither of these was published. In fact, there was no agreed policy on publishing user-oriented technology summaries and guidelines, which meant that research results, especially those relating to soil and soil fertility management, could end up buried in international journals without ever reaching the intended end-user. A project to prepare a source book on 'shelf technology', proposed by the OFR group as an aid for on-farm researchers did not materialise, because most scientists felt the technology was not yet ready.

3.4.2 Dissemination and monitoring and evaluation

The concept of technology dissemination continued to follow the conventional pathway, whereby extension would take over at the end of the development process. There were therefore few systematic attempts at dissemination, perhaps with the exception of the ILCA project about alley farming combined with livestock, although this was also basically about on-farm technology development. In the outreach projects in several countries, the on-farm tests had both testing and demonstration objectives, but there also, real dissemination was considered the task of the extension services. In essence, the role of soil and soil fertility researchers in technology dissemination remained minor and there

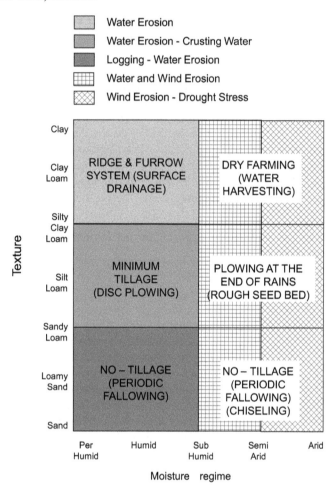

Figure 3.6 Suitable tillage systems in dependence of soil texture and moisture regime (adapted from Lal, 1985)

was no clear concept how this should be improved. There were a few cases where research interacted more closely with development projects, for example the CARDER[17] in Benin Republic.

3.5 Outputs and impact

3.5.1 Highlights of research outputs

The ecologies targeted were broadly defined as the humid and sub-humid zones of West and Central Africa. The sub-humid zone was further divided into forest fringe (where the main station was located), Derived savannah and Guinea savannah. The soils of the humid zone were mainly acidic Ultisols and

Oxisols with associated Inceptisols and Entisols, while the soils of the sub-humid zone were predominantly Alfisols, also associated with other soil classes along the toposequences. During the previous phase, a start had been made with more detailed soil mapping through the Benchmark Soils project (Chapter 2), but this work had not been completed. As a result, the information collected through multi-locational and on-farm technology testing could not be linked with reliable soil information and extrapolated to other areas. Such first-line correlation would have been useful, even though it would be overshadowed by the effect of other soil-, land use- and farmer-related factors.

Soil characterisation had now mostly turned to the Wetlands of West and Central Africa. A broad characterisation of wetland soils was produced, based on existing literature, as well as some more detailed work carried out in Nigeria and Sierra Leone. There was growing confusion about the soil classification system to be used. According to the IITA Annual Report of 1980, 'an approximate correlation of the hydromorphic soils among the three international classification systems has been prepared by IITA', but it had not been published.

This period could in some respects be called a time of 'involution' of technology development[18]. It was characterised by cyclical refinement and repackaging of several technologies under the artificial (and deteriorating) conditions of the Ibadan station, without exposing them effectively to the real farm. And when some technologies did get tested under farmers' conditions, it was mostly with a high degree of researcher supervision. Exceptions were the early on-farm alley cropping tests in the yam belt discussed in Chapter 2, and some of the new on-farm research which started around 1985 in Nigeria and Benin.

There had been a final upsurge of work on erosion and on land clearing and management, the latter with large and sometimes very complex trials, which would have to be conducted for several years to yield meaningful results. The complex combinations of land clearing methods with subsequent (and sometimes changing) cropping systems and tillage methods made it very difficult to draw firm conclusions on the effects of the many interacting factors.

The erosion studies did not produce much new insight beyond what was reported earlier, but the land clearing research showed that, for large-scale mechanised land clearing, the shear blade was the best option, provided it was combined with no-till. In case conventional tillage was chosen, the shear blade had to be combined with a root rake to fragment tree roots, but the advantage of shear blade clearing would be lost with continued mechanised tillage. After the departure of the principal soil physicist in 1987, soil physics work strongly diminished and practically ended in the course of the next episode (Chapter 4).

Summing up, however, the work on soil physics carried out since the late 1960s contributed significantly to the concept of conservation agriculture.

A lot of research was carried out on plant nutrients in both the acidic and non-acidic zones, which produced some interesting results, in particular on P-responses and on the effects of different forms of P, including rock phosphate. The work on liming confirmed that moderate amounts of lime of 0.5 to 1 t/ha once every three years was sufficient to correct acidity and Al-toxicity of Ultisols in south-east Nigeria. Much less work was done on micronutrients, although the low N-use efficiency in many trials suggested that there might often be nutrient imbalances, including micronutrient deficiencies or toxicities.

In respect of cropping systems, the running recommendation for intensive production was a cycle of three years of first season maize and second season cowpeas, followed by one year *Mucuna* fallow. Fallowing had been found indispensable under practically all conditions in several long-term experiments and observations.

The research on herbaceous and woody perennials was still work in progress, with generally positive reports on the performance of live mulch, cover crops and alley cropping. New species and varieties were being screened to optimise the technologies, but the conditions under which the technologies were tested only vaguely resembled real farmer conditions.

A listing of some detailed findings from this episode will be in order:

LAND MANAGEMENT

- The accumulated data on factors affecting soil erosion factors are yet to be converted into erosion hazard maps;
- Alfisols and soils with a high sand and silt fraction are sensitive to compaction, increasing with tillage method in the order: traditional farming < no-till < tractor ploughed < bare fallow; paraploughing combined with narrow strip tillage was the best remedy;
- Mechanised clearance resulted in long-term physical degradation at the Ibadan station, aggravated by mechanical tillage while the initial advantage of shear blade over tree pusher clearing was undone by mechanical operations in following years;
- No significant effect of tillage method was found in cassava on an Alfisol and an Ultisol, but mulching was needed on the Ultisol; ploughing was better for yams, probably because of its large tubers;
- In the Southern Guinea Savannah, manual tillage and disc-ploughing gave better maize yield than no-till; both mulching and residue burning gave higher maize yield than removal of crop residues; and
- Tied ridging in Kamboinse (Burkina Faso) gave better utilisation of rain water by crops which do not penetrate the compacted subsoil with a mixture with *Stylosanthes* penetrating deeper, possibly aiding the associated maize to access sub-soil moisture.

INLAND VALLEYS

- River overflow valleys and small inland valleys each cover 4–10% of the total land area in humid and sub-humid West Africa;
- The 'catchment factor' strongly affects the duration of wet conditions in the valleys; it varies from 100:1 to 5:1 and is higher in the drier regions, which allows rice growing in some inland valleys in the Sudan savannah;

- Iron toxicity in inland valleys can be reduced by intercepting seepage flow in the upper slope and draining the field at the end of the cropping season; and
- Improved bunds and waterways had a large effect on rice yield in Bida and Makeni wetlands.

FERTILITY MANAGEMENT

- N-use efficiency in maize varied across sites, with yield often attaining a ceiling value far below the environmental potential, a phenomenon that went largely unexplained;
- Calcium cyanamide showed promise as an N-source, because of lower leaching losses and the additional benefit of its weed control properties;
- Maize responded to P in many locations across vegetation zones and soil classes, usually up to 40–60 kg P_2O_5/ha (17.5-26.2 kg P/ha), and occasionally up to 90 $P^2)^5$ kg/ha (39.3 kgP/ha) with yield depression occurring at higher rates in acid soil due to 'induced Zn-deficiency';
- P-response of maize in a number of sites in the Derived and Southern Guinea savannah depended on whether the fertiliser contained S, confirming that sulphur is needed in many savannah locations;
- Sustained sole maize production with high fertiliser rates, especially with acidifying N-sources, caused serious degradation at the Ibadan station, while Mn-toxicity was suspectedly caused by declining pH;
- In a highly acidic Ultisol at Onne, 85% of maximum maize yield was attained with pH > 4.8, Al-saturation < 38% and Ca-saturation > 45%, requiring 0.5–1.0 t-ha of $CaCO_3$ once every three years (RCMP, 1987);
- Large amounts of N from fertiliser and mineralisation of SOM were lost by leaching from Ultisols at Onne;
- In most trials in the savannah, groundnuts and soybean responded to P up to 30–60 kg P_2O_5/ha (13.1–26.2 kg P/ha). Yield sometimes declined at higher rates, probably by nutrient imbalances;
- In Bas-Zaire, groundnut responded strongly to 500 kg/ha of burnt limestone; in on-farm trials, response to fertiliser depended on the presence of lime;
- Mixtures of cassava with groundnuts, cowpeas or maize were mostly more favourable in terms of combined yield than with soybean, except for very vigorous maize;
- Increasing P-rates may reduce the advantage of cassava-legume mixtures because of the higher ability of cassava to extract P and the stronger response of cowpeas to P; and
- The combined yield advantage of cereal-grain legume intercropping is not due to direct N-contribution of the legume to maize.

AUXILIARY CROPS

- Pigeon pea is not suitable as a supplementary crop for the maize+cassava mixture;

- Among *Mucuna, Pueraria* and *Psophocarpus* only *Mucuna* could significantly reduce *Imperata*, the other two did not develop rapidly enough;
- The maize yield increase following *Mucuna* or *Pseudovigna* fallow did not compensate for the 'lost' maize season during the legume fallow in Benin;
- The number 1 (technical) challenge with live mulch was to find a proliferous, but non-climbing species. *C. verrucosa* was the latest candidate;
- On an Apomu soil, unfertilised maize did better in alleys than in non-alley plots, with an N-effect of the alleys equivalent to 40 kg N/ha while in an Egbeda soil this equivalent advantage was maintained when 45 kg/ha N was applied to the alleys; similar results were obtained in on-farm trials in the Derived and Southern Guinea savannah of Nigeria;
- Tilled alley plots did considerably better than no-till, probably due to root competition in the no-till plots;
- In an on-farm trial in Alabata, both maize and cowpeas were depressed by *Leucaena*, while *Gliricidia* alleys had no effect, while in Gandajika, Zaire, there was no effect of the alleys on maize yield, with yield depression in the rows close to the hedges; in acidic soil in M'Vuazi, Zaire *Leucaena* developed poorly and had no effect on crop yield after three years;
- Better distributed N release of *Senna* prunings may explain that species' favourable performance as a hedgerow;
- In plantains, the effect on plantain yield of mulch was in the order *Flemingia* < *Chromolaena* < *Pennisetum;* to produce sufficient mulching biomass *in-situ* more than 40–50% of the area was needed;
- Hedgerow prunings are best applied close to planting of the crop. N-utilisation was found to be low at less than 10% of the released N;
- Fertiliser-P would be needed in the savannah for good nodulation and indigenous rhizobia may not be effective, depending on the hedgerow species;
- Results of tests and screening of herbaceous and woody leguminous species for their properties and suitability as auxiliary crops are summarised in the next chapter; and
- Statistical methods used for the analysis of farmer variability were not sufficiently powerful.

3.5.2 Uptake and impact

The uptake of the Institute's soil and soil fertility management technologies was minimal or absent at this stage, perhaps with the exception of fertiliser recommendations, but there is no record of such adoption. More systematic on-farm testing was going to be undertaken in coming years. At this point, it will be useful to consider the ecologies and the types of farming these technologies were actually targeting.

In respect of the farming systems the technologies were meant for, three categories can be distinguished: (i) technologies targeting smallholder farmers, (ii) technologies targeting medium- to large-scale farms and (iii) scale-neutral technologies. None of the Institute's soil and land use technologies were exclusively suitable for its main target group, the African smallholders, but scale-neutral technologies would be expected to suit both smallholders and larger operators. In this category, we find results on nutrient management as well as alley cropping which was being tested with smallholders across Nigeria and was beginning to be tested outside Nigeria as well.

Herbaceous legumes for fertility improvement and for the control of difficult weeds could also be useful both for small and for larger farmers, but the need to give up a season to establish a green manure crop for future gains could hamper adoption by smallholders. Also, some trials showed that the yield gains did not necessarily compensate for the foregone crop. The recommended cropping pattern of three years maize-cowpeas followed by one year of *Mucuna* fallow, would in theory also be scale neutral, but the central role of sole cowpeas in the rotation make it very risky for smallholders in view of its extreme sensitivity to insects (e.g. Mutsaers, 1991).

Live mulch had so far been tested only with a tall, erect, short season crop: maize, grown in both seasons at Ibadan. Short stature crops like cowpeas or vegetables would be overgrown, as even maize was without growth retardants. Compatibility with cassava had not been tested yet. Suitability of live mulch for smallholders was therefore questionable, but that should not prevent its testing under their conditions. At its present stage of development, it could already be attractive for larger operators, growing one or two maize crops per year.

The work on land clearing, tillage systems and erosion control plainly targeted medium- to large-scale mechanised farming, in spite of the Institute's choice for the African smallholder as its target. Some of the research in these areas was carried out in collaboration with ADPs, which were promoting a mechanised medium-scale farming model for which these technologies might have been suitable, but there is no record of actual successful establishment of such farms. The target group therefore remained largely hypothetical in West Africa.

3.6 Emerging trends

In this episode RCMP research went further afield, both within and outside Nigeria. Outreach projects were started in several countries in West and Central Africa funded by donors who would sometimes bring in their own research partners with their own objectives. The outreach projects to a large extent worked autonomously and set their own project goals. In spite of their name, their work on soils and land use was not always consistent with RCMP's core projects. As a result, even though the outreach projects may have served a useful purpose for their host country, they did not benefit enough from RCMP expertise, with the exception of alley cropping, nor did they contribute as much to the body of knowledge as would have been possible. One of the reasons was the lack of a firm framework underlying the research on soils and land use, carried out across sub-Saharan Africa, in which the work in the outreach projects could have fitted. It is symptomatic for the situation that a document on agro-ecological characterisation prepared in 1986 in support of the development of a new Medium Term Plan barely mentions the Institute's own work on soil characterisation (Goldman, 1986).

On-station research continued essentially in the same mode, often under high inputs and pest and weed control, resulting in yields which were generally much higher than in farmers' fields, and using crop rotations which were rarely if ever used by farmers, such as the maize-cowpea rotation. Research on land clearing and tillage systems, a high-profile and extensively published line of work of the first 20 years, eventually came to an end by 1990, with the departure of the lead scientist, which typifies the rather personalised nature of the research agenda at the time. Alley cropping research continued at the forefront, with organic matter dynamics as a new component.

Meanwhile, under the influence of the international FSR movement, new on-farm research activities were undertaken with the farmers' own needs rather than the technology as point of departure, even though the Institute's technologies remained the researchers' main resource. The RAMR project in Benin Republic and the Ayepe OFR site in south-west Nigeria were in this category. Furthermore, the large AFNETA, which was launched at an international meeting in 1986, was also going to adopt the same FSR/OFR approaches. AFNETA was expected to find out once and for all if and where alley farming would have a future, by testing the technology with farmers in their fields across Africa.

Standard statistical methods used at the time were not adequate to explain differences in yield levels among farmers and how these differences affected the performance of technologies in their fields, except in clear-cut cases such as the one from north-west Cameroon. Only much later were statistical tools introduced which made it possible to relate (potential) yield to individual factors from the results of large numbers of on-farm trials (Chapter 6).

Not everyone was impressed by the promise of the new FSR/OFR approaches, however, and some of IITA's leadership of the mid-1980s remained convinced that developing technology on-station and demonstrating it to farmers was the way to go. The toposequence project of the Director General set up at the Ibadan station (Section 3.3) and the idea of the Deputy Director General for Research to set up a Village Improvement Project near the Institute with a plethora of IITA technologies, are examples of past ideas lingering around.

At the end of this period, a new Medium Term Plan 1989–1993 was published (IITA, 1988), the outcome of a Strategic Study which had been started in 1985. Its main elements were as follows:

- Primary emphasis was to be put on the humid and sub-humid lowlands of West and Central Africa;
- Focus would be on the smallholder family farmer, which probably reflected concern that this had not really been the case in the past;
- Research would be decentralised by the creation of satellites in key ecological zones, one located in the humid zone, two in the savannah zone, one in an inland valley and one at mid-altitude in East or Southern Africa to extend

cassava research into that area; and
- A 'farming systems orientation' was going to be adopted, involving resource management, commodity improvement and social sciences.

The Plan further stipulated as the first priority 'the synthesis of technologies developed [...] into systems to be tested on-farm by IITA crop-based systems working groups and collaborating national scientists' and to set an accelerated process in motion to develop and test on-farm appropriate technologies for the acidic soils in the humid zones. Alley farming should receive special emphasis as 'the most promising sustainable system to emerge from the Institute's research on resource management'.

Two major elements of the Plan were made operational immediately:

- Three Crop-based Systems Working Groups were created bringing together resource management and crop improvement researchers; and
- AFNETA, the international alley farming network, sponsored by IITA, ILCA, the World Agroforestry Center (ICRAF) with support from Canada, Denmark and the United States, started a full-fledged program with the appointment of a full-time coordinator in 1989.

Notes

1 An ad-hoc 'basket' for studies on post-harvest techniques, women cassava groups, the economics of maize research, amongst other, created for internal management reasons.
2 The Wetlands Utilisation Research Project, in collaboration with three Dutch research institutions, funded by The Netherlands; the Nitrogen and Phosphorus Utilization Project with International Fertilizer Development Center (IFDC), also funded by The Netherlands; the Nutrient-use efficiency project with Buntehof, Germany; The studies on the dynamics of soil organic matter and soil fertility under different fallow and cropping systems, especially in alley cropping, the first collaborative project with KU Leuven and IITA; the Land Clearing project at Okomu, Nigeria, with the United Nations University, and in Kwara State, Nigeria, with Iloirin Agricultural Development Project (ADP); soil erosion studies in Nigeria with Munich and Ghent University and in Cameroon with GTZ; and the soil and water management project in Burkina Faso, funded by Semi-Arid Food Grains Research and Development (SAFGRAD).
3 The testing and Liaison Unit in Cameroon (USAID); the On-Farm Research project in Cameroon with IRA (IDRC); the FSR/OFR project in Rwanda (World Bank); the Applied Agricultural Research program in several areas of Zaire (USAID); the On-Farm Research Project (RAMR) in Benin Republic (Netherlands and the Near East Foundation).
4 Since one-third of the total wetland area covers an estimated 4-10% of total 'inventory area' (suitable for agriculture?), total wetland area would occupy 12-30% of total (IITA, 1984), which seems rather too high.
5 This appears to happen more frequently – scientists leaving and the programs they were leading discontinuing. One would expect to have more strategic reasons for investing in certain research themes besides priorities or interests of individual scientists.
6 The nutrient contents of NPK compound fertiliser are given in % N, % K_2O and % P_2O_5.

7　For the correlation between the USDA Soil Taxonomy and the WRB classifications, see Annexe II.

8　The effect on millet yield was not given.

9　Sols ferrugineux are equivalent to Alfisols.

10　It is not clear what caused the increase of bulk density in the forested control.

11　Results for Alabata where the test was also conducted were not given for undisclosed reasons.

12　Probably because of the departure of the lead scientist.

13　'Modified', because it also corrects for a long dry season.

14　This effect could not be replicated in some other trials, though (Hauser, pers. comm.)

15　No detailed RCMP Annual Reports were published for 1988 and 1989.

16　Recherche Appliquée en Milieu Réelle.

17　Centre d'Action Régionale pour le Développement Rural.

18　A term coined by Clifford Geertz (1963) to describe the development of Javanese sawah-based agriculture in the early 20th century into an ever more complex and intensive system of production.

References

AFNETA, 1992. *The AFNETA Alley Farming Training Manual. Volume 1, Core Course in Alley Farming; Volume 2, Source Book for Alley Farming Research.* IITA, Ibadan.

Andriesse, W, 1986. Area and distribution. In: A.S.R. Juo and J.A. Lowe (eds), *The Wetlands and Rice in Subsaharan Africa.* IITA, Ibadan.

FSP, 1986. *Annual Report 1985. Farming Systems Program,* IITA, Ibadan.

Geertz, Clifford, 1963. *Agricultural Involution. The Processes of Ecological Change in Indonesia.* University of California Press, Berkeley, CA.

Goldman, A., 1986. *Agroecological characterization and its relation to research issues in West and Central Africa.* Internal Report, IITA, Ibadan.

Grimme, H. and A.S.R. Juo, 1985. Inorganic nitrogen losses through leaching and denitrification in soils of the humid tropics. In: B.T. Kang and J. van der Heide (eds), *Nitrogen Management in Farming Systems in Humid and Subhumid Tropics.* Institute for Soil Fertility, Haren, The Netherlands.

Gunneweg, H.A.M.I., A. Evers and A. Huizing, 1986. A model to assess proposed procedures for water control, application and results for two small inland valleys. In: A.S.R. Juo and J.A. Lowe (eds), *The Wetlands and Rice in Subsaharan Africa.* IITA, Ibadan.

Hauser, S., C. Nolte and R.J. Carsky, 2004. What role can planted fallows play in the humid and sub-humid zone of West and Central Africa? *Nutrient Cycling in Agroecosystems,* 76, 297–318.

Heide, J. van der, A.C.B.M. van der Kruijs, B.T. Kang and P.L. Vlek, 1985. Nitrogen management in multiple cropping systems. In: B.T. Kang and J. van der Heide (eds), *Nitrogen Management in Farming Systems in Humid and Subhumid Tropics.* Institute for Soil Fertility, Haren, The Netherlands.

Hekstra, P., W. Andriesse, C.A. de Vries and G. Bus, 1983. *Wetlands Utilization Research Project, West Africa. Phase I, The Inventory* (4 volumes). ILRI, Wageningen.

Hildebrand, P.E., 1984. Modified stability analysis of farmer-managed on-farm trials. *Agronomy Journal,* 76, 271–274.

Hulugalle, 1994. Long-term effects of land clearing methods, tillage systems and cropping systems on surface soil properties of a tropical Alfisol in S.W. Nigeria. *Soil Use and Management,* 10, 25–30.

IITA, 1983. *Annual Report 1982.* IITA, Ibadan.

IITA, 1983–1985. *Annual Reports 1982–1984*. IITA, Ibadan.

IITA, 1984. *Annual Report 1983*. IITA, Ibadan.

IITA, 1985. *Annual Report 1984*. IITA, Ibadan.

IITA, 1988. *Medium Term Plan, 1989–1993*. IITA, Ibadan.

Juo, A.S.R. and J.A. Lowe (eds), 1986. The Wetlands and Rice in Subsaharan Africa. *Proceedings of an International Conference on Wetlands Utilization for Rice Production in Sub-Saharan Africa*, 4-8 November 1985. IITA, Ibadan.

Kang, B.T. and A.S.R. Juo, 1983. Nitrogen and potassium response of cassava. *IITA Annual Report 1982*. IITA, Ibadan.

Kang, B.T., Wilson, G.F. and Lawson, T.L., 1984. *Alley Cropping: A Stable Alternative to Shifting Cultivation*. IITA, Ibadan.

Kosaki, T. and A.S.R. Juo, 1986. Iron toxicity of rice in inland valleys: a case from Nigeria. In: A. S. R. Juo and J. A. Lowe (eds). *The Wetlands and Rice in Subsaharan Africa*. IITA, Ibadan.

Lal, R., 1983. *No-till Farming. Soil and Water Conservation and Management in the Humid and Subhumid Tropics*. Monograph no. 2, IITA, Ibadan.

Lal, R., 1985. A soil suitability guide for different tillage systems in the tropics. *Soil & Tillage Research, 5*, 179–196.

Lal, R., 1993. Conversion of tropical rain forest and agricultural sustainability in the humid tropics: A case study at Okomu in southern Nigeria. In: Juha I. Uitto and Miguel Cliisener-Godt (eds), *Environmentally Sound Socio-economic Development in the Humid Tropics. Perspectives from Asia and Africa*. United Nations University, Tokyo.

Leihner, D., 1983. *Management and Evaluation of Intercropping Systems with Cassava*. Centro Internacional de Agricultura Tropical, Cali, Colombia.

Lowe, J., 1992. *Nematological Research at IITA, 1969–1988*. Plant Health Management Research Monograph No. 2. IITA, Ibadan.

Mutsaers, H.J.W., 1991. *Opportunities for second season cropping in southwestern Nigeria*. RCMP Monograph No. 4. IITA, Ibadan.

Mutsaers, H.J.W., H.C. Ezumah and D.S.O. Osiru, 1993. Cassava-based intercropping: a review. *Field Crops Research*, 34, 431–457.

Mutsaers, H.J.W. and P. Walker, 1990. Farmers' maize yields in S.W. Nigeria and the effect of variety and fertilizer: an analysis of variability in on-farm trials. *Field Crops Research*, 23, 265–278.

Ngambeki, D.S. and G.F. Wilson, 1984. *Economic and on-farm evaluation of alley farming with* Leucaena leucocephala, *1980–1983*. Activity Consolidated Report, IITA, Ibadan.

Pleysier, J.L., Y. Arora, A.S.R. Juo and T.L. Lawson, 1983. Calcium cyanamide as a nitrogen fertilizer. *IITA Annual Report for 1982*IITA, Ibadan

RCMD, 1993. *Annual Report 1992. Highlights of scientific findings*. Resource and Crop Management Research Monograph No. 15. IITA, Ibadan.

RCMP, 1986–1988. *Annual Reports 1985–1987*. IITA, Ibadan.

RCMP, 1991. *Annual Report 1990. Highlights of scientific findings*. Resource and Crop Management Research Monograph No. 7. IITA, Ibadan.

RCMP, 1992. *Annual Report 1991. Highlights of scientific findings*. Resource and Crop Management Research Monograph No. 12. IITA, Ibadan.

Versteeg, M.N. and V. Koudokpon, 1993. Participative farmer testing of four low external input technologies to address soil fertility decline in Mono département (Benin). *Agricultural Systems, 42*, 265–276.

Wong, M. T. F., A.C.M.B van der Kruijs, and A.S.R. Juo, 1992. Leaching loss of calcium, magnesium potassium and nitrate derived from soil, and fertilizers as influenced by urea applied to undisturbed lysemeters southest Nigeria. *Fertilizer Research*, 31, 281–289.

4 Towards farming systems
1989–1994

4.1. Scope, approaches and partnerships

During this episode, the three-pronged approach of the Medium-Term Plan 1989–1993 (Chapter 3) was going to be implemented: (i) focusing on the smallholder family farmer, (ii) decentralisation by the creation of satellites in key ecological zones and (iii) a 'farming systems orientation' permeating all of the Institute's research. As described in the 1988 Annual Report, technology development would continue to follow the conventional research-extension pathway:

> a three-stage process that begins with testing on the experiment stations ... followed by transfer of promising technologies to actual fields ... information from the trials returns to the scientists ... for further rounds of improvement before the technology is ready for release to farmers through the National Agricultural Research Systems (NARS).

Decentralisation was a major organisational change, which would involve the gradual transfer of research activities to three satellite stations: an expanded Onne station in the deforested humid Ultisol zone of Eastern Nigeria, a strengthened farming systems capability at the Cotonou substation in the Lixisol[1] zone of coastal Benin Republic and a new Humid Forest Eco-regional Centre at M'Balmayo, Central-South Cameroon, for the forested humid Acrisol/Ferralsol zone. Soil and soil fertility research in Onne and Cotonou continued to bear its original signature, with Onne working mainly on soil fertility management issues of Acrisols and Cotonou focusing on demand-driven On-Farm Research (OFR), backed-up with on-station research as needed. In the new Humid Forest Station (HFS) a comprehensive research program was initiated, starting with soil and farming systems surveys, to be followed by the elaboration of a program of on-station and on-farm soil and land use research.

Outreach programs in several countries continued to carry out on-station and on-farm activities on fertility management and cropping systems according to the perceived needs of their localities. Apart from individual exceptions, the linkages with the soils and agronomy group at the 'Centre' continued to remain rather weak, until most of the projects were closed down in 1993.

In 1991 the Resource and Crop Management Program (RCMP) was renamed Resource and Crop Management Division (RCMD), with two programs: a Resource Management program and a Crop Management Research Program. The Resource Management program's task was to continue studying the physical and socio-economic resource base and developing new technologies (IITA, 1991). It harboured the soil research team, whose composition had changed significantly, from soil physics and soil chemistry to soil biology and agro-climatology/modelling, with soil fertility maintaining its former strength. The character of the work is now better described as 'soil and soil fertility research' than the previous 'soil and land use' research. The Crop Management Program, consisting of agronomists and social scientists, encompassed the three former 'crop-based systems working groups' at the Ibadan Station. They were now renamed Humid Forest Systems, Savannah Systems and Inland Valley Systems Groups. Their focus was on cassava (*Manihot esculenta*)-based, maize (*Zea mays*)-based and rice (*Oryza sativa*)-based systems, respectively (RCMP, 1992).

In the late 1980s, soil and soil fertility research presented a complex picture, consisting of several, more or less coordinated elements, most of them inherited from earlier periods:

- On-station research on soil fertility and cropping systems at the Ibadan and Onne stations and some multi-locational testing;
- On-station and on-farm testing of soil fertility and cropping systems technologies in outreach projects in Cameroon, Rwanda and Zaire (currently Democratic Republic of Congo);
- Farming systems surveys and on-station technology testing in Cameroon;
- OFR in south-west Nigeria in collaboration with the University of Ibadan, using the Institute's technology toolbox;
- OFR in Benin Republic, backed up by experiments at the Cotonou station when needed;
- On-farm testing of alley cropping in Nigeria; and
- Alley farming research initiated by the Alley Farming Network for Tropical Africa (AFNETA) in collaboration with NARS in several African countries.

These disparate elements had to be incorporated into coherent 'holistic' eco-regional programs, which would form the framework for the next Medium Term Plan (1994–1998). It was inspired by the new CGIAR-wide agro-ecological thinking, initiated by Technical Advisory Committee (TAC) in response to the centres' increasingly complex tasks (CGIAR/TAC, 1992). The first step towards that goal was another reshuffle of the RCMD in 1992, into three eco-regional programs: the Moist Savannah Program, the Humid Forest Program and the Inland Valley Systems Program. Resource management no longer functioned as a separate program, but was integrated, together with crop management, in the three ecoregional programs, supported by a new Agro-ecological Studies Unit, the successor of the former Agroclimatology Unit.

The three new programs then redefined their research goals and the tools to be used. The first step would be a systematic characterisation of their target zones as a basis for the development of technology with an increased likelihood of adoption. Sustainability of cropping systems became a key concept for which a working definition was adopted, namely,

> a cropping system [that] has an acceptable level of production of harvestable yield which shows a non-declining trend from cropping cycle to cropping cycle and is resistant in terms of yield stability to normal fluctuations of stress and disturbance over the long term
>
> (RCMD, 1993).

The research model, incorporating some of the ideas from the international Farming Systems Research (FSR) movement, would no longer follow the linear research-extension-farmer approach but involve an iterative phase of on-farm testing and adaptation, together with farmers in their fields. It was explicitly stated that in order 'to ensure the adoptability of results [there had to be] strong interaction between researchers and client farmers at all stages of technology development' (RCMD, 1993). And since NARS were directly responsible for linkage with farmers, this would imply stronger collaboration with NARS scientists (Figure 4.1). This was the model which had been used in the Ayepe OFR site in south-west Nigeria, operated in collaboration with the Agronomy Department of the University of Ibadan and the Government Extension Service. The site was handed over to the university in 1991, with a Ford Foundation grant allowing the university to continue the work and provide a training ground and research opportunities for its students.

In spite of the formal RCMD policy to move a significant part of technology development and testing on-farm, much of the work on soils and soil fertility continued to be conducted at the research farms or in researcher-controlled experiments in farmers' fields. Also, there remained a distinct separation of roles, with the NARS being responsible for on-farm testing and dissemination, and IITA taking the lead in technology generation. The research agenda continued to be set by the scientists (Section 4.3) on the basis of what they saw as the needs of the end user, identified by regular contacts, field visits and off-station trials (Figure 4.1).

Meanwhile, characterisation of the biophysical and socio-economic environment assumed a more prominent position focusing on a number of Benchmark areas, for more systematic targeting of the technology development process. This time, however, the characterisation was carried out by mixed teams of physical and social scientists in order to assess farmers' needs for technology in a more integrated manner.

At the regional level, RCMP's Crop-based Systems Groups initiated international collaborative associations, viz: the Collaborative Group on Maize-Based Systems Research (COMBS), created in 1989, and the Collaborative Group for Root and Tuber Crops Improvement and Systems (CORTIS),

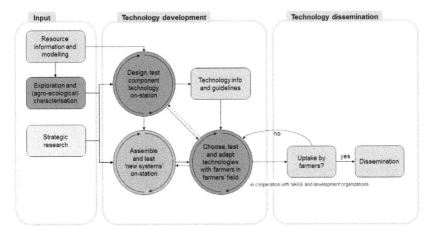

Figure 4.1 IITA's technology development and dissemination model, 1989–1994, indicating a shift towards closer involvement of farmers and more collaboration with National Agricultural Research Systems (NARS)

Shading indicates relative emphasis on the various components with darker ones indicating a relatively higher emphasis

created in 1991. Both of them emphasised on-farm technology testing and promoted research on the incorporation of herbaceous legumes into farmers' cropping systems.

Two other regional initiatives were launched or expanded during this episode, the Collaborative Study for Cassava in Africa (COSCA) and the AFNETA. The aim of COSCA was the Continent-wide assessment, in collaboration with NARS, of the status of cassava production, including the agro-ecological characterisation of cassava growing areas. After its launch in 1986, AFNETA was upgraded to an international collaborative program in 1989, to study the potential of alley farming across the humid and sub-humid zones of sub-Saharan Africa (SSA). It would eventually be transferred to the World Agroforestry Centre (ICRAF) in 1994.

The outreach programs, embedded in national research systems in several countries (Cameroon, Zaire – Rwanda, Burkina Faso), one of the pillars of the IITA structure in previous years (Chapter 3), continued for some years, but closed down between 1990 and 1994. This proliferation and geographical spread of collaborative projects of a very different nature was symptomatic of the lack of coherence of the Resource and Crop Management programs of the time (Figure 4.2).

4.2. Characterisation of soils, farms and farming systems

Stimulated by the new CGIAR-wide agro-ecological thinking, a program was set in motion during this episode to more systematically delineate and

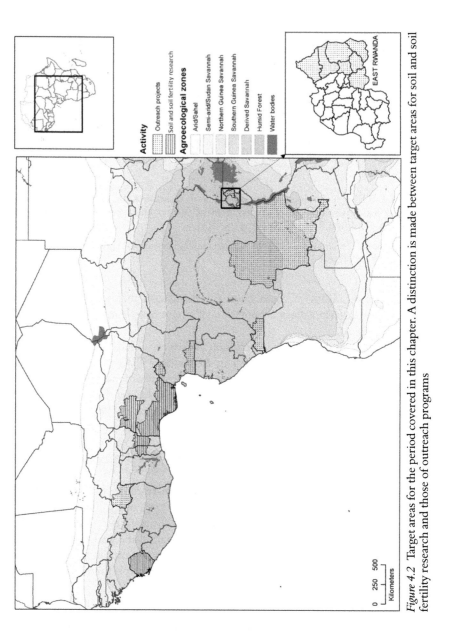

Activity

Outreach projects

Soil and soil fertility research

Agroecological zones

Arid/Sahel

Semi-arid/Sudan Savannah

Northern Guinea Savannah

Southern Guinea Savannah

Derived Savannah

Humid Forest

Water bodies

EAST RWANDA

0 250 500
Kilometers

Figure 4.2 Target areas for the period covered in this chapter. A distinction is made between target areas for soil and soil fertility research and those of outreach programs

characterise the agro-ecological conditions and the agricultural systems in the humid and sub-humid zones of West and Central Africa. All three systems groups worked on this characterisation from the late 1980s, each of them initially using more or less its own approaches. The Savannah System Group, however, had the ambition to develop a more generic methodology which could be applied throughout the mandated zones. New tools were introduced using modern computerised database, mapping and modelling techniques and new ideas were developed for more systematic technology targeting. The Agro-ecological Studies Unit's Geographic Information System (GIS) and modelling capabilities underlined the interest in modern quantitative methods for the delineation of more or less homogeneous zones on the basis of ecological and socioeconomic criteria. The aim of system characterisation was to align the technology development process with the productive potential and limitations of these zones. The other two Programs continued to use their own methods, which predated the new methodology, but with time would absorb insights contributed by the savannah group.

The Agro-ecological Studies Unit, which was to play a key role in the new characterisation drive, first developed a Resource Information System (RIS) consisting of a GIS and a database component, as a flexible tool for the storage, analysis and mapping of physical and socioeconomic features at different scales. Databases were compiled on climate, soils, vegetation, population, etc. from various sources. This included the FAO Soil Classification system, which meant a break with the past when all soil characterisation had been based on the Soil Taxonomy. The intention was further to incorporate the Institute's own accumulated data on soils, crops, farming systems and socioeconomic features into the RIS database (IITA, 1992). A start was also made in using dynamic crop modelling to calculate potential or attainable crop yield as an important environmental parameter.

4.2.1 Characterisation in the Humid Forest Zone

Once the Humid Forest Station in Cameroon had been created, characterisation of the soils at the station was carried out to assess its representativeness for Cameroon's humid forest zone. In a simultaneous reconnaissance survey of 200 sites in Southern Cameroon, five soil profile classes were identified, on the basis of soil colour and surface texture (Table 4.1).

Detailed soil sampling was carried out in 121 sites, only four of which did not belong to any of the five classes. Chemical analysis showed that the 'charge finger print' of the soils in each of the groups was very similar, with a low permanent negative charge of about 2 $cmol_c/kg$, due to dominance of kaolinite and oxides in the clay fraction, increasing to 4–5 $cmol_c/kg$ at pH 5.5. This meant that fertility management for cations would be similar for all groups and needs to take account of potentially high losses due to leaching. Furthermore, Al-saturation increased and exchangeable Ca decreased strongly with depth, which explains the limited root growth below the surface soil.

Table 4.1 Characteristics of five soil profile classes identified in south Cameroon (Menzies and Gillman, 1996)

Soil profile class	Munsell colour	Texture		Principal parent material
		Surface[a]	Subsoil[a]	
Akonolinga	10YR 4/6	FSL-FSCL	MC-MHC	Mica, schist, gneiss
Ebolowa	10YR5/6	MC	MC-HC	Granite, granodiorite
Sangmelima	2.5YR8-5YR	CL-CL	MC-MHC	Graniti rock complexes
Yaounde	10YR 3/3	C	C	Migmatite, gneiss
Mbalmayo	7.5YR 5/6	SL-SCL	MC-HC	Chlorite, greenschist

Note:
a 'C' = clay; 'FSL' = fine sandy loam; 'FSCL' = fine sandy clay loam; 'HC' = heavy clay; 'MC' = medium clay; 'MHC' = medium heavy clay.

The main difference between the soil groups was in their P-sorption curves, which determine the P-fertilisation rates needed to attain a given P-concentration in the soil solution. In most soils the amount of P needed to attain a solution concentration of 2 mM phosphate is very high: between 80 and 250 kg P/ha. The S-sorption curves also differed but most soils were considered to have adequate S-reserves (Menzies and Gillman, 1997). The soils in the HFS belonged to two of the profile groups, which represented 50% of the soils in Cameroon's humid forest zone (RCMP, 1992).

In 1992, a further resource use survey was carried out in 20 villages chosen on the basis of a geographic grid for soils and native vegetation. Using the Smith-Weber-Manyong methodology (see Section 4.2.2) as a framework, the survey sought to identify the factors which would affect adoption of new technology. The intention was to use this type of survey also in other parts of the humid forest zone. Although the findings may have influenced the research program, they were never formally published (Hauser, pers. comm.).

Finally, in 1994 the Humid Forest team and the Cameroonian NARS defined and characterised a wider benchmark area of 15,000 km² in the forest margins of Cameroon. It covered a resource-use gradient extending from hunting and gathering to intensive peri-urban production, later called the Forest Margins Benchmark, and was intended to become the program's intervention and extrapolation area.

4.2.2 Characterisation in the Northern Guinea Savannah

In the late 1980s the Savannah Group concentrated its activities in an area of relatively high potential, the Northern Guinea Savannah (NGS). The phenomenal expansion of intensive maize growing in part of the NGS since the 1970s and the decline of fallowing were leading to a loss of Soil Organic Matter (SOM) and increasing evidence of S and Zn deficiencies. Initially, the group worked on a more precise physical characterisation of maize growing environments in the NGS as a tool for technology targeting.

The physical environment of agriculture had usually been characterised by breeders on the basis of climatic factors and altitude for the choice of testing sites for their varieties. A multivariate analysis of Genotype × Environment (G × E) interaction using 26 international maize trials showed that statistical clustering of the testing sites did not reflect the definition of 'maize growing environments' very well. It did not consider soil characteristics, which appeared to affect the performance of varieties in some environments, along with management practices (RCMP, 1991). The Agro-ecological Unit tested the CERES-Maize model as a potential tool to explore the likely effects of environmental and crop factors on crop productivity, an ambitious objective considering the model's state of development at the time. The model was capable of simulating overall yield potential of maize under different environments, but it could not account for the site-specific behaviour of different varieties. In other words, it did not effectively capture G × E interaction (RCMP, 1991, 1992).

A number of constraints were known to affect crop production in the NGS: *Striga*, nematodes, fusarium wilt, maize streak and S and Zn deficiency, and multivariate methods were tried to estimate their relative influence on crop yield in farmers' fields. Principal component analysis of the data from a number of diagnostic trials and surveys showed that cation exchange capacity, exchangeable Ca and Mg, organic matter, pH and texture explained 75% of soil variability in the area (RCMP, 1992). All these variables taken together are almost synonymous with 'soil fertility', which is not of much help when trying to identify specific leverage points for the improvement of crop yield (e.g. Chapter 3 and Carsky et al., 1998a).

In a further study in Northern Nigeria in 1991 and 1992, no correlation was found between maize yield and organic carbon, total N, pH, available P or micronutrients in farmers' fields, but there was a significant correlation on sandy loam with nitrate-N at 0–15 cm depth between two and five weeks after planting, but less so on more sandy soils. As nitrate-N generally peaked early in the rainy season, early planting was thought to be a wise farmer strategy (Weber et al., 1995a). This work pointed to the need for a better understanding of the soil conditions in farmers' fields in order to synchronise crop management with nutrient dynamics and choose suitable soil amendments. More balanced cropping systems with grain legumes in the rotation were thought to be needed, but current production and consumption of legumes were found to be very low. In order to increase their role in the system, their attractiveness as cash crops and their productivity would have to be increased (RCMP, 1992).

Simultaneously with the studies on the physical production factors, the Savannah Systems Group worked on the development of a methodology for the agro-socio-economic characterisation of agricultural systems. In 1992, a broad survey of the major farming systems was carried out in six countries. In Benin Republic, for example, five systems were distinguished, viz. oilpalm (*Elaeis guineensis*)-based, maize/cassava- or maize/yam (*Dioscorea* sp.) -based, yam-based, cotton (*Gossypium hirsutum*)-based and sorghum (*Sorghum bicolor*)-based. Later the number was increased to 13 (Manyong et al. 1996). In each system key constraints were identified, related to the resource base, such as soil acidity, or to land use, such as nematodes and *Striga*. The role of grain legumes in the systems was thought to need strengthening, but the yield of cowpeas (*Vigna unguiculata*), currently the most important grain legume, would have to increase five or six-fold for the crop to become attractive (RCMD, 1993). A methodology for the agro-socio-economic characterisation of farming systems, called the Smith-Weber-Manyong methodology here, was then worked out (Smith, 1992; Smith and Weber, 1994), whereby four contrasting systems were distinguished:

- Population driven system – expansion phase;
- Population driven system – intensification phase;
- Market-driven system – expansion phase;
- Market-drive system – intensification phase.

In Nigeria, two large benchmark areas (10,000 to 20,000 km^2) were defined in the NGS covering the range of conditions occurring in the zone, one in Bauchi and one in Kaduna State (IITA, 1995). Agricultural growth in the Bauchi area was characterised as 'population-driven intensification', with low soil productivity, export of soil fertility through crop produce and crop residue, poor road access and high prices of inputs and low prices for produce. High population pressure resulted in only 15% of the fields having been fallowed and less than 5% had been manured. On-farm tests were started on rotation and intercropping with grain legumes (cowpea, soybean, groundnuts) and on the potential for Multi-Purpose Trees (MPT) (with ICRAF).

In the Kaduna/Zaria area, agricultural growth was defined as 'market-driven' intensification, due to better market access and favourable benefit/cost ratios. Half the area was planted to maize and rice and inland valleys were used in both wet and dry season. Fertiliser was widely used.

4.2.3 Characterisation of inland valleys[2]

The main findings of the earlier work were reported at a 1985 Symposium (Juo and Lowe, 1986) and details were published in the 'grey literature' Reports (Hekstra et al., 1983). An 'updated version' of these reports was published as a book in 1993 by one of the partner organisations of the Wetlands Utilisation Research Project (WURP) (Windmeijer and Andriesse, 1993).

In 1990 the Inland Valley Systems group developed a five-year research plan, which consisted of a sequence of three activity clusters: (i) inventory and classification, (ii) diagnosis and modelling and (iii) development of improved technologies (Izac et al., 1991). A similar sequence had also been followed in the earlier inland valley activities, so the question may be asked what additional information this program would generate. The answer provides insight into the changing philosophy underlying the emerging agro-ecological approaches.

The Inland Valley Strategy document explains that past regional characterisation activities had mainly been based on secondary data, and that the field data and experimental results from that period were specific for the field sites (Bida and Makeni), without a clear connection between the two. They could therefore not be extrapolated to other areas and yielded fragmented insights in the inland valley agro-ecosystems (Izac et al., 1991). The new approach would be more systematic and comprehensive and consisted of three characterisation levels, similar to the Smith-Weber-Manyong methodology:

1   *Level I.* Identification and mapping of broad agro-ecological and -economic zones in West and Central Africa on the basis of climate, soil, population density and per capita income;
2   *Level II.* The distribution and characteristics of inland valleys would be determined In a stratified sample of areas in each zone, with the help of SPOT (*Satellite pour l'Observation de la Terre*) and LANDSAT satellite images; and
3   *Level III.* Detailed on-the-ground characterisation would then be carried out for a stratified sample of valleys in each zone by year-around observations on physical and socioeconomic features and detailed characterisations using cluster analysis.

This very ambitious characterisation program was expected to be completed in about three years' time. Among other things, the collected information would enable an analysis and hopefully an explanation of the under-utilisation of valley bottoms in West and Central Africa, in spite of their apparent potential.

The level-I mapping of broad agro-ecological zones, at a scale of 1:5,000,000, was based on three parameters: Agro-Ecological Zone (AEZ), length of growing period and dominant soil group, according to the FAO classification (Thenkabail and Nolte, 1995). It was completed in 1991 and distinguished 11 large zones across West and Central Africa. Since it was found that the Bida research site was representative of only a relatively small part of the West and Central African inland valley ecologies, the site was closed down in 1992.

Next, sample areas for level II characterisation were chosen in two of the zones. One sample consisted of the Kabala and Moyamba areas in Sierra Leone, with Ferralsols, a 7-month growing period, high population density and low income. The other was the Kaduna/Minna area in Nigeria, with Lixisols, a 5–6-month growing period, high population density and low income. Characterisation involved satellite imagery-based mapping and ground truthing resulting in 1:20,000 maps allowing estimation of the proportions of land in inland valley bottoms, fringe and uplands, the intensity of land cultivation, the accessibility of the valleys, stream frequency and drainage density, as well as the shape of the valleys. It is not clear whether the full level-II characterisation in Sierra Leone and northern Nigeria was published.

In 1992, a start was made with level III characterisation (RCMD, 1993), which was meant to provide the basis for the design and testing of appropriate technology. In an earlier study in Bida and Makeni, a methodology had been developed for the detailed characterisation of inland valleys, whereby weeds had been identified as a major yield-depressing constraint (RCMP, 1992). This methodology was now used in level III characterisation, whereby special attention was given to soil fertility and weed management.

In 1993, the Inland Valley Program was scrapped as a separate program and its activities were expected to be taken over by the Forest and Moist Savannah Programs, which in the end did not happen.

4.2.4 Indigenous soil classifications

Farmers may be expected to have a soil classification of their own and know what soils are most suitable for which crops. In south-west Nigeria, for example, cocoa (*Theobroma cacao* L.) plots are almost invariably found on the heavier Egbeda-type of A Lixisol. In the past no attention had been given to indigenous classifications, neither in the very comprehensive study by Smyth and Montgomery (1962) of soils in south-west Nigeria, nor in the Institute's early Benchmark Soils Project.

An exploratory study was therefore carried out in 1991 on indigenous soil terminology and classification in northern (Hausa), central (Kulere and Nupe) and south-west Nigeria (Yoruba), in order to examine their logic. In all cases, the basic elements for characterisation were texture, colour and soil moisture, and all systems used criteria for land suitability (Warren, 1992). The collected information was necessarily scant considering the available time. Soil samples were taken in some of the sites in order to correlate measured soil properties with the farmers' assessment. It was anticipated that 'an expanded co-authored monograph comparing and contrasting the findings of the soil analyses with the indigenous soil categories [would] be completed in late 1992'. Unfortunately this monograph did not materialise and no further follow-up was planned. Especially farmers' criteria for soil suitability for agricultural purposes would be valuable, and could be easily recorded as part of all soil survey activities.

4.2.5 Modelling attainable crop yield

The foregoing characterisation work focused on the biophysical and socioeconomic factors affecting productivity, the aggregate effects of which are expressed in crop yields. This does not in itself provide insight in the gap that is likely to exist between those yields and their attainable level under optimum management. The yield gap is a yard stick to measure the possible gains from better management and its estimation is now looked upon as an important tool in development-oriented research. During this episode, a first attempt was made to introduce crop modelling for that purpose.

Crop yield is the result of the interaction of a crop's growth processes with the physical environment, modified by management. In yield gap analysis, three levels of crop yield are considered: actual, attainable and potential yield.

By *potential yield* is meant the yield which would be possible under the solar radiation and temperature regime of a specific location, everything else being optimally available. *Attainable yield* is the yield that would be possible if those yield factors which are controllable by farmers are applied in optimal quantities. Under rainfed conditions where moisture deficits occur, for instance, attainable yield will be lower than potential yield by an amount corresponding with growth reduction due to those deficits. The meaning of *actual yield* is evident. In on-farm yield studies attainable yield is often defined as the maximum possible yield under the prevalent moisture and soil conditions, everything else being done by the farmer in the best possible way. The most interesting *yield gap* is that between actual and attainable yield, because it quantifies the room for improvement by appropriate inputs and adequate management. Field trials sometimes include a treatment which is assumed to measure attainable yield, for instance in fertility trials where all nutrients that may play a role are supplied in optimum quantities. Quite often, however, one is groping in the dark, not knowing what the attainable yield would be.

Simulation models have therefore been proposed to calculate rather than try to measure attainable yield, as a reference level for the success of new technology. The Agro-ecological Unit used the CERES-Maize model to simulate attainable maize yield in the Derived Savannah (DS), Southern Guinea Savannah (SGS) and NGS[3] at different levels of N fertilisation. Measured yields were those obtained between 1992 and 1995 in multi-locational trials with unknown N-levels. Figure 4.3 shows the relationship between measured and simulated yields for three N-rates. Only the simulated yields for 90 kg N/ha come close to measured yields, but the slope of the regression lines deviate strongly from unity at all fertiliser levels, which means that the model could not account for the variability across trial locations. The conclusion that the model did reasonably well in simulating maize yield under different conditions is therefore not warranted on the basis of these results. Further, more rigorous tests would have been needed before the model could confidently be used for yield gap analysis, but it appears that no further systematic modelling work was carried out along these lines.

4.3 Technology development

Technology development had been the RCMD's stock-in-trade all along and did not stop to wait for the results of this new characterisation wave. In fact, research stayed on its earlier course for some time, in yet further refinement of technologies inherited from the previous phases, including soil and soil fertility management, herbaceous legumes and alley farming. Most research projects on land clearing, tillage, erosion and intercropping, however, came to an end by 1990 and little new research was undertaken on mineral fertilisers. The image of fertiliser use as a key technology for African farming had become tarnished, both by the environmental effects of excessive use in Asia and the poor availability or unaffordability to African farmers. Also, research had shown that in sub-Saharan Africa (SSA), agriculture relying primarily on inorganic fertiliser was

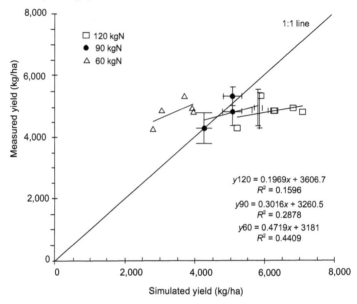

Figure 4.3 Observed and simulated maize yields in multilocational trials (adapted from Jagtap et al., 1999)

not sustainable. Attention thereby shifted to herbaceous and woody legumes for fertility improvement and maintenance, focusing on the processes involved in nitrogen fixation and dry matter accumulation, the build-up of SOM and the role of the soil fauna mediating such processes.

There was also more attention to the characteristics needed for herbaceous legumes to fit into existing farming systems, stimulated by the shift that was occurring towards more OFR, especially in the savannah areas. The earlier optimism about the potential for fertility management of traditional and exotic, herbaceous and woody legumes, however, would gradually be tempered by a growing awareness that there would have to be immediate benefits for the farmers to adopt them, not just the promise of a medium- and long-term benefit of higher yields through improved soil fertility (IITA, 1994).

4.3.1 Soil and soil fertility management

Research on soil physics and chemistry, once a spearhead of IITA's research program, was much reduced during this episode. Some field research continued on soil erosion and land management, and a project on the restoration of degraded soil was undertaken at the Ibadan station and in farmers' fields in Benin. Also, some new research on soil chemistry and soil fertility was started at the HFS in Cameroon. Otherwise, the emphasis shifted further towards the use of auxiliary, mainly leguminous, plants as components of cropping systems and on organic matter dynamics associated with those legumes.

Table 4.2 Soil losses[a] under different tillage systems on an Alfisol and an Ultisol (RCMP, 1991). No statistical information is included in the original document

Soil type (country)	Soil loss (t/ha)		
	No-till	Ploughed	Bare fallow
Alfisol (Nigeria)	0.1	2.5	140
Ultisol (Cameroun)	42	150	640

Note:
a Presumably these are annual soil losses (not indicated in RCMP, 1991).

Soil erosion control

Long-term run-off plots set up in 1986 in Cameroon (Acrisol) and Nigeria (Lixisol) showed much greater soil loss due to erosion in the Acrisol (Table 4.2). The trials also compared no-till and conventional tillage and showed no difference in maize, groundnut or cowpea yields in either soil, in spite of the higher fertility status after five years in the no-till plots (higher organic carbon and major nutrient content). Yields declined after 5–7 years of continuous cropping in both soil types, but they were higher in the ploughed plots. Also, ploughing the no-till plots after five years resulted in an upsurge of maize yield, from 1.9 to 4.3 t/ha, much higher than any of the yields in previous years, probably by a release of nutrients from decomposition of the accumulated OM (RCMP, 1991). These trials probably contributed to the database on the effects of soil management and to the quantification of erosion. No further details or publications were found on this work.

Serious soil degradation may occur as a result of over-utilisation of fragile soils. No reliable figures were available in 1989 about the land areas affected by such degradation in humid and sub-humid Africa, but indirect indications, such as sediment load of rivers suggested that soil erosion was considerable (Hulugalle, 1989). At the field level, parts of the experimental farm at Ibadan were highly degraded due to more than 15 years of intensive cropping (Chapter 3). It was therefore thought pertinent for the Institute to look into methods for the rehabilitation of severely degraded soil, whereby dense plantings of leguminous perennials were chosen as a potential remedy. Some experimental work with such species carried out at the Ibadan station and in farmers' fields in Benin is reviewed in the section on 'Integration of perennials in alley cropping systems' below.

Soil fertility management

At the HFS in Cameroon, experimental research on soil fertility management was yet to be built up, for which the reconnaissance surveys carried out in this episode would contribute content. Emerging topics were the conservation of organic matter, nutrient leaching losses due to the use of soluble fertiliser in excess of the cation exchange capacity, liming and other soil amendments, strategies to increase the content of basic cations in the nutrient-poor subsoil, and practices to overcome P-fixation (Gillman, pers. comm.).

Studies were started at the M'Balmayo site on the effects of *in-situ* burning of heaped forest vegetation, which had been observed to result in poor crop growth on the ash patches. Soil compaction was only moderate and it was thought that nutrient imbalances could be the cause (RCMP, 1991; RCMD, 1993).

Later Menzies and Gillman (2003) found that poor plant growth after burning massive amounts of biomass was not caused by the heat but by the presence of the ash. Soil pH increased significantly up to 15 cm depth and by around 1 unit at 30 cm depth below the spot where massive amounts of wood had been burned. Similar research with more realistic (lower) quantities of wood (Hauser 2006) showed that burning had a severe negative effect on *(in-situ)* maize yields with less severe reductions in each following season. Yields were lower when ash was removed than when it was retained after burning. Thus ash balanced some of the negative effects on maize yet this effect declined rapidly in the next seasons. Important for farming was that delaying seeding by 14 days after burning still caused yield losses. Soil pH increases were short lived and did not reach deeper than 40cm.

No experimental work on soil acidity and Al-saturation in the humid zone was reported in this episode. At the HFS, an idea was floated to apply lime locally in planting holes for trees meant for afforestation or alley cropping, to overcome sub-soil Al-toxicity and get the seedlings off to a good start. This work was never undertaken nor was it considered a strong idea, because native trees can do without liming and the general thinking was to avoid planting exotics (Hauser, pers. comm.).

In the NGS, detailed analyses in farmers' fields identified problems associated with intensive crop production, such as nematodes, weeds and nutrient imbalances (Carsky et al., 1998a, Weber et al., 1995a,b) and stressed the importance of diversified cropping systems and possible improvement by the introduction of legumes.

Zinc deficiency, which is widespread in the savannah zone (Kang and Osiname, 1985), may often be the cause of erratic crop responses to the application of macro-nutrients. Station research by the Institute for Agricultural Research (Samaru) showed a significant response of maize to Zn when available soil content was below 2 mg/kg. Symptoms of Zn deficiency were often observed in maize plants in the NGS and on-farm trials with and without (S and) Zn showed a wide variety of responses to Zn (RCMP, 1992)[4]. It was concluded that more analysis was needed to differentiate between responsive and non-responsive fields. No further details have been found so far on this work. This is another example of an important question emerging (the geographical distribution of Zn-deficiency), apparently without follow-up.

Soil organic matter

Soil organic matter (SOM) had long been known to play a key role in tropical soils and loss of physical and chemical soil fertility was closely associated with the decline of SOM. Although it had been routinely measured in most on-station fertility trials, no distinction had been made between different types or fractions of SOM. More systematic process studies were deemed necessary in the late 1980s on SOM dynamics, nutrient cycling and soil faunal activity, in particular in alley cropping and cover crop systems. The aim was a better understanding of the determinants of SOM accumulation and breakdown, to allow optimum management of this important soil component (RCMD, 1993). The structure and functions of soil organic matter, thus far treated as a more or less nondescript soil component, were therefore scrutinised more closely. SOM-analyses carried out with data collected in a series of alley cropping trials in Benin, Togo and Ivory Coast showed a significant effect of biomass additions (maize stover and hedgerow prunings) on particulate organic matter (POM) (Vanlauwe et al., 1999), which in turn had a significant effect on maize N-uptake.

In another study, the effect of biomass additions on N- and P-uptake by maize was examined in pot trials with soil collected from different experiments across the moist savannah. A strong linear relationship was found between N-uptake and the amount of N contained in the POM-fraction. This was less so in soils from the humid forest zone, which were also included in the tests, assumedly because of protection of part of the POM from decomposition by aggregation with the silt and clay fractions. Total maize P-uptake was again linearly related to the Olson-P content, which in turn was not much influenced by biomass additions (Vanlauwe et al., 2000). From these studies, it was concluded that POM-N content would be an 'important indicator of the soil-related N-supply of savannah soils'.

The Rothamsted Carbon Model was tested for its suitability to predict changes in soil organic C (SOC) under different cropping and fallow systems, using data from a long-term trial at the Ibadan station with *Leucaena leucocephala* and *Senna siamea* alleys (see Section 4.3.3). Note that the usual conversion factor between SOM and SOC is 1.724 (SOM = 1.724 SOC) though Diels et al. (2004) use both SOM and SOC but do not mention a conversion factor.

In the initial run the model predicted an increase in SOC for the alley cropping systems, from 14.4 t/ha[5] to 20.9 t/ha after 16 years and a practically stable content for the tree-less control, with crop residue retained. The measured contents after 16 years were very different, however: 11.8 t/ha average for the alley systems and 7.7 for the control. Also, the *Leucaena* alleys were predicted to accumulate more SOC than the *Senna* alleys, while, in fact, the reverse was the case, probably because of the slower decomposition rate of the latter, which was not accounted for in the model. The model only correctly predicted a small positive effect of fertiliser on SOC. Simulation results could only be brought in line with measured data by 'adjustment' of the model parameters. Much basic research would be needed on various SOM processes before reliable model predictions could be expected (Diels et al., 2004).

Management of wetlands soils and their fertility

Some experimental work continued in the Bida wetlands until the research sites there were closed down in 1992, because they were only representative of a relatively small part of the West and Central African inland valley ecologies (Chapter 3).

A test compared farmers' unlevelled paddies with levelled paddies as a measure for 'improved water management'. The latter involved substantial topsoil movement resulting in considerably lower yield at the fringes and higher yield at the valley bottom (RCMP, 1991).

In Bida, mounds or ridges are usually made in the valleys for crops grown during the dry season. No differences in physical soil properties were found to result from either method, both creating a favourable loose growing medium (RCMP, 1991). No yield effects were reported. Hence, the question remains why farmers do build these (often large) mounds.

4.3.2 Cropping systems

Cropping systems research carried out during this episode consisted almost entirely of work with herbaceous and perennial legumes. Trials were conducted on legumes as live mulch or cover crops and leguminous (and a few other) tree species as hedgerows for alley cropping, while new species were screened for their suitability for these technologies.

A large, long-term on-station project compared different cropping systems including farmers' own, mimicking a real-life situation in respect of crop association, cycle length and inputs. The outcome of this trial, showing doubtful advantages of either alley cropping of *Mucuna pruriens* fallow compared with natural fallow, probably contributed to the eventual demise of alley cropping as a major alternative for shifting cultivation in the humid and sub-humid tropics.

Integration of herbaceous legumes

On-station research with herbaceous legumes continued, but the emphasis shifted back from live mulch to cover crops, stimulated by the success of *Mucuna* for green manure and weed control in Benin Republic (Section 4.4). Some work continued on live mulch, a technology which suffered from lack of exposure to the real farm and was bogged down by the technical problem of preventing the live mulch from smothering the crop.

Crotalaria verrucosa, a fast growing non-climbing herbaceous legume, had been tested earlier as a potential live mulch species, because of its non-climbing habit (Chapter 3). Live mulch had apparently been given up tacitly and the species was now tested for green manure and weed suppression, interplanted into maize. When planted not later than two weeks after planting the maize, it did not affect maize yield and reduced weeding needs to only one round. In another test with maize+cassava, the yields of both crops were similar with and without *C. verrucosa* , cassava yielding 28.4 and 28.1 t/ha respectively. The residual effect on the next crop consisted of less weed pressure and 'reduced fertiliser-N demand' (RCMP, 1992; RCMD, 1993).

A new round of on-station tests was conducted with pigeon pea (*Cajanus cajan*) interplanted into maize+cassava, the dominant pattern of the forest savannah transition zone. In earlier tests (Chapter 3), pigeon peas had been found to be unsuitable as a full intercrop with maize+cassava, because at normal densities it was suppressed by the crops. The idea this time was to grow the pigeon pea as an undergrowth during the first year and let it take off after the cassava harvest, as a short fallow, similar to *Mucuna*. This required pruning the pigeon peas once or twice after the maize harvest. Pruning up to 30 cm was tolerated if the crop mixture was planted in the second growing season after the maize harvest (RCMP, 1991), but should be pruned to 50–75 cm when planted in the first season planting. Vigorous cassava growth also increased pigeon pea mortality. The yield of maize planted after one year of the ensuing pigeon pea fallow (following maize+cassava+pruned-pigeon pea) was 23–31% higher than after maize+cassava without pigeon peas (Adekunle, 1998).

Weeds and low fertility having been identified as major constraints in inland valleys across different ecologies, legume technology was chosen as potential remedy for both. In 1992, 12 legume species were screened at the Ibadan station for their suitability for inland valley conditions, viz. *Calopogonium mucunoides, Mucuna, Crotalaria retusa, Stylosanthes gracilis, Vigna radiata, Vigna luteola,* pigeon pea, *Sesbania rostrata, Crotalaria calycina, Pueraria phaseoloides, Clitoria ternatea* and *Senna occidentalis.* The preliminary conclusion was that pigeon pea, *Vigna radiata* and *Sesbania rostrata* showed most promise. Pigeon pea had the added advantage of being an important legume grown for its edible grain in some areas in West and East Africa (RCMD, 1993). *Sesbania* had also been tested earlier on as a green manure crop for inland valleys in Bida, planted between late February and early May, well ahead of rice planting[6]. It produced most biomass when pruned to 25 cm (RCMP, 1991).

Pseudovigna argentea, a non-climbing herbaceous legume was tested at the Ibadan station for suitability in a live mulch system, with the results shown in Table 4.3. There was no N-effect in the live mulch treatments, in spite of the low overall yield level. In an earlier trial (Chapter 3), a small to zero nitrogen effect was also observed in the live mulch treatments, but at a higher

Table 4.3 Maize yields under live mulch with *Pseudovigna argentea* and zero tillage at the Ibadan station (RCMP, 1992; RCMD, 1993)

Treatment	Weeding regime	Maize grain yield (t/ha)			
		Fertiliser-N application (kg N/ha)			
		1991		1992	
		0	90	0	90
Pseudovigna	Unweeded	1.5	'no N-effect'	NA	1.4
	Weeded	2.3	'no N-effect'	NA	NA
Zero tillage	Unweeded	NA	NA	NA	1.8
	Weeded	0.5	2.0	NA	

Note:
'NA' = 'not available'; the figures were mentioned in the text, not tabulated

overall yield level. This phenomenon has remained unexplained. The parasitic nematode population in the soil was lower under the *Pseudovigna* live mulch but the incidence of maize stem borers was higher (RCMP, 1992).

A variety of traditional and non-traditional legumes were screened both by IITA scientists and by collaborating NARS, participating in COMBS and CORTIS, for their possible contributions and chances for adoption (RCMP, 1992). Most of this work was reviewed in papers presented at the 1999 Cover Crop Conference in Cotonou. Table 4.4 presents a partial summary of findings. Some of the results will also have been included in LEXSYS, the legume expert system (see below).

Integration of perennials in alley cropping systems

In the late 1980s and early 1990s, high expectations persisted for alley cropping, now renamed 'alley farming', because of its perceived dual potential, for cropping and livestock fodder. Doubts started to creep in, however, due to variable results, especially in on-farm trials. Major objectives of the early 1990s were to make a realistic comparison of alley cropping with other systems, determine the adaptation of alley cropping to different climatic and edaphic conditions and screen new potential hedgerow species. Furthermore, the study of multi-purpose trees (MPT) broadened beyond the confines of alley farming by looking at their wider potential, for example for the rehabilitation of degraded soils.

At a more fundamental level, the attention turned to processes below the surface, in particular root competition and organic matter dynamics.

ALLEY CROPPING IN THE SAVANNAH ON LIXISOLS

A test was carried out from 1989 to 1996 with the three most commonly used hedgerow species *Leucaena, Gliricidia sepium* and *Senna* on a degraded and a non-degraded coastal Ferralic Nitisol (*terre de bare*) in Benin to assess their potential to restore and maintain soil fertility. All plots received 150 kg/ha of 'cotton fertiliser' (14% N, 23% P, 14% K, 5% S, 1% B). Maize yields in successive years are shown in Table 4.5 (Aihou et al., 1999). The interesting finding here was that in the medium-term both *Leucaena* and *Senna* restored and maintained soil fertility, with *Senna* doing best in both respects. *Senna* hedges contributed much more N and biomass on the degraded than on the non-degraded soil, which was explained from the higher subsoil fertility of the degraded soil and the ability of Senna to access that fertility.

In another researcher-managed test with four hedgerow species, a single maize crop was grown annually in the alleys in replicated trials along a north-south transect in Togo (Tossah et al., 1999) on widely different soil types (Table 4.6). Fertiliser was applied at 45 kg N, 26 kg P and 50 kg K. In the southern site, the maize yields were high and the differences among hedgerow species and between hedgerows and no-hedgerows (control) were not significant. In the

Table 4.4 Partial summary of findings about properties of herbaceous legumes, various sources (Chapters 3 and 4, Carsky et al., 1999)

Legume species	Suitable for:				Notes
	Intercrop cover	Live mulch	Imperata control	Acidic soils	
Aeschynomene histrix	++	—			Upright, persistent, suicidal germination of Striga in NGS
Mucuna pruriens	+++	—	+++	+	Fire-sensitive, volunteers in next crop
Pueraria phaseoloides				++	
Psophocarpus palustris		++		—	For live mulch needs growth retardant
Canavalia ensiformis	+++				
Centrosema pubescens		++		++	For live mulch needs growth retardant
Stylosanthes guineensis	++				Penetrates hardened soil layer in NGS
Arachis repens		++			
Vigna unguiculata				+	
Desmodium ovalifolium				—	
Lablab purpureus					Well-adapted to drier savannah climate
Crotalaria verrucosa		+++?			Non-climbing
Crotalaria juncea	++				Tested in Rwanda
Pseudovigna argentea		+?			Reduces nematode damage
Cajanus cajan	++	—	—		Erratic results; suitable for inland valleys
Vigna radiata	++				Suitable for inland valleys
Sesbania rostrata	++				Suitable for inland valleys

Key:
+ = favourable; — = unfavourable; empty cell = unknown

Table 4.5 Mean maize and pruning N-yields on a degraded and a non-degraded 'terre de bare' in Benin, 1990–1996, recalculated from graphs (Aihou et al., 1999)

| Year | Maize grain yield (kg/ha) | | | | | | | |
| | Non-degraded soil | | | | Degraded soil | | | |
	Leucaena	Gliricidia	Senna	Control[a]	Leucaena	Gliricidia	Senna	Control[a]
1990[b]			2,181				400	
1994	1,981	1,441	2,396	1,675	539	519	1,244	36
1995	1,120	830	1,607	633	202	306	695	10
1996	3,220	1,882	3,080	1,353	1,400	674	1,981	104
SED[c]				415	within site: 15.9; between sites: 19.0			
N yield[d] (kg/ha)	87	25	49		76	47	155	
SED[c]								
Fresh wood[d] (t/ha)	10.0	2.2	6.6		8.6	6.6	13.9	
SED[c]					within site: 1.96; between sites: 2.47			

Notes:
a Oil palm fallow.
b Pre-trial yield.
c 'SED' = 'standard error of the difference'.
d Annual mean at first pruning.

Table 4.6 Mean maize yields[a] in alley cropping trials along a transect in Togo, 1991–1996 (Tossah et al., 1999)

Site/year	Maize yield (kg/ha)				
	Albizia	Leucaena[c]	Gliricidia	Senna	Control
Glidji (south)	Soil type: Rhodic Eutrustox/Rhodic Ferralsol				
1991/92[b]			2,475		
Mean 1995/96	3,040	–	3,230	3,485	3,030
Amoutchou (centre)	Soil type: Oxiaquic Ustipsament/Haplic Arenosol				
1991/92[b]			1,365		
Mean 1995/96	NA	1,810	2,790	1,005	1,015
Sarakawa (north)	Soil type: Kanhaplic Haplustult/Ferric Acrisol				
1991/92[b]			2,045		
Mean 1995/96	2,790	–	3,845	3,765	2,135
SED[d] (1995/1996)			396		

Notes:
a　Rounded estimate from graphics.
b　Pre-trial yield.
c　Not planted in Glidji and Sarakawa.
d　'SED' = 'standard error of the difference'.

sandy soil of Amoutchou, *Gliricidia* outperformed the others with *Leucaena* in second place and *Senna* doing no better than the control. In the northern site, *Gliricidia* and *Senna* did better than *Albizia lebbeck*. The variation in performance of *Senna* is most pronounced, in particular the low yield in Amoutchou, a site with sandy soils and unfertile subsoil, compared with the high yield in the other two sites (Table 4.6). The pruning and N-yield of the *Senna* hedgerows was also lowest in Amoutchou, which led the researchers to look at the relation between the amounts of N contributed by the prunings and the yield increment relative to the control, i.e. the slope of the response line in Figure 4.4 (a), which is a measure of the agronomic efficiency of the hedgerow-supplied N.

The results are striking, although in line with expectations: the N-use efficiency was lower as total soil N-content, measured at the start of the season, was higher, going to practically zero efficiency at around 0.05% N (Figure 4.4 (b))[7]. At the high end of the scale, the best efficiency barely reached 10 kg maize grain/kg N, a common finding in most alley cropping (and green manure and live mulch) trials. It was suggested that the efficiency of pruning-N is directly related to total soil N-content, although other factors may have played a role, since a similar linear relationship existed with the yield of the control, which is a measure of overall field conditions.

An implication of this work, in the authors' words (IITA, Project 1, 1999) was that: 'Top and subsoil characteristics [are] important modifiers of the

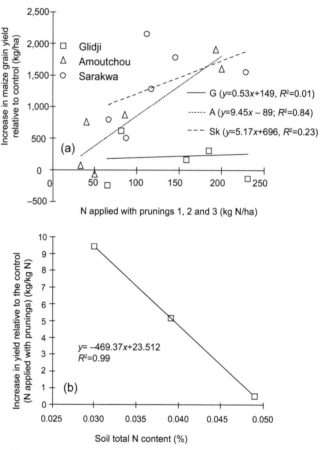

Figure 4.4 Efficiency of hedgerow-supplied N: (a) yield increments with supplied N relative to control at different sites; (b) slopes (i.e. N-use efficiency) as a function of mean total soil N-content (adapted from Tossah et al. 1999)

functioning of alley cropping systems and should be taken into account when a decision is made on whether to use alley cropping and when the hedgerow species are being selected.' Another property of the hedgerow species to be considered in the savannah was the ability to recover from bush fire, which affected this trial in Amoutchou and Sarakawa. In the former, *Leucaena* took more time to recover than the other species.

The long-term experiment at the Ibadan station with *Leucaena* and *Senna* alleys, which started in 1986 and would run for 20 years, compared the performance of these hedgerow species with and without NPK fertiliser[8]. An annual crop of maize followed by cowpeas was grown (Vanlauwe et al., 2005). Pruning yields of both hedgerow species declined steadily, most drastically for *Leucaena,* in part due to tree mortality. Even so, N- contribution in 2002 was still 200 and 160 kg/ha for *Leucaena* and *Senna* respectively. Figure 4.5 shows the evolution of maize

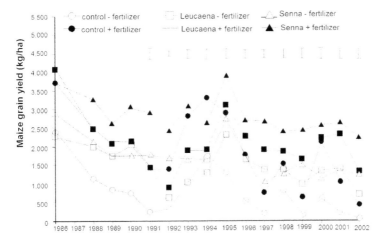

Figure 4.5 Maize grain yields, with error bars (from 1992), for the different treatments in an alley cropping trial in Ibadan, 1986–2002 (adapted from Vanlauwe et al., 2005)

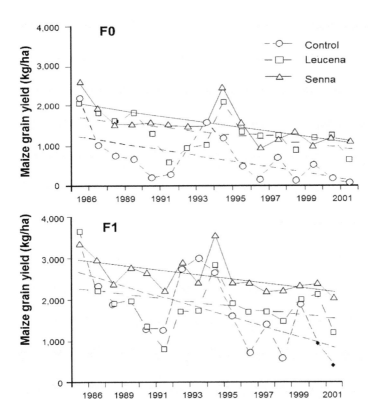

Figure 4.6 Trend line of maize yields in an alley cropping trial in Ibadan in the treatments without (Fo) and with (F1) fertilizer, 1986–2002; same data as Figure 4.5 (adapted from Vanlauwe et al., 2005)

yield for all the treatments. If these results are plotted separately for the fertilised and unfertilised treatments and the year-to-year fluctuations are filtered out by fitting trend lines, fairly clear patterns emerge (Figure 4.6).

In the absence of fertiliser, maize yield in the control fell below 1 t/ha after just a few years and declined to practically zero, while both hedgerow species maintained yield above 1 t/ha for about 10 years, with a small, advantage for *Senna*. It is unlikely, however, that the yield increment compensated for the extra work of maintaining the hedgerows. With fertiliser *Leucaena* initially did worse than the control without hedgerows, but the latter declined more steeply, possibly because of the lower fertiliser rate after 1993. *Senna* did best, maintaining maize yield above 2 tons throughout. The expected benefit from the alleys in terms of additional N-uptake by the maize, over and above that from fertiliser-N, only occurred in the case of *Senna* in the last six years, which implies that the N fixed by the hedges must have been partly lost and partly taken up by the hedges themselves, since there was no build-up of SOM.

As usual in this environment, cowpea yields were extremely variable, from 0–1,256 kg/ha, mainly due to pest problems, Differences were mostly not significant, except in a few years when the *Leucaena* plots did worse than the others.

In a trial at the Ibadan station, comparing weed incidence in *Leucaena* alleys in no-till plots, weed biomass weight initially was not different, but later weed 'density' was lower in the alley cropped plots. The floristic composition also shifted from perennial grasses and *Euphorbia heterophylla* in the no-till plots to shade-tolerant *Synedrella nodiflora* and no perennial grass in the alley plots. In another alley cropping trial with alternate year cropping, volunteer seedlings of *Leucaena* could be controlled by two weedings per season. In a test on speargrass (*Imperata cylindrica*)-infested land near IITA, *Gliricidia sepium* was found to be better at reducing spear grass *(Imperata)* density and biomass than *Leucaena leucocephala* (RCMP, 1991, 1992; RCMD, 1993). In later years (IITA, 1999), similar results were obtained with mulches from *Leucaena, Gliricidia* and *Senna siamea*, the latter two being more effective in suppressing weed growth than *Leucaena*.

Minirhizotron equipment was acquired to study root interaction between hedgerows and crops. In a first study at Ibadan, it was estimated that less than 50% of *Leucaena* roots were located in the top 20 cm of the soil profile (RCMD, 1992). Further investigations were conducted on the horizontal proliferation of hedgerow roots (Hauser, pers. comm.), but no further publications were found.

In spite of the sometimes impressive nitrogen content of hedgerow prunings, in most experiments their effect in terms of N-use efficiency was quite low (Chapter 3). This was at least partly attributed to lack of synchronisation between N-release from the prunings and N-demand by the crop (IITA, 1995). In following years, attention would increasingly turn to the efficiency of N-use originating from symbiotic N-fixation and its improvement.

In a study in 16 established *Leucaena* and *Senna* alleys of 2–5 years old in the forest area of south-west Nigeria (Ayepe), seedlings of 32 indigenous woody species were found to colonize the hedgerows, some at a very low frequencies, others up to a frequency of about 3 seedlings per 50 metres of hedgerow. One of the occurrences of average frequency was Iroko (*Milicia excelsa*), a valuable hardwood timber species. The authors recommended planting double hedgerows to enhance the environment for woody volunteers, thereby helping to conserve useful species through alley cropping in the forested area (Akinnifesi et al., 1995).

ALLEY CROPPING IN THE ACRISOL/FERRALSOL ZONES

In a long-term test with five hedgerow species (Chapter 3), *Gmelina arborea* and *Acacia mangium* were written off for alley cropping because the former had a (strongly) negative effect on maize and cassava yield, caused by allelopathic effects on root growth of both crops (Hauser, 1993a), while the latter did not survive pruning. Of the remaining three species, *Dactyladenia barteri* and *Flemingia macrophylla* alleys gave better maize and cassava yield than *Senna*. In all cases yields declined with time and the inclusion of a fallow period was deemed necessary (RCMP, 1992). *Flemingia* tended to dry up during the fallow and this, combined with decreasing pruning yields, made this species less suitable, leaving only *Dactyladenia* and perhaps *Senna* as suitable species. *Dactyladenia* was shown to have the most favourable rooting pattern in this acid soil with abundant roots in deeper horizons (RCMP, 1991). *Senna* and *Gmelina* could be suitable fallow species for soil fertility regeneration, because of their potential for increasing soil organic matter, basic cations and cation exchange capacity (RCMD, 1993).

Research on the importance of mulch and how to produce it in association with plantain had been going on for years, without reaching firm results. The fastidiousness of plantains in respect of soil fertility has limited that crop's niche in the forest zone to recently cleared (secondary) bush fallow, where a rapid decline of the crop would set in after one or two cycles. Shortening fallows are causing the crop's disappearance from farming systems in the humid zone, except as a back yard crop. Some work had been carried out at Onne on the possibility to substitute fallow by cut-and-carry mulch, by alley cropping, or by maintaining strips of fallow vegetation between rows of plantains, as a source of mulch.

Disappointing results with *Flemingia* as a live cover or as a hedgerow species for plantains (Chapter 3) apparently led to abandoning this species. In a trial with strips of the natural vegetation between plantain rows, started in 1987 (Chapter 3) it was found that with heavy fertiliser application, there was no sign of decline in plantain yield in 1992 (IITA, 1994). Within the hedges, 120 plant species were found growing. Later on plantain yield also started declining, however (Hauser, pers. comm.).

Another trial was set up in the late 1980s, comparing different hedgerows species with 'cut-and-carry' elephant grass (*Pennisetum purpureum*) mulch, produced in a separate fertilised field (80 t/ha fresh weight!). Best results were obtained with *Dactyladenia* pruned twice: 14.5 t/ha of fresh bunch weight, compared with 16.5 t/ha for the mulched plots (RCMP, 1991). Plantain-cum-*Dactyladenia* alley cropping was proposed as a sustainable system for plantain production, whereby the lower plantain yield compared to elephant grass mulch was thought to be redeemed by weed suppression and soil improvement by the *Dactyladenia* hedgerows (IITA, 1994). In any case, bringing in large amounts of elephant grass mulch from outside was hardly a serious option.

Alley cropping 'management tests' were conducted on-station by AFNETA collaborators across Africa. Averaged over ten trials a 32% lower cassava yield was reported under alley cropping and 12% lower cotton yield over four trials (Table 4.7). Maize yield increased by 180% averaged over 44 trials, while common beans, cowpeas and rice had a 12–16% better yield under alley cropping. All yields were calculated relative to total area, i.e. including the area occupied by the hedgerows (RCMP 1992; Woomer et al., 1995).

SOIL REHABILITATION BY LEGUMINOUS TREES

Severe soil degradation had occurred in some of the treatments of the large land clearing trial at the Ibadan station, due to ten years of continuous cropping with tractor-mounted machinery, resulting in soil compaction, hard setting and erosion (Chapter 3). A trial was started in 1989 to see whether the soil could be improved by a medium-term planted fallow. A variety of annual and perennial legumes were compared with natural regrowth and continuous maize+cassava cropping (Kang et al., 1997). Of the 15 planted fallow species, only six survived: the perennials *Acacia auriculiformis*, *Acacia leptocarpa*, *Leucaena leucocephala* and *Senna siamea* and the herbaceous legumes *Psophocarpus palustris* and *Pueraria*. In 1993, the plots were divided in two, with one half cleared, the vegetation burned and maize+cassava planted, while the other half continued under (planted) fallow. The crop yields in the cleared halves are shown in Table 4.8. The LSD values might suggest significant differences between some of the extremes, but, in fact, only the difference between the average of the fallow treatments and continuous cropping is significant[9]. When looking across fallow types, years and crops, however, natural regrowth (presumably *Chromolaena odorata*) consistently scored comparatively well.

None of the soil parameters were significantly improved in the unburned plot halves after four years of fallow by any of the treatments (some selected values are shown in Table 4.8). Apparently, neither planted nor natural fallow (*Chromolaena*) could significantly improve chemical or physical conditions in this highly degraded Lixisol in four years' time. All analytical values went up strongly after burning the fallow vegetation, however, which was probably responsible for significantly higher crop yields after four years fallow than in the continuously cropped plots (Table 4.8). One or two years after burning, the differences with the uncleared halves had become negligible again, with the exception of P, which remained higher (figures not shown here). All maize yields were pathetically low, while cassava yields of around 10 t/ha were similar to that of the lowest quartile of farmers' yields in the area. Apparently, four years of fallow was not enough to restore fertility in severely degraded soil.

Table 4.7 The ratio of alley faring control monocrop yields (FYI) and the increase in yield per unit of ally farming inputs (RETURN) from AFNETA field sites. (Woomer et al., 1995)

Crop	Tree	n	FYI	Return
Cassava	(overall)	10	0.68 (0.08)	–1.0 (0.35)
	Anthonotha macrophylla	1	0.80	0.80
	Calliandra calothyrsus	1	0.23	–2.80
	Cajanus cajan	7	0.79 (0.15)	–0.53 (0.47)
	Harungana madagascariensis	1	0.29	–3.04
Cotton	(overall)	4	0.80 (0.06)	–0.29 (0.09)
	Senna spectabilis	2	0.754	–0.22 (0.02)
	Leucaena leucocephala	2	0.854	–0.36 (0.08)
Maize	(overall)	4	2.83(1.65)	0.24(0.30)
	Faidherbia albida	1	3.98	1.19
	Acacia auriculiformis	3	3.45 (2.07)	0.09 (0.03)
	Albizia lebbeck	4	2.28 (1.70)	0.16 (0.11)
	Acacia mangium	2	3.33 (2.74)	0.09 (0.04)
	Calliandra calothyrsus	2	2.02 (0.72)	0.12 (0.05)
	Senna siamea	3	4.06 (2.82)	0.11 (0.03)
	Senna spectabilis	2	1.72 (0.45)	0.07 (0.03)
	Flemingia macrophylla	2	2.49 (1.38)	0.23 (0.13)
	Gliricidia sepium	5	2.84 (0.95)	0.48 (0.42)
	Gliricidia sepium (hyb)	1	3.42	0.66
	Leucaena diversifolia	2	3.40 (3.59)	0.11
	Leucaena leucocephala	14	2.52 (1.76)	0.14 (0.17)
	Prosopis juliflora	1	3.62	1.24
	Tephrosia candida	2	2.54 (1.78)	0.25 (0.03)
Beans	(overall)	3	1.13 (0.06)	0.02 (0.01)
	Acacia saligna	3	1.13 (0.06)	0.02 (0.01)
Rice	(overall)	6	1.14 (0.43)	0.01 (0.06)
	Senna siamea	2	0.92 (0.36)	–0.02 (0.05)
	Gmelina arborea	2	1.25 (0.67)	0.01 (0.09)
	Gliricida sepium	2	1.25 (0.46)	0.03 (0.06)
Cowpea	(overall)	9	1.16 (0.26)	0.02 (0.04)
	Anthonotha macrophylla	1	1.27	0.02
	Calliandra calothyrsus	1	1.60	0.10
	Senna siamea	2	1.00 (0.03)	0.00
	Gmelina arborea	2	1.17 (0.43)	0.01 (0.04)
	Gliricida sepium	2	1.14 (025)	0.01 (0.02)
	Harungana madagascariensis	1	0.96	–0.01

Table 4.8 Maize and cassava yields after burning four-year-old fallow vegetation, and selected chemical soil properties before burning (Kang et al., 1997)

| Fallow species | Maize grain yield (t/ha) | | Cassava fresh root yield (t/ha) | | Chemical soil properties[a] | | | | | |
| | | | | | pH | | Organic C (%) | | P–Bray–I[a] (mg/kg) | |
	1993	1994	1993/94	1994/95	PB	AB	PB	AB	PB	AB
P. phaseoloides	0.73	0.69	16.0	11.8	5.2	6.5	0.9	1.7	4.4	23.2
P. palustris	1.17	1.19	14.0	12.4	5.3	7.0	1.3	2.1	3.0	25.3
S. siamea	0.73	0.99	11.0	7.3	5.5	7.4	1.1	2.3	5.7	26.6
L. leucocephala	0.98	0.82	9.0	8.9	5.5	7.7	1.1	1.7	2.6	29.4
A. leptocarpa	0.65	0.82	10.0	10.0	5.1	7.0	1.3	2.2	4.0	20.2
A. auriculiformis	0.90	0.50	8.0	6.3	5.2	7.5	0.9	2.2	3.3	45.4
Natural regrowth	0.88	1.02	13.0	12.3	5.5	7.5	1.0	1.5	2.7	35.7
Continuous cropping	0.43	0.36	6.0	9.3	5.5	5.7	0.8	1.0	2.1	11.9
LSD[b] (5%)	0.72	0.59	5.0	NS	NS	0.7	0.4	0.8	3.5	24.8

Notes:
a 0–15 cm, measured in 1993 before (PB) and after (AB) burning the fallow vegetation.
b 'LSD' = 'least significant difference'.

Finally, an interesting observation was the strong border effect by the fallow species with the highest biomass production, *Senna* and *Acacia auriculiformis*, which was thought to explain the low crop yields in the cropped halves of those fallow plots. This border effect confirms the concern about possible faulty conclusions from alley cropping trials with small plots (Chapter 3).

Tephrosia candida was tested at Onne as a potential perennial species for the restoration of a degraded Acrisol. The species controlled weeds effectively and improved soil properties more effectively than the *Chromolaena* and the natural grass fallow, as indicated by better maize growth after a one year fallow[10]. The results also showed that *T. candida* re-cycled calcium and added nitrogen to the soil from N-fixation (RCMP, 1992; no figures given).

In the savannah area of north-west Benin (Atacora Province), preliminary observations were carried out in 1993 in collaboration with farmers on the soil protective and ameliorative effects of four woody legumes (*Leucaena leucocephala, Acacia auriculiformis, Gliricidia sepium and Senna siamea*) and a grass (*Vetiveria zizanioides*. Vetiver grass was still being promoted in Benin on a limited scale in 2014 by GRIPE, an NGO (Versteeg, pers. comm.), but no results seem to have been published.

COLLECTION AND SCREENING OF MPTS

Pentaclethra macrophylla (oil bean) was chosen as a new species for testing its suitability as a hedgerow, because of its rapid growth and good coppicing properties, while also producing edible seeds when left unpruned (RCMP, 1991).

Gmelina, although not suitable for alley cropping in the acidic humid zone because of root competition, could be useful for soil fertility regeneration. This also applies to *Senna siamea* although it is less unfavourable as a hedgerow species than *Gmelina* (RCMD, 1993). *Acacia auriculiformis* may be suitable for soil regeneration in the sub-humid zone and was tested for that purpose on-farm in Benin (see Section 4.4 on OFR).

A collaborative ethnobotanical survey was carried out in south-east Nigeria and a large collection of potential MPTs for the Acrisol zone from the survey was established in arboretums at Onne, Ibadan and M'Balmayo (RCMD, 1992). The accessions were chosen for their potential for soil enrichment, food/ fruit, medicine, ornament and other indigenous uses. The collection formed the material for a PhD thesis, but around 1995 alley cropping was no longer considered an option for the forested humid zone, so no further field testing was carried out, (S. Hauser, pers. comm.).

More potential tree species for alley farming were screened in collaboration with national institutes in many countries, as a back-up for the expanding regional and ecological coverage through AFNETA. Studies were carried out at IITA and by national research institutes on the growth habits and performance of a range of (potential) hedgerow species, their biomass production, decomposition of their prunings, nitrogen use efficiency and effects on weeds. Table 4.9 presents a tentative summary of findings on ecological adaptation of several species from various studies. More details may be found in the Proceedings of the Alley Farming Conference held in 1992 (Kang et al., 1995)

Table 4.9 Tentative ecological adaptation of perennial leguminous and non-leguminous species; RCMP, 1992, 1993; IITA, 1995; Chapters 3 and 4 of this document; AFNETA, 1992, Tossah et al., 1999)

Species	Ecological adaptation			Vigour in suitable ecology	Notes
	Humid forest, acidic soils	Moist savannah	Semi-arid zone		
Leguminous					
Acacia auriculiformis		++		+	Poor coppicing
Acacia mangium	+++				Does not tolerate pruning
Albizia lebbeck	++	++	++?	++	
Cajanus cajan		++		+	Sensitive to termites, degenerates after 2 years; frequent replanting necessary
Calliandra calothyrsus		+++		+++	Edible for ruminants
Flemingia macrophylla	+++	+++		+	Slow decomposition; degenerates after some years
Gliricidia sepium	+	++++		+++	Good at control of Imperata and other weeds
Leucena leucocephala	+	+++		+++	Profuse seeding, can become weed; shallow root system in acidic soil; high tree mortality; slow recovery from bush fire
Pentaclethra macrophylla	+++			+++	Good coppicing, edible seeds when unpruned
Prosopis spp.			++		
Senna siamea	+++	++++		+++	Non-nodulating, shallow rooting in acid soils?, good weed control
Sesbania sesban		++		++	Regular replanting necessary
Tephrosia candida	+++			++ ?	Suitable for regeneration of degraded land

Species	Ecological adaptation			Vigour in suitable ecology	Notes
	Humid forest, acidic soils	Moist savannah	Semi-arid zone		
Non-leguminous					
Alchornea cordifolia	+		+++		Root competition, depresses maize and cassava yield
Dactyladenia barteri	++++			++	Slow decomposition, good weed suppression, rooting in sub-soil
Gmelina arborea	+++	++	+++	++	Competes with cassava, suitable for soil regeneration
Moringa oleifera		+++ ?	+++		

Key:
+ = favourable (a higher number of crosses indicates a higher suitability); empty cell = unknown (cf Table 4.4)

Systems comparison

During the late 1980s, there was growing concern about the artificiality of much on-station research, which often used cropping patterns which hardly resembled those of the farmer and applied heavy fertiliser rates and pesticides. In 1989, it was therefore decided to set up a large, long-term experiment at the Ibadan station to put the Institute's two sterling land use systems to a realistic test: *Leucaena* alley cropping and *Pueraria* green manuring, compared with the farmers' natural fallow-based system. *Pueraria* was chosen rather than *Mucuna*, because the latter would dry up during the dry season. The 'system plots' were split into four sub-plots, each with a different fallow length of zero (continuous cropping) to three years, alternating with one year of maize+cassava, the dominant cropping pattern in south-west Nigeria. The soil was a relatively fertile Lixisol (Egbeda series), which had previously been under 23 years of secondary bush fallow. The initial P-status of the soil was moderate[11]. Prior to the actual trial, the land was planted uniformly to maize+cassava (Tian et al., 2005). No fertiliser was applied at any time and fallow residues were burned before planting, as farmers would do. The trial eventually ran until 2000, but the results are presented in this chapter because the trial is typical for this transitional period. Tables 4.10 and 4.11 show the evolution of maize and cassava yields over the years.

In the treatments with one year fallow, the best maize yields were obtained with the *Pueraria* cover crop system, but natural fallow caught up when one cropping year alternated with two years fallow. Alley cropping produced less than the other two systems across the board.

Table 4.10 Maize yield in a long-term trial comparing three land use systems, Egbeda soil, Ibadan station (Tian et al., 2005)

Maize grain yield (t/ha)

Year	Continuous cropping			1 year fallow			2 years fallow			3 years fallow		
	Natural fallow	Pueraria fallow	Alley cropping	Natural fallow	Pueraria fallow	Alley cropping	Natural fallow	Pueraria fallow	Alley cropping	Natural fallow	Pueraria fallow	Alley cropping
1990	1.88	1.24	1.90	fallow	fallow	fallow	fallow	fallow	fallow	fallow	fallow	fallow
1991	2.33	1.71	2.47	3.01	3.08	2.85	fallow	fallow	fallow	fallow	fallow	fallow
1992	1.21	2.08	0.71	2.12	2.94	1.41	3.52	2.77	2.39	fallow	fallow	fallow
1993	1.16	1.41	1.49	2.46	3.04	2.41	3.25	3.22	3.15	3.19	3.47	3.12
1994	1.10	1.19	1.28	2.09	2.56	2.05	2.78	3.08	2.79	2.96	3.78	2.61
1995	0.63	0.93	1.02	1.38	1.66	1.37	1.97	1.83	1.74	1.94	2.10	2.08
1996	0.32	0.66	0.29	1.44	2.47	0.90	2.64	2.36	1.34	2.65	2.34	2.12
1997	0.31	0.48	0.32	0.78	0.99	0.44	1.61	1.52	1.18	1.66	1.62	0.96
1998	0.57	0.64	0.47	2.09	1.89	0.85	1.76	2.09	1.01	2.28	2.45	1.43
1999	0.26	0.86	fallow	0.75	0.99	fallow	1.62	0.98	fallow	1.72	1.65	fallow
2000	0.21	0.80	0.50	0.80	1.54	1.11	2.12	1.35	1.59	2.26	1.30	1.34
Mean[a]	0.97	1.11	1.05	1.80	2.24	1.49	2.46	2.28	1.90	2.42	2.44	1.95

LSD[b] (5%)

Year:	1993	1994	1995	1996	1997	1998	1999	2000
Between system means	0.38	0.37	0.16	0.45	0.24	0.48	0.58	0.49
Between fallow length within systems	0.70	0.67	0.42	0.63	0.44	0.61	0.44	0.74

Notes:
a Excluding 1999.
b 'LSD' = 'least significant difference'.

Table 4.11 Cassava yield in a long term trial comparing three land use systems, Egbeda soil, Ibadan station (Tian et al., 2005)

Year	Cassava root yield (t/ha)											
	Continuous cropping			*1 year fallow*			*2 years fallow*			*3 years fallow*		
	Natural fallow	*Pueraria fallow*	*Alley cropping*	*Natural fallow*	*Pueraria fallow*	*Alley cropping*	*Natural fallow*	*Pueraria fallow*	*Alley cropping*	*Natural fallow*	*Pueraria fallow*	*Alley cropping*
1990	?	?	?	fallow	fallow	fallow	fallow	fallow	fallow	fallow	fallow	fallow
1991	5.8	1.4	5.2	8.9	3.2	6.1	fallow	fallow	fallow	fallow	fallow	fallow
1992	5.1	2.9	4.2	5.6	3.2	6.2	9.5	3.9	5.5	fallow	fallow	fallow
1993	6.6	8.5	4.6	12.0	9.4	7.9	12.7	11.1	9.8	11.5	12.5	10.7
1994	3.8	3.3	3.2	8.8	6.2	5.6	10.2	8.8	6.3	10.4	9.5	6.0
1995	8.1	6.5	7.0	8.9	7.3	6.8	9.5	8.3	7.5	11.0	10.3	8.4
1996	6.6	7.5	5.2	10.1	14.2	6.1	8.9	13.3	6.0	11.4	12.2	7.2
1997	4.0	3.5	5.2	8.8	5.9	3.6	6.9	5.7	4.0	6.8	5.9	4.9
1998	6.4	2.6	2.4	5.4	5.6	5.4	13.5	9.7	6.0	9.9	10.3	6.5
1999	7.6	8.3	fallow	12.9	13.3	fallow	17.6	15.4	fallow	17.9	18.8	fallow
2000	2.0	2.1	2.6	2.8	4.6	3.2	5.0	5.0	4.2	7.5	5.0	3.0
Mean[a]	4.84	3.83	3.96	7.92	6.62	5.66	9.53	8.23	6.16	9.79	9.39	6.67

LSD[b] (5%)

Year:	1993	1994	1995	1996	1997	1998	1999	2000
Between system means	2.9	2.6	1.6	3.0	2.3	3.8	5.8	1.1
Between fallow length within systems	3.8	3.1	1.6	3.1	3.9	4.3	7.2	1.8

Notes:
a Excluding 1999.
b 'LSD' means 'least significant difference'.

As for cassava, natural and *Pueraria* fallow gave about the same cassava yield, with alley cropping lagging behind both. Hence, in the absence of fertiliser one year of maize+cassava alternating with two years of either natural fallow (*Chromolaena*) or *Pueraria* fallow would be a reasonably stable system in terms of maize and cassava yield, although maize yield would still decline slowly. The *Chromolaena* fallow would be most economical, since it would involve no extra investment or work. This is similar to what farmers in the area were doing, the difference being that they will leave the cassava in the field for up to two years, getting about 25% higher cassava yield than in the trial, plant a second maize+cassava cycle and finally leaving the land to a 3–5 year (*Chromolaena*) fallow. These were sobering conclusions, showing essentially no advantage of either *Pueraria* fallow or alley cropping over *Chromolaena* fallow. Only in the case of continuous cropping did *Pueraria* 'fallow' perform somewhat better than the other two, but in that case *Chromolaena* would not even be able to get established.

These results are likely to have been affected by inadequate trial management, however. The *Pueraria* climbed into the cassava, covering it over the dry season, while the *Leucaena* caused severe shading in rainy years because of late pruning (Hauser, unpublished observations). On the other hand, these are precisely the challenges making the two technologies difficult to handle for smallholder farmers.

After three years, there were no significant differences for any of the measured chemical soil parameters, either between fallow systems or between fallow lengths. After six years, relatively small differences had developed, but they were hard to interpret, except that SOC, total N and P tended to increase with fallow length. Another observation was the low P-accumulation in the *Leucaena* prunings compared with the biomass in the natural and *Pueraria* fallow. The authors argued that the lower cassava yield in the *Leucaena* alleys, apart from root competition, was due to severe sub-soil P-limitation. They further assumed that overall soil-P was the main limiting nutrient responsible for the yearly decline in crop yield and recommended that P-fertilisation be considered (Tian et al., 2005).

4.3.3 Technology validation

On-Farm Research came in two flavours during this episode:

- 'Needs-driven' research, which took the farmers' needs and opportunities as point of departure, whereby IITA-developed technology could, but did not have to be proposed to the farmers; and
- 'Technology-driven' research, which intentionally tested IITA technology or technological concepts under farmers' conditions for which such technology was expected to be suitable, based on system characterisation and technology targeting research.

In future years, OFR would become more and more needs-driven and evolve into what would come to be known as 'Research for Development', but in the late 1980s and early 1990s testing IITA technology remained, explicitly or implicitly, the objective of much OFR. Furthermore, RCMD's portfolio of OFR activities narrowed, both because of the closure of most outreach projects and the departure of most scientists who had been involved in OFR in the early 1990s, as well as by an increased focus on legume-based technology.

Herbaceous legumes

In Benin Republic, the RAMR (*Recherche Appliquée en Milieu Réel*) project focused on soil fertility improvement to address the great fertility problems in their densely populated intervention area in south-west Benin. The project, which was closely linked with the Institute through an IITA agronomist attached to the project, set great store by the potential of herbaceous and woody perennials for fertility improvement, based on their experimental results and farmers' reactions to what they saw in demonstration plots.

In their early on-farm trials, maize was intercropped with pigeon peas or *Mucuna* to boost soil fertility. The legume would take over the field after the maize harvest as a 'short fallow', to be followed by maize again in the next rainy season. The results of a trial with pigeon pea as short fallow are shown in Table 4.12. Total maize yield over four seasons for the pigeon pea treatment, averaged over 25 farmers, was 610 kg/ha less than the control, so the foregone second season maize of year one was not compensated by the gain due to pigeon pea in the other seasons. The harvest from the pigeon pea itself of 310 kg/ha did not make up for the short fall. Furthermore, in fields of low fertility growth of pigeon peas was very poor.

In another trial, comparing *Mucuna* and pigeon pea, maize yield in the year after the short fallow was less than 350 kg/ha higher than the control for the pigeon pea treatment, while it was 1,400 kg/ha higher for *Mucuna* (Table 4.13). Again, the increase in maize yield of the pigeon pea plots was much less than the missed second season maize yield in the previous year (yield not given), while it was considerably higher for the *Mucuna*.

Table 4.12 Maize yield in on-farm tests with pigeon pea 'short fallow', 25 farmers in south-west Benin; Versteeg and Koudokpon (1993)

Season	Maize grain yield (kg/ha)							
	1988[a]			1989[b]			Total	
	Control	With pigeon pea	LSD[c] (5%)	Control	After pigeon pea	LSD[c] (5%)	Control	With pigeon pea
1st season	1,755	1,700	NS	1,540	1,915	205	3,295	3,615
2nd season	1,015	–	–	550	635	NS	1,565	635
Total maize	2,770	1,700	–	2,090	2,550	–	4,860	4,250

Notes:
a 1988 is the year with pigeon pea.
b 1989 is the year after pigeon pea fallow.
c 'LSD' = 'least significant difference'.

Table 4.13 Maize yield, average of local and improved variety, in on-farm tests (the number of farmers is not given) after pigeon pea and *Mucuna* 'short fallow'; Versteeg and Koudokpon (1993). No statistical information is reported

Season	Maize yield (kg/ha)		
	Control	*After pigeon pea*	*After* Mucuna
1st season	1,520	1,745	2,170
2nd season	549	655	1,300
Total	2,069	2,400	3,470

What attracted farmers' attention most, however, was the smothering effect of *Mucuna* on *Imperata*, a major problem weed in the area. In the following years, large numbers of farmers would adopt *Mucuna,* primarily because of its spectacular ability to overgrow *Imperata*-infested land, and several development organisations including World Bank-supported Agricultural Development projects (*Centre d'Action Régionale pour le Développement Rural* – CARDER) and Sasakawa Global 2000 (SG2000) promoted its adoption.

A similar observation was made in Cameroon where cover crops were initially disliked by farmers because of the difficulty of controlling them in crops. After a few years, farmers came around to like the *Pueraria* because the soil was found to be practically ready for planting. They only appreciated the weed control and labour saving in the fallow phase (Hauser, pers. comm.).

In the NGS of Nigeria, four legumes were selected for on-farm testing: *Aeschynomene histrix, Centrosema pubescens, Stylosanthes* and *Senna.* Of these, *Aeschynomene* performed best, as it stimulates suicidal germination of Striga, has good fodder value, erect growth type and slow early growth. It has no insects or diseases in common with cowpea (RCMP, 1992). The species attracted much attention and was tested in farmers' fields in savannah areas by several research groups in West Africa, with mostly favourable results for the reduction of the Striga seed bank (Weber et al., 1995b), as well as for its soil improving ability and fodder value as reported at the Cover Crops conference of 1999 (Carsky et al., 2000a).

Alley cropping

A total of about 50 farmer-managed alley cropping trials with *Leucaena leucocephala and Senna siamea* hedgerows were established between 1986 and 1989 in the villages Ayepe and Ohosu in the humid forest area and Alabata in the Derived savannah of southern Nigeria. All hedgerows were sown into the dominant maize+cassava association. Although soil conditions were still relatively favourable in Ayepe and Ohosu, it was thought that alley cropping could stabilise the system and prevent its degeneration into grass savannahs as happened elsewhere in the humid forests areas of the world. It was further hypothesised that the technology would only be adoptable if farmers would perceive an additional benefit. They were therefore offered hybrid oilpalm seedlings or plantain suckers as a 'bonus' to plant with the hedgerows, the idea being that these perennials would also benefit from the hedgerows. In a survey carried out in Ayepe in 1989, it was found that establishment and management of the hedgerows varied from poor to reasonable (Shokunbi, 1989). Hedgerow development was slow due to shading by the crops and weed competition especially after the maize harvest, when the cassava stayed in the field for up to another 18 months, and heavy hedgerow stand losses occurred in several of the fields. No benefit of the hedgerows could therefore be expected during the first cropping cycle. In 1990, only one farmer had started a new crop cycle after the cassava from the initial cycle had been harvested.

Regeneration of degraded soil

Apart from its successful work with *Mucuna*, the RAMR project in Benin also promoted *Acacia auriculiformis* as a potential restorative species for severely degraded soil (Versteeg and Koudokpon, 1993). In spite of the vigorous growth of *Acacia* on the degraded land, farmers generally opted for *Mucuna* because it had an impressive effect on maize yield after only one season, while the expected benefit from *Acacia* would take a few years to materialise. In later years, the promotion of *Acacia* by different organisations would be more successful (Chapter 5).

4.4 Technology delivery and dissemination

4.4.1 Databases, technology digests and guidelines

Publication of digested research findings in a user-oriented form was much better than before, but remained a weakness for some key research areas, such as live mulch and nutrient responses. An important cause, apart from scientists' understandable reluctance to make premature claims about their work, was the absence of a clear and dynamic policy on the production of digested research findings, which is essential for further on-farm testing and dissemination of technologies by national institutions.

A major shortcoming was the lack of consolidation of the vast body of data on soils, soil fertility and nutrient responses, collected all over Africa. The soils database, as far as it physically existed, was not structured in such a way that it could provide guidance for users to tackle soil fertility issues in their own area. A clear underlying framework for the research data, such as originally intended by the Benchmark Soils project, was missing. Even in the Institute's own experiments, it is rarely clear on what basis routine nutrient combinations were chosen.

The live mulch technology, one of the hopes of the 1980s, practically disappeared from the active research program in the 1990s before ever reaching the on-farm testing stage. Its benefits and potential were elaborated upon in many scattered publications, but there was never a consolidated publication, describing the accumulated experiences which future researchers could turn to. A start was made, however, on the creation of a general database for herbaceous legumes, to be incorporated into the Legume Expert System (LEXSYS, see Section 4.4.2).

A favourable case was the work on tillage systems by the soil physics group, which was published for the general user in 1985 (Lal, 1985; Chapter 3). An updated version, however, incorporating the findings since 1984, was never produced, as the principal scientist departed in 1987.

The results of early characterisation work up to about 1985, by the Wetlands Project were also well published, although not by IITA but by ILRI, the Dutch partner in WURP (Windmeijer and Andriesse, 1993). The methodology developed since for a comprehensive three-level characterisation of the wetlands of West and Central Africa was also well documented, with concrete examples (Thenkabail and Nolte, 1995). The actual field studies themselves, however, remained unpublished.

There were further positive cases. One was a good summary of the Institute's technical research on rice-based production in inland valleys published in a 1992 RCMP Monograph (Carsky, 1992), with some excursions to the work by other organisations. A strength of the report is its emphasis on N-use efficiency, a weakness its failure to relate the findings on soils and soil fertility to the Soil Taxonomy classification used in all previous characterisation work by the Institute. Another publication was the *Training Manual on Alley Farming* produced in 1992 by AFNETA, at the service of its expanding African network (Tripathi and Psychas, 1992). It can be seen as a state-of-the-art document with contributions by several of the main actors in alley cropping/farming research.

Proper OFR methodology was increasingly perceived as essential for on-farm technology testing. The Collaborative Group on Maize-Based Systems Research, for instance, recognised the methodological problem of farmer participation in on-farm experimentation experienced by its teams (RCMP, 1992). A *Field Guide for On-Farm Experimentation*, describing the technicalities of all phases of on-farm research, especially farmer-managed experimentation, was therefore elaborated and finally published in collaboration with the International Service for National Agricultural Research (ISNAR) (Mutsaers et al., 1997)[12].

4.4.2 Decision support

The Legume Expert System (LEXSYS)

The body of knowledge on herbaceous legumes, 'from 800 literature citations' (Carsky et al., 2000b) was consolidated in the form of a computerised 'expert system', Legume Expert SYStem (LEXSYS), actually a searchable database which allowed researchers to select legumes on the basis of quantitative and qualitative specifications (COMBS, 1993; RCMD, 1993). It is not clear to what extent experimental results with various legumes by IITA and its research partners had been incorporated.

In order to choose a species through LEXSYS, a wide range of plant- and environment-related properties may be specified, such as tolerance to drought and acidity, cycle length, weed suppression and many others. The user has to put together a desirable species profile for the target cropping system and access the database for the combination of properties corresponding with the profile (Figure 4.7).

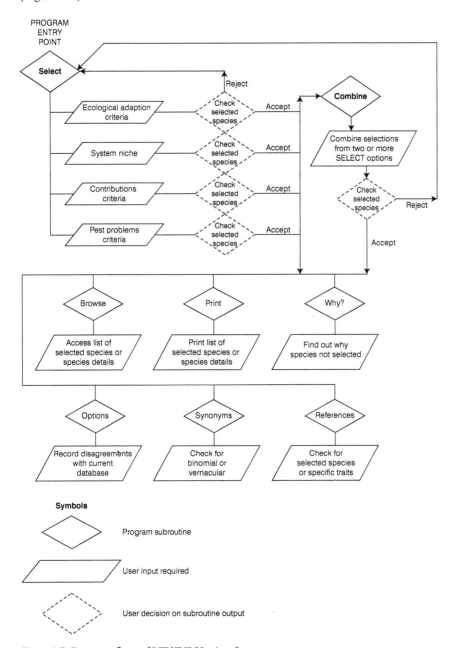

Figure 4.7 Program flow of LEXSYS Version 2

A preliminary version of LEXSYS was tested at the Crops Research Institute in Kumasi, Ghana, and the National Cereals Research and Extension Project in Cameroon to select potential legumes for the local cropping systems. In both cases, a small number of potential species was selected for multiplication and testing, but no subsequent trial results have been reported.

An expert system for alley cropping

An abortive attempt was made in 1992 in collaboration with the University of Hawaii to develop a 'decision aid' for the design of new alley farming systems. The attempt was given up rapidly because it was concluded that insufficient data were available to model the alley cropping system, in particular on the suitability of potential hedgerow species for the farmers' intended purposes as well as on the species' requirements for water, nutrients and management. It also became clear that for application to specific sites, the required soil data might not be widely available and it was concluded that nutrient cycling data would have to be collected in all (on-going and future) alley farming trials (RCMP, 1992). There is no further mention of modelling alley cropping in following years. Much later, Mutsaers (2007) made a simple spreadsheet model to simulate long-term alley cropping processes, using soil and nutrient cycling data from IITA experiments, including those of Tables 4.10 and 4.11. There has been no follow-up on this work.

4.4.3 Dissemination and monitoring and evaluation

Successful legume technologies such as *Mucuna* short fallows, were actively disseminated by development partners in Benin, viz. the World Bank-supported CARDER development projects and SG2000. The numbers of farmers in Benin testing *Mucuna* grew rapidly, from about 100 in 1989 (Versteeg and Koudokpon, 1990) to 3,000 in 1993, as estimated by the extension service (Carsky et al., 1998b).

Otherwise there was little systematic dissemination of soil and soil fertility technology. At a meeting of the RCMD with Dr Borlaug for SG2000 in the early 1990s, soil fertility issues and common nutrient deficiencies were discussed in general terms, but no specific recommendations for fertility were proposed, besides the *Mucuna* technology.

In 1992, a survey was carried out of about 50 farmer-managed alley cropping trials in southern Nigeria, which found that most of the alley cropping fields had gone into fallow after the cassava had been harvested and some farmers had uprooted or burned down the hedgerows. The combination with plantains was not viable because of heavy competition by the hedgerows. The oil palms, however were well taken care of by the farmers and generally developed vigorously (RCMD, 1993).

4.5 Outputs, impact and emerging trends

In many ways, this episode marked the end of an era, which was dominated by the search for new soil and soil fertility management technologies on the research stations, in many cases without exposing them to farmers' conditions during the early phases of their development. The realisation of this defect led to a renewed thrust in the study of existing farming systems and their constraints and opportunities, an assessment of the potential niches for alley cropping and a growing numbers of on-farm experiments, while some of the earlier on-station work continued.

4.5.1 Highlights of research outputs

There were worrying signs about the performance of alley cropping as a technology for permanent cropping under the conditions of the real farm. The worries were reinforced by the results of a long-term cropping systems test at the Ibadan station, which showed better results for (herbaceous) *Pueraria* or natural *Chromolaena* fallow than for *Leucaena* hedgerows. Furthermore regular fallowing was shown to be needed to maintain (low) maize and cassava yield without fertiliser, irrespective of the cropping system. The best strategy for more or less sustainable production would be one full cropping year of maize+cassava followed by two years of natural or *Pueraria* fallow, a sobering conclusion, because it was similar to the farmers' practice of one or two maize+cassava cycles alternating with relatively short natural fallows. Another sobering result from the acidic Acrisol zone was that the most suitable species for alley cropping turned out to be *Dactyladenia*, which had been used for ages by farmers in South-East Nigeria, because of its effect on soil fertility, its use as yam stakes and browse and the quality of its wood (Kang et al., 1991).

Because of this kind of result, as well as observations in Cameroon, there was growing interest in the role of *Chromolaena* in the humid zone, and an issue emerged whether it should be considered primarily as a 'problem weed', as farmers do, to be controlled by shading and slashing (IITA 1995) or as a favourable species for short duration fallows (e.g. Mutsaers et al., 1981; Slaats, 1995). Many years of research on herbaceous legumes had resulted in a few species standing out, in particular *Mucuna*, which was being adopted by many farmers in Benin because of its effect on soil fertility and its capacity to suppress *Imperata*. Pigeon pea, once considered a good candidate as an auxiliary species in several guises (intercrop, short fallow, hedgerows) had performed disappointingly in practically all respects.

Some further results from this episode were in the areas of characterisation, alley cropping and SOM. A generic methodology was developed for the characterisation of agricultural systems, their constraints and potential, at regional and local level. A start was also made with the introduction of modern quantitative characterisation methods, including GIS and dynamic simulation of crop growth. Attempts to quantify the effects of soil fertility constraints under farmer conditions mainly showed the complexity of the interaction of various factors, which was difficult to disentangle. More balanced cropping systems with a greater role for grain legumes were thought to be indicated.

The importance of micro-nutrient deficiencies emerged again from several on-farm studies in the Guinea savannah, in particular Zn and S, which could probably explain the often disappointing responses to major nutrients. No systematic attempts were, however, undertaken to further quantify their magnitude and geographical spread.

In spite of the rising doubts about the future of alley cropping, several studies were conducted to further refine the technology and identify the mechanisms of the interaction of hedgerow and crop growth under different soil conditions. *Senna*, a non-nodulating species, turned out to perform mostly best in non-acid areas and *Dactyladenia* in the acid soil zone. Use efficiency by crops of hedgerow-N was usually low, possibly because of poor synchronisation between N-release and crop demand. In the forest-savannah transition zone the hedgerows were found to be colonised by many forest species.

There is often no clear relationship between total SOM and crop N-uptake and yield. It was found, however, that biomass additions to the soil had a significant effect on POM in savannah soil, which in turn had a significant effect on maize N-uptake. POM-N content could therefore be an indicator of the soil-related N-supply in savannah soils.

4.5.2 Uptake and impact

Whether a technology is taken up by its intended users will depend to an important extent on whether it fits into their system and satisfies a felt need in a profitable way. In other words, was the technology adequately targeted? In the largely technology-driven research system of the previous episodes, the question of technology targeting therefore was very important. With the gradual change towards a more needs-driven research during this episode, the emphasis shifted from finding suitable conditions for a particular technology to identifying appropriate technologies for a particular system, based on its physical and socioeconomic conditions. The Smith-Weber-Manyong methodology for agro-socio-economic characterisation of farming systems (Section 4.2.2) was an important step towards this last goal. The methodology was to be used in the context of the benchmark areas, which were in the process of being chosen by the end of this episode.

The ambitious Inland Valleys characterisation work was also to result in a better focussed experimental program, but with the transfer of most activities to the West African Rice Development Agency (WARDA, now called AfricaRice) in 1993, there was no direct follow-up by the Institute, apart from some (intended) involvement in level II characterisation.

The RAMR project in Benin developed an approach to technology targeting, called the 'problem-oriented group approach' (Versteeg and Koudokpon, 1993); for example, groups were formed around the problem of soil degradation whereby three potential technologies were proposed and farmers joined sub-groups to test the technology that they liked. Significant adoption of *Mucuna* resulted from this approach. The RAMR team also looked into possible use of *Mucuna* grain for food, to increase the attractiveness of the species for farmers (Versteeg et al., 1998).

The LEXSYS decision aid was another targeting tool, for a specific class of technologies: herbaceous legumes. At the same time, it can be seen as a repository of experimental results obtained both by IITA and by others for a wide range of legumes. Some exploratory tests were carried out in collaboration with COMBS members, but it was too early to pass judgement on its usefulness.

Technology uptake and impact

The most successful technology thus far was *Mucuna* in Benin Republic, which was adopted by numerous farmers as a short-term fallow, initially for fertility improvement, but eventually mostly for *Imperata* control. The technology was heavily supported by the RAMR team with advice and free seed supply, and no figures were available yet about 'autonomous' adoption by farmers themselves.

Between 1978 and 1991, altogether 471 alley cropping plots had been established on-farm through IITA and ILCA, from the humid forest to the SGS, 159 of which were still cropped in 1991 (Whittome et al., 1995). From an analysis of the data, it was concluded that alley cropping was most suited to areas where:

1 Maize is the dominant crop;
2 Annual rainfall is 'sufficient to avoid serious water competition' (>1200 mm);
3 Land is scarce and fertility declining, due to high population density (>30/ km^2);
4 Fuelwood or yam stakes are scarce; and
5 Land is individually owned.

The Institute's GIS facility was used to map the potential area in West and Central Africa for alley farming with *Gliricidia* or *Leucaena* (Figure 4.8). The suitable area appears to be quite restricted, but it should be remembered that *Gliricidia* and *Leucaena* are not suited to highly acidic soils. Also, the much larger

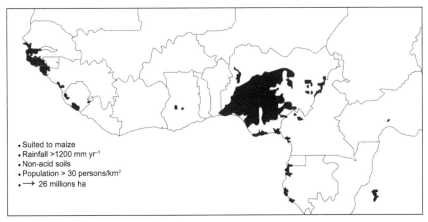

- Suited to maize
- Rainfall >1200 mm yr⁻¹
- Non-acid soils
- Population > 30 persons/km²
- → 26 millions ha

Figure 4.8 Potentially suitable area for alley cropping with *Leucaena* and *Gliricidia* (adapted from Whittome et al., 1995)

area of suitability in Nigeria compared to other countries is not well explained. Although the methodology is solid, the results must be seen as tentative.

In another study of essentially the same set, carried out between 1987 and 1990, Dvorak (1996) concluded that alley cropping will only be adopted when the crop is sole maize with yields below 2 t/ha and there is an immediate sufficiently large yield benefit. The last condition, especially, was rarely satisfied, at least partly because hedgerow maintenance was inadequate. The author stresses the 'unforgiving' nature of the technology, with heavy penalties for late pruning, especially in crops like yams and cassava.

4.6 Emerging trends

The Institute's soil and land use research so far can be characterised by two features:

- It was strongly station-based, with on-farm testing taking second position, often at the end of a long station research process; and
- Its aim was the development of 'International Public Goods' (IPG), i.e. prototype technologies, suitable for more or less well-defined classes of ecologies and farmers.

This was going to change drastically in the years to come. The lack of adoption of soil and land use technology as determined from post-OFR surveys was becoming a serious concern and it was felt necessary to transfer a greater part of the research process to the real farm, in order to enhance the chances of adoption. The Savannah Systems Group, especially, had already made a start in that direction. The decline in core funding and the growing influence of donors through project funding and their call for concrete results were also important factors steering research into farmers' fields. The challenge would be how to generate IPG while carrying out farmer-oriented research under the specific conditions of farmers' fields in specific areas. This would be tackled through the ecoregional-cum-benchmark approach.

There was also a narrowing down of the range of technologies tested to 'auxiliary crops': herbaceous and woody legumes as components of cropping systems for fertility improvement and weed control. In coming years soil fertility research was going to be transformed increasingly into Integrated Soil Fertility Management (ISFM), combining soil amendments, fertiliser, auxiliary crops and cultural practices, according to the specific conditions of the farmers in the pilot research areas.

Auxiliary crops are not uncommon in some traditional systems, so with the right conditions assembled, farmers might be willing to try them. This was demonstrated for example by the 'three success stories' of Kang et al. (1991) where farmers themselves have conceived systems where perennials are used at least partly because of their fertility enhancement properties. It was the merit of Smith, Manyong and Weber to systematically explore the conditions which must be satisfied for farmers to be receptive to such novelties and to use that knowledge in developing and targeting new technologies.

Whereas during the 1980s the emphasis in herbaceous legume agronomy had shifted from temporary legume cover crops as short fallow, to permanent covers as live mulch, in the 1990s there was another change, back to the former: legumes as intercrops, rotation crops and short fallows. The interest in live mulch dwindled, probably because of continuing technical complications of incomplete weed suppression, difficult crop establishment through the live mulch and the need for growth retardants to prevent vigorous legumes from smothering the crops.

The results with alley cropping had been disappointing and the validity of positive experimental results, obtained from relatively small research plots, was found to be questionable. Especially at Onne, it was shown that *Dactyladenia* and *Senna* roots were occupying the entire no-tree control plot soil and thus potentially reducing the control yields, while nutrient export to alley cropped plots would have increased their yield (Hauser 1993b). All hope was not lost, however, and work on the technology would continue in the next episode, focusing on the optimisation of nutrient flows and the build-up and decomposition of soil organic matter in alley cropping systems.

In most field experiments carried out over the past decades, there was no reference yield level against which to compare the results. It was often suspected that yields were held down by factors other than those being investigated, which made results on nutrient responses tenuous. A methodology would be needed to assess the gap between actually measured and attainable or potential yield. A start was made to assess maize yield gaps in the savannah using the CERES-Maize model, but the issue was not pursued with sufficient vigour and seems to have been abandoned after these initial attempts. It would be revitalised and tackled with new tools several years later, when most research was transferred to the real farm.

Implementation of the ecoregional concept, promoted across centres by the CGIAR Technical Advisory Committee (CGIAR/TAC, 1992), took off in 1994. Two of the ecoregional initiatives concerned the Institute's activities in soil and land use research, viz.:

- The Global Initiative on Alternatives to Slash-and-Burn, apparently high-jacked by ICRAF, with the Institute being responsible for benchmarks in the Congo Basin area; and
- An Ecoregional Program for the Humid and Subhumid Tropics of sub-Saharan Africa (EPHTA), partly led by IITA, through its existing research centres and sub-centres.

EPHTA would be implemented by three consortia of national and international institutions, one each for the Humid Forest, the Moist Savannah and Inland Valleys. At IITA the first two coincided with the existing RCMD Programs of the same name, while the Inland Valley Consortium was going to be led by AfricaRice (WARDA), with IITA contributing to characterisation.

The intention was to choose a number of benchmark areas representing the range of conditions of the Humid Forest and Moist Savannah, and pilot research sites within benchmarks, where the actual farmer-oriented research would be carried out.

Apart from the gradual transition into an eco-regional mode as the Institute's organisational principle, the entire research portfolio was also reshuffled in 1994 into 16 research projects, cutting across the ecoregions and benchmarks, resulting in what was called a soft matrix of ecoregions and projects (Chapter 5).

By the end of this episode, several RCMD scientists had left the institute and most of the research programs initiated in earlier phases had come to an end. A lot of the findings of the first almost 30 years had not been capitalised, most not even properly documented, and some squandered, viz. the soil characterisation and mapping work, now tacitly replaced by the FAO classification.

The views about inland valleys had evolved. They were now seen as intrinsic parts of the wider farming systems in an area, whereby it was no longer thought feasible or desirable to develop all of them into productive entities. The 1992 RCMD Annual Report, for instance stated: 'Different kinds of improved technologies will be needed for [...] different categories [of inland valleys]. Some categories may be better suited for resource and biodiversity conservation.' The Inland Valley Systems Program, one of RCMD's earlier three programs was transferred to WARDA in 1993, but IITA continued to contribute expertise for characterisation and GIS.

And finally, a new collaborative concept was adopted for the research-development continuum, whereby national and international researchers and farmers would be partners in a joint venture with a collegial balance which corresponded with their level of involvement at every stage (IITA, 1994).

Notes

1   Henceforth, we will use the WRB soil classification, contrary to earlier chapters when the Institute's convention of using the USDA Taxonomy was followed; the tentative correlation is given in Annexe II.
2   The Institute's work on Inland Valleys was supported by the Netherlands-funded WURP project from 1981 until 1987 and characterisation work had already started in 1980.
3   For definitions of the savannah zones, see Chapter 5.
4   No details were found on this work.
5   Based on an equivalent soil mass of 1700 t/ha.
6   This is common practice in rice-growing areas in Indonesia.
7   The data points in Figure 4.3 (b) are measurements and the line fitted through these indeed resulted in a very close relationship.
8   The rates were quite high in the first eight years: 120 kg N, 90 kg P and 30 kg K/ha. N and P were reduced to 60 and 30 kg/ha from then on.
9   For individual differences a multiple range test has to be used.
10 *Chromolaena* does not grow well at Onne, however, because of the low Ca-status (S. Hauser, pers. comm.), and one year is a short time for natural fallow to establish well.
11 The pre-trial Bray1 P-content of 12 ppm went down to 2-3 by 1993.
12 It was based on an earlier version published in 1987.

References

Adekunle, A.A., 1998. Incorporation of pigeonpea (*Cajanus cajan* L.) into maize/cassava intercrop and its effect on the development of component crops. PhD thesis, University of Ibadan.

Aihou. K., N. Sanginga, B. Vanlauwe, O. Lyasse, J. Diels and R. Merckx, 1999. Alley cropping in the moist savanna zone of West Africa: I. Restoration and maintenance of soil fertility on "terre de barre" soils in Benin Republic. *Agroforestry Systems*, 42, 213–227.

Akinnifesi, F.K., H.J.W. Mutsaers and H. Tijani-Eniola, 1995. Association of pioneer woody species in *Leucaena leucocephala* and *Senna siamea* hedgerows in a forest ecosystem in southern Nigeria. In: B.T. Kang, A.O. Osiname and A. Larbi, *Alley Farming Reserarch and Development: Proceedings of the International Conference on Alley Farming*, IITA, Ibadan.

Carsky, R.J., 1992. *Rice-based Production in Inland Valleys of West Africa: Research Review and Recommendations*. Resource and Crop Management Research Monograph No. 8. IITA, Ibadan.

Carsky, R.J., S. Nokoe, S.T.O. Lagoke and S.K. Kim, 1998a. Maize yield determinants in farmer-managed trials in the Nigerian Northern Guinea Savanna. *Experimental Agriculture*, 34, 407–422.

Carsky, R.J., S.A. Tarawali, M. Becker, D. Chikoye, G. Tian, and N. Sanginga, 1998b. *Mucuna – Herbaceous Cover Legume with Potential for Multiple Uses*. Resource and Crop Management Research Monograph No. 25, IITA, Ibadan

Carsky, R.J., A.C. Eteka, J.D.H. Keatinge and V.M. Manyong (Eds), 2000a. Cover crops for natural resource management in West Africa. In: R.J. Carsky, A.C. Etèka, J.D.H. Keatinge and V.M. Manyong (eds), *Cover Crops for Natural Resource Management in West Africa*. Proceedings of a workshop organized by IITA and CIEPCA, 26–29 October 1999, Cotonou, Benin.

Carsky, R.J, G.K. Weber and A.B.C Robert, 2000b. LEXSYS: A computerized decision-support tool for selecting herbaceous legumes for improved tropical farming systems In: R.J. Carsky, A.C. Etèka, J.D.H. Keatinge and V.M. Manyong (Eds). *Cover Crops for Natural Resource Management in West Africa*. Proceedings of a Workshop Organized by IITA and CIEPCA, 26-29 October 1999, Cotonou, Benin.

CGIAR/TAC, 1992. An ecoregional approach to research in the CGIAR. In: *Expansion of the CGIAR System*. FAO/TAC Secretariat, FAO, Rome, Italy.

COMBS, 1993. *LEXSYS-Legume Expert System. Decision Support for the Integration of Legumes into Farming Systems*. Collaborative Group on Maize-based Systems Research, WAFSRN/RESPAO, Zaria and IITA, Ibadan.

Diels, J., B. Vanlauwe, M.K. van der Meersch, N. Sanginga and R. Merckx, 2004. Long-term soil organic carbon dynamics in a subhumid tropical climate: ^{13}C data in mixed C_3/C_4 cropping and modeling with ROTHC. *Soil Biology & Biochemistry*, 36, 1739–1750.

Dvorak, K.A., 1996. *Adoption Potential of Alley Cropping*. Resource and Crop Management Research Monograph No. 23. IITA, Ibadan

Hauser, S., 1993a. Effect of *Dactyladenia (Acioa) barteri, Senna (Cassia) siamea, Flemingia macrophylla* and *Gmelina arborea* on germination of maize seeds and cassava cuttings. *Agriculture, Ecosystems and Environment*, 45, 263–273.

Hauser, S., 1993b. Root distribution of *Dactyladenia (Acioa) barteri* and *Senna (Cassia) siamea* in alley cropping on Ultisol: Implication for field experimentation. *Agroforestry Systems*, 24, 111–121.

Hauser, S. 2006. Soil temperatures during burning of large amounts of wood, effects on soil pH and subsequent maize yields. TROPENTAG 2006, 11–13 October, Bonn, Germany. Available online at www.tropentag.de/2006/abstracts/full/54.pdf

Hekstra, P., W. Andriesse, C.A. de Vries and G. Bus, 1983. *Wetlands Utilization Research Project, West Africa. Phase I, The Inventory* (4 volumes). ILRI, Wageningen.

Hulugalle, N.R., 1989. *Regeneration of Degraded Alfisols and Associated Soil Groups in the Sub-humid and Humid Tropics of West Africa: A Proposed Long Term Experiment at the International Institute of Tropical Agriculture (IITA)* RCMP Discussion Paper 89/2, IITA, Ibadan.

IITA, 1991–1995. *Annual Reports 1990–1994.* IITA, Ibadan.

IITA, 1999. Project 1. Short fallow systems to arrest resource degradation due to land-use intensification. *Annual Report 1998.* IITA, Ibadan

Izac, A-M.N., M.J. Swift and W. Andriesse, 1991. *A Strategy for Inland Valley Agroecological Research in West and Central Africa.* Resource and Crop Management Research Monograph No. 5. IITA, Ibadan.

Jagtap, S., F.J. Abamu and J. Kling, 1999. Long-term assessment of nitrogen and variety technologies on attainable maize yields in Nigeria using Ceres-maize. *Agricultural Systems*, 60, 77–86.

Juo. A.S.R. and J.A. Lowe (Eds.), 1986. The wetlands and rice in Subsaharan Africa. *Proceedings of an International Conference on Wetlands Utilization for Rice Production in Sub-Saharan Africa*, 4–8 November 1985. IITA, Ibadan.

Kang, B.T. and O.A. Osiname, 1985. Micro-nutrient problems in tropical Africa. In: Paul L.G. Vlek (ed.) *Micro-nutrients in Tropical Food Crop Production.* Development in Plant and Soil Sciences series vol. 14. Martinus Nijhoff/Dr. W. Junk, Dordrecht, The Netherlands.

Kang, B. T., M.N. Versteeg, O. Osiname and M. Gichuru, 1991. Agroforestry in Africa's humid tropics: three success stories. *Agroforestry Today*, 3(2), 4–6.

Kang, B.T., A.O. Osiname and A. Larbi (eds), 1995. Alley farming research and development. *Proceedings of the International Conference on Alley Farming*, IITA, Ibadan.

Kang, B.T., F.K. Salako, I.O. Akobundu, J.L. Pleysier and J.N. Chianu, 1997. Amelioration of a degraded Oxic Paleustalf by leguminous and natural fallows. *Soil Use and Management*, 13, 130–136.

Lal, R., 1985. A soil suitability guide for different tillage systems in the tropics. *Soil & Tillage Research*, 5, 179–196.

Manyong, V.M, J. Smith, G.K. Weber, S.S. Jagtap, and B. Oyewole, 1996. *Macrocharacterization of Agricultural Systems in West Africa: An Overview.* Resource and Crop Management Research Monograph No. 21. IITA, Ibadan.

Menzies, N.W. and G.P. Gillman, 1997. Chemical characterization of soils of a tropical humid forest zone: a methodology. *Soil Science Society of America Journal*, 61, 1355–1363.

Menzies, N.W., and G.P. Gillman, 2003. Plant growth limitation and nutrient loss following piled burning in slash and burn agriculture. *Nutrient cycling in Agroecosystems*, 65, 23–33.

Mutsaers, H.J.W., 2007. *Peasants, Farmers and Scientists. A Chronicle of Tropical Agricultural Science in the Twentieth Century.* Springer, Dordrecht, The Netherlands.

Mutsaers, H.J.W., P. Mbouemboue and Mouzong Boyomo, 1981. Traditional foodcrop growing in the Yaounde area (Cameroon). I. Synopsis of the system. *Agro-Ecosystems*, 6, 273–287.

Mutsaers, H.J.W., G.K. Weber, P. Walker and N.M. Fisher, 1997. *A Field Guide for On-Farm Experimentation.* IITA/ISNAR/CTA, Ibadan.

RCMD, 1993. *Annual Report 1992. Highlights of Scientific findings*. Resource and Crop Management Research Monograph No. 15. IITA, Ibadan.

RCMP, 1991. *Annual Report 1990. Highlights of Scientific Findings*. Resource and Crop Management Research Monograph No. 7. IITA, Ibadan

RCMP, 1992. *Annual Report 1991. Highlights of Scientific Findings*. Resource and Crop Management Research Monograph No. 12. IITA, Ibadan.

Shokunbi, O.A., 1989. Field visits to established alleys (1986–1988). Unpublished student report, IITA, Ibadan.

Slaats, J.J.P., 1995. *Chromolaena odorata fallow in food cropping systems: an agronomic assessment in south-west Ivory Coast*. Doctoral thesis, Wageningen Agricultural University. Wageningen, The Netherlands.

Smith, J., 1992. *Socioeconomic Characterisation of Environments and Technologies in Humid and Sub-humid Regions of West and Central Africa*. Resource and Crop Management Research Monograph No. 10. IITA, Ibadan.

Smith, J. and G.K. Weber, 1994. Strategic research in heterogeneous mandate areas: an example from the West African Savanna. In: J. Anderson (ed.) *Current Policy Issues for the International Community*. CAB International, Wallingford.

Smyth, A.J. and R.F. Montgomery, 1962. *Soils and Land Use in Central Western Nigeria*. Ministy of Agriculture and National Resources, Government Printer, Ibadan, Nigeria.

Thenkabail, P.S. and C. Nolte, 1995. *Mapping and Characterizing Inland Valley Agroecosystems of West and Central Africa: A Methodology Integrating Remote Sensing, Global Positioning System, and Ground-Truth Data in a Geographic Information Systems Framework*. Resource and Crop Management Research Monograph No. 16. IITA, Ibadan.

Tian, G., B.T. Kang, G.O. Kolawole, P. Idinoba and F.K. Salako, 2005. Long-term effects of fallow systems on crop production and soil fertility maintenance in West Africa. *Nutrient Cycling in Agroecosystems*, 71, 139–150.

Tossah, B.K, D.K. Zamba, B. Vanlauwe, N. Sanginga, O. Lyasse, J.Diels and R. Mercia, 1999. Alley cropping in the moist savanna of West Africa: II. Impact on soil productivity in a north-to-south transect in Togo. *Agroforestry Systems*, 42, 229–244.

Tripathi, B.R and P.J. Psychas, 1992. *AFNETA Alley Farming Training Manual, Vols. I and II*. IITA, Ibadan.

Vanlauwe, B., S. Aman, K. Aihou, B.K. Tossah, V. Adebiyi, N. Sanginga, O. Lyasse, J. Diels and R. Merckx, 1999. Alley cropping in the moist savanna of West-Africa. III. Soil organic matter fractionation and soil productivity. *Agroforestry Systems*, 42, 245–264.

Vanlauwe, B., K. Aihou, S. Aman, B.K. Tossah, J. Diels, O. Lyasse, S. Hauser, N. Sanginga and R. Merckx, 2000. Nitrogen and phosphorus uptake by maize as affected by particulate organic matter quality, soil characteristics, and land-use history for soils from the West African moist savanna zone. *Biology and Fertility of Soils*, 30, 440–449.

Vanlauwe, B., J. Diels, N. Sanginga and R. Merckx, 2005. Long-term integrated soil fertility management in south-western Nigeria: Crop performance and impact on the soil fertility status. *Plant and Soil*, 273, 337–354.

Versteeg M. and V. Koudopkon. 1990. *Mucuna* helps control *Imperata* in southern Benin. *West African Farming Systems Research Network (WAFSRN) Bulletin*, 7, 7–8.

Versteeg, M.N and V. Koudokpon, 1993. Participative farmer testing of four low external input technologies, to address soil fertility decline in Mono Province (Benin). *Agricultural Systems*, 42, 265–276.

Versteeg, M.N., F. Amadji, A. Eteka, A. Gogan, and Y. Koudokpon. 1998. Farmers' adoptability of *Mucuna* fallowing and agroforestry technologies in the coastal savanna of Benin. *Agricultural Systems*, 56, 269–287.

Warren, D.M., 1992. *A Preliminary Analysis of Indigenous Soil Classification and Management Systems in Four Ecozones of Nigeria.* RCMD Discussion Paper 92/1. IITA, Ibadan.

Weber, G.K., V. Chude, J. Pleysier and S. Oikeh, 1995a. On-farm evaluation of nitrate-nitrogen dynamics under maize in the Northern Guinea Savannah of Nigeria. *Experimental Agriculture*, 31, 333–344.

Weber, G.K., K. Elemo, A. Awaro, S.T.O. Lagoke, and S. Oikeh. 1995b. *Striga hermonthica (Del.) Benth in cropping systems of the northern Guinea savanna.* Resource and Crop Management Research Monograph No. 19. IITA, lbadan, Nigeria.

Whittome, M.P.B., D.S.C. Spencer and T. Bayliss-Smith, 1995. IITA and ILCA on-farm alley farming research: lessons for extension workers. In: B. T. Kang, O. Osiname and A. Larbi (eds), *Alley Farming Research and Development*. Proceedings of the International Conference on Alley Farming, IITA, Ibadan.

Windmeijer, P.N. and W. Andriesse (Eds.), 1993. *Inland Valleys in West Africa: An Agro-ecological Characterisation of Rice-Growing Environments.* Publication 52, ILRI, Wageningen, The Netherlands.

Woomer, P., O. Bajah, A.N. Atta-Krah and N. Sanginga, 1995. Analysis and interpretation of alley farming network data from tropical Africa. In: B.T. Kang, O. Osiname and A. Larbi (eds), *Alley Farming Research and Development*. Proceedings of the International Conference on Alley Farming, IITA, Ibadan.

5 Ecoregions, benchmarks and projects
1995–2001

5.1 Scope, approaches and partnerships

At the beginning of this episode decentralisation of IITA's research activities had advanced, with specialised centres now catering for the research needs of the Humid Forest (the Humid Forest Station in Cameroon), the Degraded Forest (Onne), the Forest-Savannah transition zone and the Southern Guinea Savannah (SGS) (Benin and Ibadan), the Northern Guinea Savannah (NGS) (Kano) and the Wet Mid-Altitude Zone (Namulonge, Uganda). This division of tasks was consistent with the eco-regional approach advocated since the late 1980s by the CGIAR, as a way to systematically address the research needs of the various ecologies which make up humid and sub-humid West and Central Africa (TAC, 1996). The Institute's mandated zone was further expanded to include parts of East and Southern Africa, in particular the mid-altitude zones, where the emphasis initially was on crop breeding.

The goals of the Institute's resource and crop management research were reformulated during the 1995 External Program Review as:

> the promotion of the development of sustainable production systems for sub-humid and humid African agriculture through improved resource management, with special attention to problems of environmental degradation.

These goals were in essence the same as those formulated at the institute's inception and the grand target remained robust production systems with wide applicability to the mandated zone, now with the explicit condition of sustainability. In other words 'International Public Goods' (IPG) type of technology.

The Review implicitly and explicitly criticised the lack of impact of natural resource management research so far, by expressing the hope that:

> the high standards achieved in IITA's […] commodity improvement research will also be achieved in sustainability-oriented research such as on resource management and its socio-economic studies

and that 'in the future, impact will also be obtained through IITA's research on natural resources management' and recommended strengthening the research on resource and crop management and reinforcing partnerships with National Agricultural Research Systems (NARS), non-governmental organisations (NGOs) and the private sector.

In the first few years, research activities by the Resource and Crop Management Division (RCMD) still took place at the research centres of Ibadan, M'Balmayo and Cotonou, but with time the emphasis shifted to on-farm research until, by the early 2000s, it formed an important part of the research portfolio. The goals, however, as expressed in the 2001–2010 Strategic Plan, remained essentially the same: the development of technology for intensified production in the humid forest and moist savannahs of sub-Saharan Africa, whereby peri-urban agriculture, perennial crops and crop-livestock integration would receive increased attention (IITA, 1999a).

The eco-regional approach sprang from the recognition that the great variation of agricultural conditions in most tropical regions demanded technology adapted to specific agro-ecologies, rather than the widely applicable technology which triggered the Green Revolution in Asia. Eco-regions were delineated and benchmark areas were chosen to capture the variation in conditions within eco-regions. Research would then be carried out in pilot sites within each benchmark, along a gradient capturing the variation within a benchmark. The work would be carried out in close collaboration with the NARS in whose mandated area the Benchmarks were located.

IITA was designated as the CGIAR's lead institute for the Ecoregional Program for the Humid and Subhumid Tropics of sub-Saharan Africa (EPHTA), which was launched in 1996 and would cover two agro-ecological zones (AEZs): the Humid Forest and the Moist Savannah Zones. Three sub-zones were distinguished in each of them, whereby those in the Moist Savannah were characterised by the length of their growing period[1] (Jagtap, 1995), and those in the Humid Forest zone by the stage of deforestation (Section 5.2). In each sub-zone, one Benchmark Area was chosen, centred on a NARS station that would act as main collaborator and host (IITA, 1997; Douthwaite et al., 2005)[2]. Although EPHTA's name suggested that its target area was all of sub-Saharan Africa, in actual fact it was limited to West and Central Africa.

Research in the pilot research sites would address important researchable issues, identified in principle by all relevant stakeholders working together. This would imply a major turn-about, from mainly technology-driven to demand-driven, from 'reductionist' to more 'holistic' Integrated Natural Resource Management (INRM) research, from disciplinary to inter-disciplinary, and from station- to on-farm research. New research with this signature would be initiated, while research inherited from earlier episodes would be phased out.

In 1995, soil and soil fertility management research went through a crisis, due to organisational and financial problems, an exodus of researchers and the failure to achieve impact. Attempts were therefore made to better match research with farmers' needs, in accordance with their resource endowment and the forces driving the development of their farming systems. As a result the following research activities emerged in the following years:

1 Systematic characterisation of soil resources and agricultural production systems in the moist savannah and forest areas and identification of their constraints and needs for innovations;
2 Development and targeting of new cropping technology, crop mixtures and rotations for the high potential Guinea savannah zone, in particular the introduction and optimisation of legumes in cereal-based systems;
3 Improvement of productivity of food crop systems and reduction of weed pressure in the humid areas through fertility enhancing technologies, including herbaceous and woody legumes; and
4 Development of innovations for perennial crop production: plantains (*Musa* spp.), cocoa (*Theobroma cacao*), coffee (*Coffea* spp.) in the humid forest zone.

Perennial crops as research targets were new for the Institute, but one of the most important small farmer perennials, oil palm (*Elaeis guineensis*), remained absent from the list. The other orientations were a reflection of the urgent need for impact of the work on soils and land use.

In the following years, the shift of soil and soil fertility research from mainly on-station to mostly on-farm was completed and the role of research stations diminished accordingly. Researchers formed mixed teams with scientists from NARS and worked together with development partners and farmers in all phases of the research-development process. Platforms were constituted where the dialogue among development partners took place and where the roles of the partners were discussed.

During the programming of EPHTA and its World Agroforestry Center (ICRAF)-led sister program 'Alternatives to Slash and Burn' (ASB) it was realised that research for the development of the humid area should no longer ignore perennial crops, in particular cocoa and oil palm, both as sources of revenue and as stabilising elements in the exploitation of the forest ecology. Similarly, in the savannah area research should address the association of crop and livestock production as a condition for sustainability, and in both zones attention had to be given to the potential of inland valleys for 'off-season' production (IITA, 1998, 1999a).

The chart in Figure 5.1 illustrates the research process, with shades of grey indicating the relative emphasis of the different research components during this episode. It shows that research was still very much in the exploratory stage, with renewed surveying and characterisation, to lay the foundation for future technology development. Most experimental research, although increasingly carried out in farmers' fields, remained to a large extent researcher-managed, aiming at better understanding of the processes involved in different land use practices and systems. A crucial stage in the research-development process is, of course, the assessment of adoption or non-adoption by farmers of the experimented and demonstrated technology. In many cases, the life of the research projects undertaken in this episode have been too short to really ascertain genuine technology adoption.

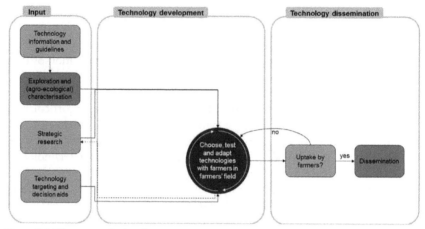

Figure 5.1 The research-development continuum, 1995–2001

The shades of grey indicate the relative emphasis during this episode with darker components receiving relatively more emphasis

Partnership with leading NARS in the Benchmark areas was an important principle of eco-regional research, as well as an inevitable consequence, because of the distributed nature of the work. It helped to bring national scientists back into the farmers' fields, where many of them had carried out on-farm research, during the Farming Systems Research (FSR) era. International funding for such activities had dried up in the early 1990s (e.g. McHugh and Kikafunda-Twine, 1995) because the FSR approach had not delivered on its promises. Only RAMR (*Recherche Applique en Milieu Réel*; Chapter 4) continued to be supported by the Netherlands, with technical support by IITA and the Royal Tropical Institute (KIT).

Declining core funding also forced the establishment of new collaborative projects with advanced institutions, financed by their home country governments[3]. These projects funded most of the soils and land-use work in the Savannah Benchmarks carried out in Benin and Nigeria by the Institute and its national partners.

In the Humid Forest Zone, collaboration was initiated with other CGIAR Centres under the umbrella of the eco-regional programs, viz. the IITA-led EPHTA, the ASB program, led by ICRAF, and the System-Wide Livestock Research Program by the International Livestock Research Institute (ILRI).

On the downstream side there was collaboration with World Bank and Sasakawa Global-2000 (SG-2000) programs in West Africa which were attracted by IITA technology, notably crop varieties and the successful *Mucuna pruriens* cover crop technology in Benin. Other partners included the Catholic Resource Centre (CRC) in northern Nigeria which collaborated in some on-farm technology testing.

These collaborative arrangements were a reflection of the growing need for strong partnerships resulting from several interrelated trends: (i) the adoption of eco-regional concepts, (ii) the gradual shift from station to on-farm research,

(iii) the wider geographical spread of research sites and (iv) donors opting out of core funding and into financing special projects.

Administratively, IITA's research programmes were carved up into 16, later reduced to 14 multidisciplinary, systems-based projects, each with a multidisciplinary team of researchers. Nine of the projects in principle had some soil and land-use components, viz. (IITA, 1996):

1   Agroecosystems Development Strategies: Delineation and characterisation of different agro-ecosystems, analysis of the likely development paths of their production systems and planning of appropriate research interventions in support of that development;
2   Short Fallow Stabilisation: This project intended to repeat the successes of *Mucuna* in Benin Republic, by testing and promoting *Mucuna* and other species as short fallows in other countries (e.g. *Lablab purpureus* in the Nigerian savannah, *Senna spectabilis* and *Calliandra calothyrsus* in the Cameroonian forest zone);
3   Farming Systems Diversification: The project intended to help farmers identify and exploit possibilities for diversifying their product range and cash income, e.g. through fruit tree plantings, fisheries, livestock, off-season production in inland valleys;
4   Integrated management of Striga and other Parasitic Plants: Improvement of soil conditions, e.g. through crop rotation with legumes, was one of the approaches to control these pests; and
5   Improving dissemination of IITA's research results (1996 and 1997 only).

Then there were several projects around major crop-based systems, with agronomic and soil fertility elements:

1   Improvement of Plantain- and Banana-based Systems;
2   Cassava productivity in the Lowlands and Mid-altitude Agro-ecological Zones;
3   Improvement of Yam-based Systems;
4   Improvement of Maize-Grain Legume Systems in the Moist Savannah; and
5   Cowpea-Cereal Systems Improvement in the Dry Savannahs.

This reorganisation marked a shift towards more 'systems thinking', but was not a direct corollary of the eco-regional concept. In order to reconcile the two, all of the research was conceived as a matrix of projects and benchmarks, with some of the cells remaining empty. Most soil and soil fertility work took place under the first three projects, which cut across the forest and savannah benchmarks (IITA, 1995)[4] (Figure 5.2). The last mentioned project, which addressed 'dissemination' as a separate topic, was scrapped in 1998, transferring monitoring of dissemination to the subject matter projects.

The first three projects were reshuffled again in 2000 to accommodate new research on perennials and peri-urban vegetable production, adding vagueness to fragmentation:

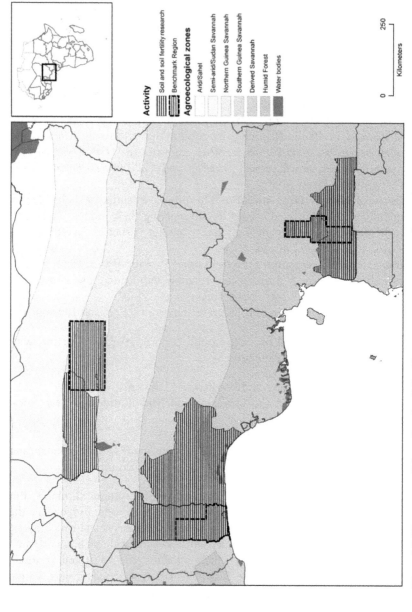

Activity

Soil and soil fertility research

Benchmark Region

Agroecological zones

Arid/Sahel

Semi-arid/Sudan Savannah

Northern Guinea Savannah

Southern Guinea Savannah

Derived Savannah

Humid Forest

Water bodies

0   250

Kilometers

Figure 5.2 Target area of the soil and soil fertility management program during the period covered in this chapter

1   Protection and enhancement of vulnerable cropping systems;
2   Improvement of high-intensity food and forage crop systems; and
3   (Development of) integrated annual and perennial cropping systems.

Soon after the publication of the Strategic Plan, however, the Institute was put on a different course, away from the eco-regional approach towards more direct involvement with the farmer through involvement in various development-oriented projects and the adoption of a Research for Development (R-for-D) approach. This was endorsed later by the 2007 External Program Review and the CGIAR Science Committee (CGIAR, 2008), which recommended that the centre should: develop an overall strategy to contribute to R-for-D and strengthen the alignment between such a strategy and the Center's [core] projects and individual grant projects. The Institute's research on soils and soil fertility management has since continued to be inspired by the goal of contributing directly to development and most of its research has moved on-farm.

5.2 Characterisation of soils, farms and farming systems

Renewed characterisation of farming systems was a major feature of this episode, which can be looked at as a resumption of the characterisation work of the past, presumably better structured this time through the eco-regional-benchmark concept and using more systematic agro-socio-economic methodology developed in the early 1990s (Chapter 4). The objectives were three-fold:

1   Selection and delineation of representative benchmarks for each of the AEZs on the basis of the farming systems, their biophysical and socio-economic characteristics, and their variation across the benchmarks;
2   Further characterisation of the farming systems in the benchmarks and their determinants, in particular their crops, natural resource use and constraints to improved productivity; and
3   Identification of different categories of farmers and their needs for improved technology.

The results of these characterisation activities were meant to guide future systematic technology development, specific for different farmer categories within and across AEZs.

5.2.1 Benchmark delineation and characterisation

During the previous episode, each of the three systems programs within RCMD had worked on the development of methodology for farming systems characterisation (Chapter 4), which would now be further worked out and used for the delineation and characterisation of six large Benchmark areas, representing the agro-ecological conditions of humid and sub-humid West and Central Africa (Table 5.1 and Figure 5.3). The Benchmarks were delineated

Table 5.1 Agroecological zones in humid and sub-humid Africa, with the location of Benchmark Areas and the host National Agricultural Research Systems (NARS)

Agroecological zone (AEZ) and sub-zone	Length of growing period (days)	Country of Benchmark area	Host NARS[a]
Humid Forest AEZ	> 270	–	–
Forest Margins	as above	Cameroun	IRAD, Yaoundé
Forest Pockets	as above	Ghana	CSIR, Kumasi
Degraded Forest	as above	Nigeria	NRCRI, Umudike
Moist Savannah AEZ	150–270	–	–
Derived/coastal Savannah	211–270	Benin	INRAB, Cotonou
Southern Guinea Savannah	181–210	Ivory Coast	IDESSA, Bouké
Northern Guinea Savannah	150–180	Nigeria	IAR, Samaru

Note:
a 'IRAD' = Institute de Recherche Agricole pour le Développement, 'CSIR' = Council for Scientific and Industrial Research; 'NRCRI' = National Root Crops Research Institute, 'INRAB' = Institut National des Recherches Agricoles du Bénin; 'IDESSA' = Institut des Savannes; 'IAR' = Institute for Agricultural Research

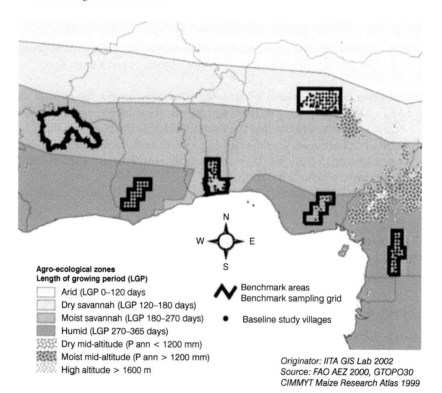

Figure 5.3 Agro-ecological zones and proposed Benchmark areas in West and Central Africa (adapted from Douthwaite et al., 2005). Note that only the Benchmark areas in Northern Nigeria, in Southern Benin Republic, and in Cameroon were activated

in 1995 in cooperation with researchers from the national research stations, earmarked to become the Benchmarks' host Institutions (Douthwaite et al., 2005). They did not cross national boundaries and had to be within easy reach of the host institute. In the Humid Forest Zone, they reflected the status of forest resources, while in the Moist Savannah they were representative for the three sub-zones with their characteristic rainfall regimes.

The next step would be a further characterisation of each Benchmark and the choice of representative pilot research sites, which in the end was only carried out for the 'Forest Margins' Benchmark in Cameroon, the Derived savannah of Benin and the Northern Guinea savannah in north-west Nigeria. No further systematic agro-economic surveying took place in the other benchmarks, which never became operational.

5.2.2 Surveys of farming systems and technology targeting in the 'Forest Margins' Benchmark

In the 'Forest Margins' Benchmark in Cameroon, a natural resource management (NRM) survey method, developed by Baker and Dvorak (quoted by Douthwaite et al., 2005), was combined with a rapid characterisation of soils and vegetation (Menzies and Gillman, 1997; Tchienkoua and Menzies, 1996). The study confirmed the existence of three relatively homogenous blocks and a north-south trend of population densities, as identified by the earlier NRM survey. Two research villages were selected in each block. A comparable characterisation of the other two humid zone benchmarks never took place.

Subsequent agro-socio-economic surveys (IITA, 1999c) drew attention to the importance of cocoa and intensive horticulture and their geographic distribution, with horticulture close to urban centres and cocoa in rural areas. Unsurprisingly, it was found that extension of cocoa (and plantain) growing would only be positive for 'carbon sequestration', an emerging global concern (IPCC, 2001), if new plantations were established in already deforested land. The importance of the 'cocoa agroforests' and village oil palm plantations, both as a source of revenue and as a more or less stable carbon store, comparable to forest, would bring these crops into the foreground as targets for research.

Detailed surveys of 225 households (Douthwaite et al., 2005) identified four categories of farming families according to their land and labour endowments (Table 5.2). A major constraint for the economically important cocoa and horticulture enterprises was found to be the lack of rural credit[5]. A more substantive but also well-known finding was that one reason why farmers cut down forest was their belief that plantains will only do well in newly cleared forest. This led researchers to demonstrate that plantains can be grown on less pristine soil if nematodes in the planting material are killed by hot water treatment and some fertiliser is added (Hauser, 2000). Another conclusion from a technology targeting point of view was that commercial tomato (*Solanum lycopersicum*) producers close to Yaoundé (land scarce, labour abundant category) could benefit from the use of peri-urban poultry manure.

Table 5.2 Livelihood patterns of four farmer categories in the Forest Margins Benchmark Area, Cameroun (after Douthwaite et al., 2005)

Land resources	Family labour	
	Scarce	*Abundant*
Scarce	Subsistence, off-farm employment, artisanal food processing	Sell much of food crop production, intensive horticulture (young household heads)
Abundant	Cocoa agroforests, hunting, fishing	Intensive cocoa and commercial food crop production

The Smith-Weber-Manyong approach, which characterises systems on the basis of population density, access to markets and land use intensity (Chapter 4) does not seem to have played a major role in these analyses, apart from their insights that farming systems are dynamic and that the development trajectory in one area may be predicted on the basis of the trajectory already gone through in another in the same general area.

Next, considerable effort was devoted to the study of various aspects of agriculture in different benchmark areas, mostly through separate surveys (IITA, 1998, 1999b, c), such as (i) deforestation and firewood; (ii) current and potential role of perennial crops in the forest (cocoa, oil palm, fruit trees), and the savannah (cashew (*Anacardium occidentale*), guava (*Psidium guajava*) and *Citrus* spp.), (iii) (peri-urban) vegetable production and marketing; cropping options for thinned timber plantations, (iv) the importance of small stock in the forest margin, (v) animal fattening in the NGS; the potential for integrated crop-livestock systems for the savannah, and (vi) the potential for dry season cropping in inland valleys.

Usually these studies were carried out by mixed international and NARS teams. They were meant to lay the foundation for future research-for-development activities on potential agri-silvicultural systems. No comprehensive accounts of these surveys have been published.

5.2.3 Surveys on farming systems, soils, fertility status and nutrient use in the moist savannah

It was apparently felt that the Smith-Weber-Manyong method developed in the previous episode was not sufficiently discriminating to adequately characterise the farming systems in the NGS Benchmark, so additional surveying was conducted. Four 'resource domains' were identified through multivariate analysis of the survey data, characterised by: (i) low, (ii) low to medium, (iii) medium to high and (iv) high resource use (Manyong et al., 1998; Douthwaite et al., 2005). The first domain corresponded with 'population-driven expansion', the second and third with 'population-driven intensification' and the fourth with 'market-driven intensification', according to the Smith-Weber-Manyong method. Three villages would be chosen in each domain, ensuring in each case good representation of the north-south rainfall gradient. Soil and soil fertility surveys were now conducted in two villages each of the Derived Savannah (DS) of Benin (Zouzouvou and Eglimé) and the NGS of Nigeria (Danamayaka and Kayawa). The information collected by the surveys was used for selecting representative farmers for on-farm fertility trials. The soils in Zouzouvou, derived from coastal sediments, all belonged to the Ferralic Nitisol group, while in the basement complex area of Eglimé the soils varied strongly over short distances along the toposequence. Typical toposequences with soil units according to the FAO system are shown in Figure 5.4.

Parent material and classifications of the soils in the four pilot villages and average values of some soil properties from 12 farmers' fields in each village are shown in Table 5.3 (Vanlauwe et al., 2002).

Next, a missing-nutrient pot experiment was carried out with soil from each farmer's field to identify nutrient deficiencies. The response to N of maize (*Zea mays*) shoot biomass yield, measured as percentage growth in the absence of N was similar in all villages and there was only a weak relationship with total soil N-content of the individual fields. The response to P was stronger in the NGS than in the Derived Savannah (DS) and stronger in Zouzouvou than in Eglimé, while the pooled data showed a more or less linear response pattern with the soils' P-Olsen figures tending to a threshold of about 12 mg/kg (Figure 5.5). This is similar to the P-Bray1 threshold of 14 found earlier for the Lixisol zone (Chapter 1). The stronger P-response in the NGS is probably due to higher P-sorption of the loess layer in that zone, while the small or absent P-response in many of the Eglimé fields will be related to habitual fertiliser use in cotton (*Gossypium hirsutum*). Responses to the cations and to S and micronutrients were minor everywhere (Vanlauwe et al., 2002).

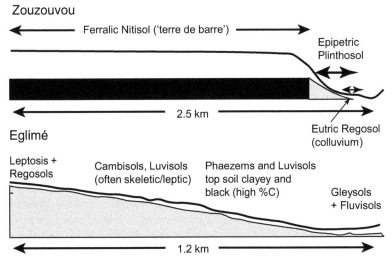

Figure 5.4 Soils along toposequences in Zouzouvou (coastal sediments) and Eglimé (basement complex) in Benin

No detailed explanation is given in relation to the various patterns used in the original document (adapted from IITA, 1999c).

Table 5.3 Average properties of soils in four villages in the Derived Savannah of Benin and the Northern Guinea Savannah of Nigeria; after Vanlauwe et al., 2002

Agroecological zone:	Derived Savannah (Benin)		Northern Guinea Savannah (Nigeria)	
Village:	Zouzouvou	Eglimé	Danayamaka	Kayawa
Parent material:	Coastal sediments	Basement complex rock	Quaternary loess over basement complex-derived soil	
Soil classification:	Ferralic Nitisol (terre de bar)	Soil associations along toposequence[a]	Soil associations along toposequence[b]	
Physical				
Sand (%)	83.4	73.6	60.6	56.2
Silt (%)	6.1	14.7	27.6	30.0
Clay (%)	10.5	11.7	11.8	13.8
Gravel (%)	0	19	6	2
Chemical				
pH (H_2O)	6.7	6.7	6.1	6.0
Organic C (%)	7.9	10.7	5.5	7.1
Total N (%)	0.62	0.78	0.46	0.53
Exchangeable K (cmol$_c$/kg)	0.15	0.38	0.32	0.32
Exchangeable Mg (cmol$_c$/kg)	0.94	1.65	0.66	0.65
ECEC[c] (cmol$_c$/kg)	4.61	9.38	4.10	5.48
P-Olson (mg/kg)	8.1	13.3	5.1	5.8
POM[c] (g/kg)	0.57	0.54	0.51	0.68

Notes:
a Acrisols, Lixisols, Luvisols and Leptisols with inclusions of Vertisols and Cambisols.
b Plinthosols, Luvisols/Lixisols, Gleysols, Fluvisols.
c 'ECEC' = 'effective cation exchange capacity; 'POM' = 'particulate organic matter'.

An agro-economic survey in 1998 of the use of organic and inorganic inputs by 200 farmers in the NGS of Nigeria showed that 90% used inorganic fertiliser, mostly at less than half of the recommended rate of 120 kg N/ha. Less than 30% used organic inputs, including animal manure, green manure and household refuse. An *a priori* classification of farmers in two groups using an N-use threshold of 30 kg/ha captured 75% of the farmers classified through discriminant analysis with a number of variables. This very simple grouping could be used for technology

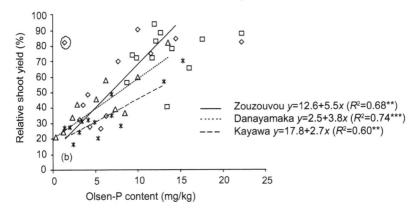

Figure 5.5 Relationship between Olson-P and relative maize biomass yield in pot trials with soils from the Derived Savannah and the Northern Guinea Savannah (adapted from Vanlauwe et al., 2002)

targeting. In a similar study in two villages with more variable conditions in the DS Benchmark in Benin the 30 kg/ha N threshold captured a smaller part of the variability in the population (IITA, 1999c) (Manyong et al., 2001).

5.2.4 Soils and soil fertility in cassava growing areas

The Collaborative Study of Cassava in Africa (COSCA), initiated in 1990, collected continent-wide information on the ecologies where cassava (*Manihot esculenta*)-based cropping was important. Although cassava is not restricted to the humid and sub-humid zone, it had its greatest expanse there. The detailed surveys included information on soils, soil fertility, cropping patterns and yields (Asadu and Nweke, 1999). It was concluded that in general the fertility conditions in cassava fields were similar to those in fields where other crops were grown and that the cropping systems, in terms of fallow length, were also not very different from other crops except yams (*Dioscorea* spp.). This refuted the widely held opinion that cassava was predominantly grown on poor soils, which are no longer suitable for more demanding crops.

Thirty-five percent of the yield differences between the sampled fields were explained by soil properties, but no clear regional patterns were found. It is not unthinkable that the great expansion in cassava growing in the past decades has pushed the crop into conditions where it was not grown traditionally. The extensive analysis of 2,300 soil samples was done exclusively by multiple regression, which is insufficiently powerful to identify structure in the data.

5.3 Technology development

Technology development interventions now followed two major routes, one continuing to aim at changing existing systems through improved short-term fallows and a second aiming at improved soil fertility management within existing farming systems. This dichotomy also attributed varying importance to the use of fertiliser, in the former as a supplement to N fixed by the fallow species and in the latter as an essential ingredient in any system. This was stimulated by concern about the declining soil nutrient stock in intensified traditional systems with little or no external plant nutrient inputs, termed 'nutrient mining' (e.g. van der Pol, 1992; Smaling et al., 1993).

Experimental research was being moved out of the stations into farmers' fields, but remained mostly under management by the researchers. Multi-locational designs were used, in some cases with a single replicate per farm, in order to capture between-farm variation within locations. Major sources of variation in soils and households were not necessarily included explicitly in the trial design, however, and trials were often allocated randomly within locations.

5.3.1 Soil and soil fertility management

Driven by the increased emphasis on sustainability of intensified production, research during this episode shifted further from mainly chemical and physical to more biological factors, related to soil organic matter (SOM) dynamics, N-fixation and unlocking tightly bound P. In particular the ability of many legumes in symbiosis with arbuscular mycorrhizal fungi (AMF) to extract P from soil sources inaccessible for most other (crop-) species came to the forefront, both in the work on short fallow legumes and in integrated nutrient management studies in farmers' fields in the DS and NGS benchmark areas. The recognition of the often low nutrient-use efficiency also led to more attention to the need for increased efficiency in order to make fertiliser use profitable.

Soil organic matter build-up, being a long-term process, would be prohibitively expensive to carry out experimental research in different benchmarks and under different land use systems to study its dynamics. Modelling was therefore chosen as a predictive tool, using soil and cropping systems parameters measured in different trials as inputs, supplemented with data from specific trials carried out under representative agro-ecological conditions (Diels, et al., 2004).

Soil fertility management

Nutrient management research in the Humid Forest and the Moist Savannah Groups went in different directions. The former group studied basic issues on nutrient dynamics in acid soils following up on earlier studies in south-east Nigeria, while the latter studied the performance of integrated nutrient management packages and locally available organic material in farmers' fields.

NUTRIENT MANAGEMENT IN THE HUMID FOREST AND MOIST SAVANNAH ZONES

In the humid forest zone, household ashes were thought to be a potential source of nutrients, with an estimated average production of about 130 kg per household per year in villages of the Humid Forest Zone. Laboratory analysis showed the following nutrient contents: 16% Ca, 2.8% Mg, 5.5% K and 0.85% P, i.e. an unimpressive total store of 21, 3.6, 7.2 and 1.1 kg of Ca, Mg, K and P (IITA, 1999b). When applied on a limited area, however, the pH-raising effect of the ashes may improve P-availability and N fixation.

Other laboratory tests showed increased leaching of Ca, Mg and K due to surface addition of KCl and gypsum. K_2SO_4 only marginally increased leaching and dolomite $(CaMg(CO_3)_2)$ not at all. Large amounts of ashes applied to the soil caused only slow nutrient movement in the soil and minor changes in soil pH at deeper layers (IITA, 1999a; Hauser 2006).

Missing-nutrient tests were carried out in so-called 'groundnut (*Arachis hypogaea*) fields' (*afub owondo*) at ten sites in the Cameroonian forest margin. The groundnut field is the dominant food crop field type in the area, consisting of an association of groundnuts, maize, cassava and plantain (Mutsaers et al. 1981). There was no response of groundnuts to leaving out any of the nutrients K, P, S or B. In other trials responses to Ca or P had been observed, but only on 'extremely acid or P-deficient soils (IITA, 1999b). The tests showed only a positive response of cassava to K and an (unexplained) negative effect of N and P. Although soil-K was generally below the 0.15 cmol/kg threshold, there was no significant correlation between the K-effect and soil-K.

In the moist savannah zone, researcher-managed on-farm tests with conventional fertility enhancement technology were carried out in two villages each in the DS of Benin and the NGS of Nigeria (Iwuafor et al., 2002). The objective was to examine to what extent expensive fertiliser could be substituted by locally available organic materials. Missing-nutrient pot tests with soil from the trial plots, carried out in advance of the trials, had shown minor responses to cations and S and a much stronger response to P, especially in the NGS, while the response in the DS was small in fields where fertilised cotton had been grown.

Researcher-managed tests (one farmer-one replicate) were carried out with 12 farmers each in the DS (Zouzouvou in Eglimé) to study the response of maize to a combination of urea-N and organic residue, compared with urea or organic residue alone, in all cases with the same total N rate. The organic residue came from dual purpose cowpeas (*Vigna unguiculata*) planted in April–May ahead of the maize[6] and redistributed in the experimental plots according to the treatments. There were five treatments: urea only, applied at 0, 90 and 135 kg N/ha; half urea and half cowpea residue at an N-rate of 90 kg/ha; and cowpea residue only, also at 90 kg N/ha. All plots received P and K at 30 kg/ha. The yields for the bottom and the top 25% of the fields over both villages are shown in Figure 5.6.

For the top 25% producers the yield effect of 90 kg N/ha applied as pure urea compared to the unfertilised control (1280 kg/ha) was considerably larger than for a mixture of half urea and half cowpea haulm (700 kg/ha). For the bottom 25%, however, it was practically the same at about 700 kg/ha, but at a very low agronomic nitrogen use efficiency of 8.5 kg maize grain/kg applied N. Even for the top 25%, the N-use efficiency for pure urea was quite low (14.2 kg/kg),

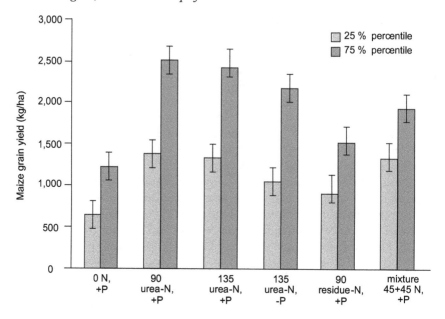

Figure 5.6 Maize yields in an on-farm fertiliser trial in the Derived Savannah agro-ecological zone of Southern Benin: treatment means and their 95% confidence intervals for the bottom and top 25% value of site mean (adapted from IITA/KUL, 1999)

though, so there must have been other major yield limiting factors, which is also clear from the fact that there was no further yield increment at 135 kg N/ha and even a yield decline in the absence of P. Rainfall in the second rainy season is notoriously erratic, which may have caused this.

Researcher-managed trials with a similar objective were conducted in the NGS, with 14 farmers in Danayamaka and 13 in Kayawa, using farmyard manure (FYM) instead of plant residues as organic material. In addition to fertiliser and FYM the trials included treatments with soybeans (*Glycine max*) with and without P (Table 5.4), in view of expected P-deficiency. Interestingly, the effect of 60 kg N in the form of FYM only was quite small in both villages, while the combination of half urea and half FYM had almost the same effect as urea alone in Danayamaka, where the overall yield level was very low. This is comparable to the 25% lowest producers in the DS, where urea combined with groundnut haulms had the same effect as urea alone and was tentatively attributed to synergism between mineral and organic N at a low yield level. Further studies were planned to find an explanation for this phenomenon, but they were not implemented. The average nitrogen use efficiency up to 60 kg/ha of urea-N was about 14 kg maize grain/kg applied N in Danayamaka and 22 kg/kg in Kayawa, The latter is quite respectable as an average and suggests that the better producers may have attained a figure close to 30 kg/kg.

As to the effect of P, the low yield at 120 kg N/ha in the absence of fertiliser-P in both sites confirmed the importance of P fertilisation in this environment.

Table 5.4 Maize grain yield[a] and soybean fresh pod yield in on-farm trials in the Danayamaka and Kayawa villages in the Northern Guinea Savannah of Nigeria, testing combinations of chemical fertiliser, farmyard manure (FYM, for maize only), and rock phosphate (soybean only)

Crops	Applied nutrients (kg/ha)			Crop yield (t/ha)	
	N	P	K	Danayamaka	Kayawa
Maize	-	30 (TSP)[b]	30 (KCl)	0.66 (0.13)[c]	0.89 (0.26)
Maize	60 (FYM)	30 (TSP)	30 (KCl)	0.91 (0.10)	1.12 (0.20)
Maize	60 (Urea)	30 (TSP)	30 (KCl)	1.46 (0.12)	2.22 (0.24)
Maize	120 (Urea)	30 (TSP)	30 (KCl)	2.07 (0.27)	2.95 (0.28)
Maize	30 (FYM)+30 (Urea)	30 (TSP)	30 (KCl)	1.38 (0.13)	1.83 (0.29)
Maize	120 (Urea)	-	30 (KCl)	1.29 (0.26)	1.80 (0.39)
Soybean	-	-	-	2.33 (0.24)	2.13 (0.20)
	-	90 (RP)[b]	-	2.34 (0.23)	2.38 (0.22)

Notes:
a Recalculated from graphics in IITA (1999b).
b 'TSP' = 'triple super phosphate', 'RP' = 'rock phosphate'.
c Numbers in parentheses are the standard errors of the mean.

Soybean fresh pod yield did not respond to rock phosphate, but nodule weight and AMF infection did, which raised the expectation that there could be an effect on the next crop. This was not verified.

LEGUMES AND PHOSPHORUS

The availability of P, an essential element for all crops, is limiting in many soils, but N-fixation by legumes is especially sensitive to low soil-P content. Earlier work had shown that legumes often respond to P-application and that P-uptake is enhanced by association with AMF, especially in P-limited soils. It was further observed that many 'wild' plants and weed species were able to grow well in low-P soil, as does cassava (Chapter 4), which may be due to their ability to explore a large soil volume by an extensive root system or through an effective AMF association. Another postulated mechanism was the excretion of organic compounds by the roots to unlock otherwise insoluble P. This mechanism was thought to be used by some legumes like pigeon pea *(Cajanus cajan)* (IITA Project 1, 1999b; Vanlauwe et al., 2000).

Various studies were carried out to gauge the variation in the ability of legumes to extract P and to elucidate the mechanisms of P-extraction from tightly bound soil sources. A study with a range of soybean accessions showed significant differences among soybean lines in their response to AMF inoculation, suggesting the possibility to breed soybean varieties with less dependency on AMF and better ability to extract tightly bound P (IITA, 1999b).

Mucuna, which normally nodulates well with indigenous rhizobia, failed to do so in some fields in the DS of Benin. In a trial in 15 farmers' fields, the response of shoot dry matter to inoculation varied inversely with the number of rhizobia already present in the soil, where that count was smaller than 5 per gram of soil, except in two fields where extractable soil-P content was less than 10 mg/kg soil. A significant relationship was further found between mycorrhizal colonisation and growth and nodulation of *Mucuna*. It was therefore concluded that *Mucuna* will fix N effectively in those fields where farmers' management practices, such as good crop rotation, combined with rhizobial inoculation, will result in a high degree of AMF infection, thereby alleviating P deficiency (IITA, Project 1, 1999b; Houngnandan et al., 2000).

The use of rock phosphate (RP) as a potential remedy for the low P-status of many tropical soils had been a recurrent topic since the institute's creation. Acidulated RP did improve the response of maize to P (Chapter 3), but the (availability and) cost posed a problem for adoption. Several other approaches were tested, including mixing of RP with decomposing plant residues and incorporation into compost, none of which had a measurable effect (IITA, 1999c).

In a test along a toposequence in Kasuwan Magani in the Nigerian NGS, the effect of RP on the growth of *Mucuna*, *Lablab purpureus* and maize as well as on a subsequent maize crop was studied (Vanlauwe et al., 2000). The RP effected a significant increase in both biomass and P-content of the legumes and doubled the yield of the maize crop following the legumes on the plateau soil (Figure 5.7)

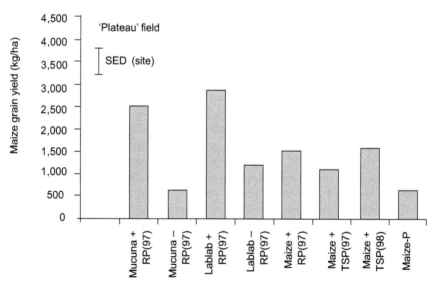

Figure 5.7 Yield of maize following previous season legumes or maize with or without rock phosphate, plateau soil, Northern Guinea Savannah, Nigeria (adapted from Vanlauwe et al., 2000)

and a smaller but still significant increase in the valley field. There was no effect in a slope field with an Olsen soil-P content of about 14 mg/kg, where the maize yield level was much higher than on the plateau (Vanlauwe et al., 2000). A dual purpose cowpea which was also included in the test did not itself respond to RP, but the subsequent maize crop also had a higher grain yield (1,000 against 600 kg/ha) (IITA, 1999b).

5.3.2 Cropping systems

It had frequently been observed that fallows were shortening in many areas in response to increasing population density, while an ageing farming population was often unable to cope with the task of clearing heavy secondary forest fallow vegetation. The initial hypothesis had been that, in order to modernise agriculture, fallowing should be replaced entirely by different fertility management techniques. Results of its long-term research, however, showed that doing away entirely with fallow was an illusion, especially in the wetter parts of humid Africa and that some sort of fallow was essential for sustainable farming in practically all environments (e.g. Chapter 2; Juo and van der Meersch, 1983; Tian et. al., 2005). Thus, research findings and actual farming trends were converging onto short fallows as the future for farming in humid and sub-humid Africa and most of the cropping systems' work in this episode was concerned with the management or improvement of such short fallow systems. Finding suitable N-fixing legumes for farmers' systems and developing management practices for their incorporation into those systems thereby acquired the status of a high priority objective.

Integration of herbaceous legumes

It was hypothesised that the improvement of short fallows would often require herbaceous legumes or other fast growing species to replace or enrich the natural fallow. In the forest zone and the DS such species would replace *Chromolaena odorata*, the dominant fallow species which normally invaded the land rapidly during the final phase of, or after, cropping. It was considered a noxious weed by farmers and scientists alike (e.g. IITA, 1999a), although in a long-term trial in Ibadan (1990–2000) *Pueraria phaseoloides* did not do any better than *Chromolaena* fallow and *Leucaena* alleys did worse (Tian et al., 2005; Chapter 4), This, however, may have been due at least partly to poor trial management (Hauser, unpublished observations). Studies in Ivory Coast had also shown the beneficial role of *Chromolaena* as a short fallow species (Slaats, 1995). In seriously degraded soils in the forest and DS zone, however, *Chromolaena* does not survive and is replaced by more noxious spear grass (*Imperata cylindrica*) while in the Guinea savannah other grasses dominate with insufficient ability to restore fertility. Under those conditions, planted short fallows, in particular with legumes, were thought to be the answer.

Shortening the fallow by introducing herbaceous legumes into the farmers' cropping system posed several major challenges, however (IITA, 1999b). The legume must restore soil fertility in a short period, whereby total crop yield in a crop-fallow cycle is maintained at the same or preferably a higher level compared to conventional fallow. Furthermore, it must control weeds carried over from the previous cropping or from the fallow itself, as in the case of *Chromolaena*. An important additional condition for adoption would be that the farmer does not have to give up a crop to get the legume established, while it would help if the legume had some additional immediate benefit, like edible grain or fodder in case there is livestock.

Much work was carried out during this and the previous episode on the characterisation of leguminous species, to be grown either sole or in mixtures, as well as on screening of different strains (accessions) of several species for flowering, biomass and seed production, nutrient accumulation and mycorrhizal symbiosis. Table 5.5 shows features which were considered important for a species to be suitable as a component of a short fallow. A large number of potential legumes for the Forest Margin Benchmark were screened in 1996–1998 at the Humid Forest Station for rapidity of early growth, in order to be competitive against *Chromolaena*. Examples of 'very fast' species (>50% cover at 2 months after sowing) were *Mucuna* spp., *Crotalaria* spp., *Centrosema* spp. and

Table 5.5 Some characteristics of herbaceous legumes with importance for their use in planted fallows (IITA, 1999b)

Species characteristics	Notes
Growth type	Annual, persistent, perennial
Earliness of cover	In view of early suppression of weeds (e.g. *Chromolaena*, *Imperata*)
Growth/canopy duration	Short duration will allow weed growth after the legume dries up (e.g. *Mucuna*)
Drought resistance	Can species survive dry season?
Weed suppression	–
Acidity tolerance	–
P-uptake capacity	Can species unlock tightly-bound P, from soil or rock phosphate?
Seed yield	Does high seed yield harm suitability as a cover crop? Does species re-sow? Can farmers collect their own seed?
Dry matter yield	–
Nutrient accumulation	Above/below ground
Useful secondary products	Are the grain or leaves edible for humans, the stover for cattle?

Clitoria spp. Very slow species were *Psophocarpus palustris, Canavalia ensiformis, Tephrosia candida, Centrosema plumieri, Chamaecrista rotundifolia, Indigofera tinctoria, Pseudovigna argentea* and *Stylosanthes guianensis* (these were all preliminary results) (IITA, 1999a). The data gathered through this kind of screening were also built into the LEXSYS database.

Mucuna was considered the most promising annual legume, in spite of its fire-sensitivity in the dry season, mostly based on on-farm experiences in Benin. It featured prominently in all research related to short fallow improvement in this episode. Several accessions were screened in the DS (Ijaiye, south-west Nigeria) and the Humid Forest Station (Mbalmayo, Cameroon), showing large differences in canopy duration, canopy density and canopy persistence (Table 5.6). The amount of N fixed, measured in Mbalmayo, was significantly lower for *Mucuna ghana* than for the other species/varieties (Hauser and Nolte, 2002). In speargrass-infested plots in Ijaiye, *Mucuna cochinchinensis* and *M. veracruz* reduced speargrass biomass most after 16 weeks[7] (IITA, 1999b; Carsky et al., 1998).

Pueraria, one of the species thought to have good potential for fertility improvement and speargrass control, was found to be very sensitive to low soil-K levels. In a test on a Ferallic Nitosol in Benin deficiency symptoms appeared at a soil K-content of 0.20 cmol$_c$/kg and below. The species was thought to be suitable as an indicator crop for low soil-K.

A range of herbaceous species was also tested with 20 farmers in two villages in the Forest Margin Benchmark area in Cameroon for their suitability as short fallow cover crops (IITA, 1999b). The species were *Mucuna pruriens var. veracruz, Pueraria, Stylosanthes,* pigeon pea, *Crotalaria juncea* soybean and cowpea. Differences in dry matter production were not significant, mainly by the 'tremendous variation [...] across fields with values ranging from 2.2 to 12.9 t/ha'. Researchers intervened in various ways which probably affected the results: early weeding to ensure establishment of some species (pigeon pea, *Stylosanthes, Pueraria* and soybean), late weeding in some farms because farmers failed to weed in a timely way.

Table 5.6 Growth habits of five *Mucuna* species/varieties with crop cover data form Ijaiye, Nigeria and N fixation data from Mbalmayo, Cameroon (IITA, 1999a). Values followed by the same letter are not significantly different, at P varying from 0.029 to 0.0985

Species/variety[a]	Early growth	Cycle	Cover at 31 WAP (%)	N fixed (kg/ha)
M. cochinchinensis	Slow	Long	98	127(bc)
M. ghana	Very fast	Short	2	85(c)
M. jaspeada	Medium fast	Medium long	58	161(ab)
M. pruriens var *utilis*	Medium fast	Medium long	41	145(b)
M. veracruz	Very fast	Long	53	171(ab)

Note:
a   The nomenclature gets confused here; these are probably all botanical varieties of *Mucuna pruriens*.

Mixtures of legume species with complementary characteristics (in vigour, persistence, drought tolerance) were tried between 1997 and 1999 in the NGS and the DS as a way both to get an early full cover and to bridge the dry season. The treatments involved mixtures of three species with contrasting characteristics: one with fast, early establishment but little persistence (*Mucuna, Lablab* or *Centrosema pascuorum*) one with slower establishment but persistence into subsequent seasons (*Aeschynomene histrix, Stylosanthes guianensis* or *Stylosanthes hamata* and a third with the ability to remain green in the dry season (*Centrosema brasilianum*) (IITA, 1999b). Results were variable, with the annual species suppressing the others in the DS site, whereas in the NGS sites the persistent or drought resistant species replaced the annual ones in the second year (IITA, 1999b), resulting in less risk for the mixture. Growing mixtures of legumes with contrasting or complementary properties could be an interesting concept to rapidly establish a continuous cover with a better chance to suppress noxious weeds such as *Imperata*[8].

MANAGEMENT OF SHORT FALLOWS IN THE FOREST AREA

Short legume fallow was compared with alley cropping and natural fallow in southern Cameroon in 'groundnut fields' under two contrasting field conditions: one site had been under secondary forest regrowth of 25 years old, while in the other site *Chromolaena* was the dominant fallow species. After clearing, a blanket groundnut/maize/cassava crop was planted in 1994, followed in 1995 by the actual trial, comparing (i) natural fallow, (ii) *Pueraria* cover crop and (iii) *Calliandra* hedges. After 2 years the fallow vegetation was cleared, the hedgerows pruned, the residues burned and the plots planted again to groundnut/maize/cassava. The amount of biomass was considerably greater for *Calliandra* hedges than for the other fallow types. Groundnut yield after natural fallow and *Pueraria* varied across sites without a clear trend, but it was depressed in the *Calliandra* plots in both sites, in spite of pruning and a large number of the trees having died after the pruning. Maize and cassava yields did not differ significantly among treatments or sites. The labour requirements for clearing and burning of *Calliandra* were considerably higher than for the other treatments, which would pose an obstacle to farmers' adoption (IITA, Project 1, 1999b, Hauser et al., 2006).

This trial was actually set up as a long-term study, which should have run for up to 20 years to really measure the effect of different types of fallow in this environment, in a similar way as the long-term trial at the Ibadan station (Chapter 4). A number of successive cropping-fallow cycles would have to be grown for long-term effects to express themselves. The large plots were therefore divided in two, one of which was cropped again for another year (sequence 2) and then put under fallow, while the other went into fallow immediately (sequence 1). These fallows lasted for two years, followed again by two crop-fallow sequences. The crop yields for the two cycles, averaged over the sites (secondary forest versus *Chromolaena* pre-trial fallow)[9], are shown in Table 5.7.

Table 5.7 Dry matter yields[a] in two crop-fallow cycles of a long term trial, Humid Forest Station, Cameroun; 1997–2001 (sequence 1) and 1998–2002 (sequence 2). Values followed by the same numbers are not significantly different (Hauser, unpublished data)

Treatments	Groundnut yield (t/ha)		Maize yield (t/ha)		Cassava yied (t/ha)	
	Sequence 1	Sequence 2	Sequence 1	Sequence 2	Sequence 1	Sequence 2
Cycle 1						
Natural fallow	0.54	0.45	1.29	0.92	5.57	6.55
Pueraria	0.52	0.45	1.35	1.05	5.83	7.26
Calliandra	0.44	0.46	0.97	1.04	5.01	7.79
Cycle 2						
Natural fallow	0.49	0.29	0.87	1.37	4.55	4.52(a)
Pueraria	0.38	0.28	1.17	1.54	4.64	4.87(a)
Calliandra	0.34	0.27	0.97	1.13	4.29	3.39(b)

Note:
a   Cassava yield was significantly higher in cycle 1 - sequence 2 than in the other cases.
    Significance of the groundnut differences was not tested (Hauser, pers.comm.).

Most differences between fallow types were not significant, only cassava yield was significantly lower with *Calliandra* hedges in the second cycle of the second sequence. Groundnut yields were lowest in sequence 2, second cycle, but significance was not tested. Thus, overall the different fallow treatments did not yet show much effect. In the end, the trial was terminated prematurely with the closure of the Humid Forest Station in 2006. No further yield data are available beyond those of Table 5.7.

Mucuna establishment and seed collection is easy and the plants die off in the dry season. This makes it easy to handle for farmers, contrary to *Pueraria* which is difficult to establish and cumbersome in seed collection and did not die in the dry season. Both legumes were tested with and without burning the biomass before seeding maize on an Acrisol in southern Cameroon. In the first four years, cumulative maize grain yield was higher after *Mucuna* and *Pueraria* fallow than after natural regrowth (Hauser et al., 2002). Burning the biomass had no negative effects on grain yields. Trials in farmers' fields confirmed these results (Hauser et al., 2008).

Some preliminary work was also carried out in southern Cameroon with pigeon pea planted into the cassava left over after harvesting the groundnut and maize in the groundnut field (IITA, 1999b). Farmers were said to have noted that 'where pigeon pea established well, weeds were suppressed and cassava harvest was easier compared to the weedy control'.

MANAGEMENT OF SHORT FALLOWS IN THE SAVANNAH AREA

Legumes as components of short fallows in the savannah zone were expected to serve two purposes: (i) improvement of soil fertility, (ii) eradication or at least reduction of problem weeds, in particular *Imperata* and possibly (iii) contributing to fodder production.

A collaborative testing program was started in 1996 in collaboration with NARS scientists to assess the suitability of short legume fallows for the local cropping systems in three savannah ecologies in Nigeria:

1   In the forest-savannah transition area (Ibadan, south-west Nigeria), first season maize was interplanted with *Mucuna, Pueraria, Mucuna + Pueraria* or cowpea; after the maize harvest the legumes took over in the short rainy season and the dry season, and were followed by the conventional maize+cassava in year 2;

2   In the SGS, in Ilorin (south-west Nigeria), first season maize was interplanted with pigeon pea, *Crotalaria ochroleuca, Crotalaria verrucosa, Centrosema pascuorum, Lablab* or *Mucuna*, which took over during the short rains and dry season and were followed by sole maize in the second year; and

3   In the NGS (Shika, northern Nigeria), pigeon pea, *Centrosema pascuorum* or *Sorghum almum* were planted as full season crops, harvested for animal fodder, and followed by sole maize in the next year.

The tests were conventional replicated researcher-managed trials in farmers' fields and different fertiliser regimes were applied in each site (IITA, 1999b).

In Ibadan and Ilorin, *Mucuna* smothered the maize, reducing maize yield by 50% and 99% respectively in the absence of fertiliser, probably because of poor development of the maize which yielded less than 1.5 t/ha in the control plots. *Lablab* also reduced maize yield by 72% in Ilorin. The positive effect of *Mucuna* and *Lablab* cover on maize yield in the following year was therefore rather meaningless. The after-effect of the other legumes on maize did not make up for the foregone second season crop (in Ibadan that would be intercropped cassava and in Ilorin relay-cropped sorghum (*Sorghum bicolor*). In any case, the tests were inconclusive at best, because the more vigorous legumes, interplanted too early, outcompeted poorly developed associated maize, which makes the technology risky for farmers.

In Shika, the after-effect of the cover crop on the yield of fertilised maize was larger than in the other two sites but only significant for *Centrosema* (4.5 t/ ha, compared with 3.1 t/ha for natural fallow). Here, the question is whether the fodder produced by the cover crops plus the maize yield advantage would compensate for the efforts involved in growing the cover crop.

In a test in the DS and degraded forest areas of Nigeria on the effect of legumes on speargrass, it was found that legumes interplanted into maize or maize+cassava took three cropping seasons to reduce rhizome weight of speargrass by more than 80%. *Mucuna* and *Lablab* were most effective in the

DS (Ijaiye, south-west Nigeria) and *Mucuna* and *Pueraria* in the degraded forest (Ezilla, south-east Nigeria). In another test in Kouti (Benin) and Ogoja (Nigeria), the *Imperata* rhizomes were almost completely eliminated after 2 years by *Mucuna* and *Mucuna+Pueraria* cover crops (IITA, 1999b).

A long-term research project was carried out along a north-south transect in Nigeria on the effect of legume fallow (species not mentioned) with or without grazing on the fallow vegetation and soil dynamics. Some interesting results from these trials were (IITA, 1999c):

- Maize yield in the DS was higher for grazed than for non-grazed plots in both natural and legume fallow system: 3.0 and 3.3 t maize/ha respectively compared with 2.1 t/ha;
- In natural fallow, soil organic carbon (SOC) was higher in non-grazed than in grazed land, while the opposite was the case in legume fallow; and
- Grazing resulted in higher available-P content of the soil in both systems.

In a parallel study along the same transect, the effect of different utilisation ratios of maize-*Lablab purpureus* residue for mulching (M) and fodder (F) on subsequent maize yield and soil chemistry was studied. The ratios ranged from 100%M-0%F to 0%M-100%F. Preliminary results for the DS showed no effect of different ratios on subsequent maize or *Lablab* yield.

Integration of perennials in alley cropping systems

Evidence had been accumulating (Chapter 4) that adoption of alley cropping/ farming remained far below expectation and that there might be only a limited niche for the technology. Nevertheless, some new field research was initiated in the humid forest zone of Cameroon where little work had so far been carried out on alley cropping. The first alley cropping trials performed poorly. Several different hedgerow arrangements were tried in farmers' fields, all of them at 40% of the 'normal' alley cropping density. The arrangements were: equidistant, clustered, hedgerows or fence. There was large and seemingly unsystematic variation in dry matter production between treatments across villages and farmers and no significant trends emerged. For the farmers, the depressing effect the shrubs (*Calliandra*) had on *Chromolaena* turned out to be their most important consideration, which made them opt for the standard alley arrangement as the best (Nolte et al., 2003).

The fact that even with changed pruning height and changed planting patterns of cassava and maize the results did not get better (Hauser et al., 2000) led to the 'formal' termination of the work on 'standard' alley cropping. One of the trials was retained as a two year fallow-one year cropping rotation with the common groundnut/cassava/maize intercrop. After three fallow/crop cycles, none of the tree-based systems had been able to out-yield the no-tree control (Hauser, 2008) and the entire approach of planted tree-based fallow management was abandoned.

5.3.3 Technology validation

In the past on-farm technology validation had been undertaken by various scientists and their collaborators, but the work had lacked an agreed methodology, which often made the results difficult to interpret or compare. There was, for example, no consensus about the way participating farmers had to be chosen and about the 'acceptable' degree of farmer management. One school of thought maintained that farmers should only be invited to participate in the trials rather than be selected, and researchers should intervene as little as possible in the non-treatment aspects, i.e. the general management of the tests. At the other extreme, the researchers would select farmers according to strict criteria and run on-farm trials in much the same way as they did at the station. The former approach (Mutsaers et al., 1997; Versteeg and Koudokpon, 1993) had been used in the 1980s and early 1990s in several 'On-farm Research villages' in south-west Nigeria and Benin and to some extent in outreach projects, but in most other cases the researchers kept more or less strict control over the trials. During the present episode, there was a trend towards stronger involvement of farmers in the trials, as partners rather than passive observers, but the trials still remained essentially under researcher control. Most of the on-farm trials were therefore discussed in Sections 5.3.1 and 5.3.2.

Herbaceous legumes

In Benin, the *Mucuna* cover crop technology had passed the testing phase and was being vigorously promoted by development organisations (see Section 5.4). There was concern among the researchers, however, about the stability of adoption by farmers. In a survey of 277 farmers, they found only 25% 'confirmed adopters', i.e. farmers using *Mucuna* twice or more for *Imperata* control or fertility improvement (Versteeg et al., 1998). Farmer interviews suggested that the attractiveness of the species would be enhanced if the grain could be made edible by detoxification, and research was therefore initiated with farmers to test preparation methods which would enhance the palatability of the grain, e.g. as flour mixed with maize.

Alley cropping

Although some experimental research on alley cropping continued during this episode, there was no longer an active dissemination thrust, especially since the demise of AFNETA in the mid-1990s, when some of its activities were absorbed by ICRAF and it essentially died there as a separate network. Disappointing adoption and consequent decline of funding were the causes (Atta-Krah, pers. comm.).

5.4 Technology delivery and dissemination

During this episode, dissemination of technology continued to be seen as primarily the responsibility of development organisations, according to the linear research-extension model. Dissemination requires information on the technology's likely performance under conditions other than those where it was developed. The Institute's mission was to develop prototype technology which could be adapted and disseminated by other organisations, and the Institute would therefore be expected to prepare technology digests and guidelines for its clients. So far the record had not been impressive. In previous chapters, some examples were given of practical manuals and guidelines, but a lot of the Institute's work had not been consolidated and operationalised in this way. During the current episode, some progress was made in that respect for herbaceous legume-based technology.

In spite of the prevalence of the linear research-extension model, researchers also started to associate themselves more directly with dissemination, through several collaborative initiatives with government and non-government development organisations. These activities were still largely technology-driven, in particular around legume-based technology, although the RAMR approach of offering a number of options to farmers (for rehabilitation of degraded soil) may have been adopted by the collaborating organisations.

5.4.1 Databases, technology digests and guidelines

From 1998 through 2001, IITA-Benin operated a Center for Information on Cover Crops in Africa (CIEPCA) which provided both information and seed for a range of herbaceous legumes and managed the LEXSYS package. Altogether eight semi-annual CIEPCA newsletters were published and sent to more than 300 readers. Cornell University collaborated by posting the Newsletter on its website until publication stopped in 2001. A comprehensive summary of research findings on *Mucuna* as a cover crop (Carsky et al., 1998) was widely distributed as a guide for further testing by NARS and other organisations.

The data on the performance of alley cropping collected in primarily on-station trials by AFNETA collaborators had been analysed earlier by Woomer et al. (1995), showing widely varying yield effects, positive for maize, moderately positive to neutral for grain legumes and negative for cassava and cotton (Chapter 4).

5.4.2 Decision support

The decision support package LEXSYS, developed in the late 1980s (Chapter 4) is a good example of what could be done. During this episode, the package was further developed and distributed to many potential users in the region and elsewhere. Feedback from the field, however, was said to be sparse (IITA, 1999a).

Alley cropping, though a 'mature technology' by now, had not been the subject of a similar effort to bring together the findings in the form of a decision aid tool. Something approaching this was the Training Manual on alley farming produced by AFNETA in the early 1990s (AFNETA, 1992).

The BNMS-I project was keenly aware of the need to convert its research findings into a form which made them useable for NARS scientists and extension personnel. Hence, the original project design included as one of its outputs: 'Decision support systems for implementation of BNMS technologies for extension and research'. In the course of the project, it turned out to be an unrealistic goal within the time span of the first phase and was therefore postponed to Phase II (BNMS-I, 2000).

5.4.3 Dissemination and monitoring and evaluation

Dissemination in fact starts as soon as farmers are exposed to innovations in on-farm tests and demonstrations, and come to the conclusion that they may be worth applying in their farms. M&E should therefore be constantly alert for indications that farmers are actively engaged in applying the innovations on their own account. This principle was beginning to be used in some of the on-farm research.

In the early-1990s, two large international organisations, SG2000 and the World Bank, had picked up the idea of using *Mucuna* as an auxiliary crop, inspired by the successes in Benin. Seed was initially provided by RAMR/IITA and multiplied and used by the extension organisations supported by these organisations. By 1996, SG2000 reported that 10,000 farmers were testing *Mucuna* in their fields in 500 m² demonstration plots (Carsky et al., 1998; Douthwaite et al., 2002). A large amount of seed of *Mucuna, Aeschynomene, Centrosema brasilianum, Centrosema pascuorum, Lablab, Stylosanthes guianensis* and *Stylosanthes hamata,* was distributed through CIEPCA to International Agricultural Research Centers (lARCs), NARS and NGOs in and outside Nigeria (IITA, 1999a).

Later on collaboration was established between IITA/BNMS, RAMR and SG2000 for the on-farm demonstration of integrated soil fertility management, consisting of fertiliser and organic matter, produced *in situ* by dual purpose legumes (cowpea, soybean) or brought in from outside as manure, combined with improved management (Iwuafor et al., 2002). In northern Nigeria, the initial advantage of the improved package compared to farmer practice declined in following years, which was explained by farmers adopting (part of) the package in their own system. Other interesting observations were that farmers' yields differed widely and that they chose different elements of the package, according to their resources (BNMS-II, 2003).

5.5 Outputs, impact and emerging trends

This episode marked the beginning of a new phase in the evolution of the Institute's soil and soil fertility research, with extensive surveying for the development of new decentralised programs, on-station research to test new ideas, on-farm research to test mostly simple innovations in farmers' fields as well as some terminal activities around research projects from the previous phases. We present an overview here of the most important findings as well as some preliminary results of monitoring and evaluation of recently tested innovations and final results on the (non-)adoption of once prominent technologies, viz., *Mucuna* short fallow and alley cropping.

5.5.1 Highlights of research outputs

The numerous surveys for benchmark characterisation contributed insights which would influence the Institute's future research programs, even though most of the findings were really restatements of well-known phenomena:

- In the Forest Margin benchmark, farming households could be meaningfully characterised in terms of their access to land resources and the availability of family labour; the perennial crops cocoa and oil palm made an important contribution to the rural economy;
- In the NGS of Nigeria, high value production (peri-urban vegetable growing and livestock) was important; a useful dichotomy of farmers was based on the amount of fertiliser-N they used on their crops, with a threshold of 30 kg/ha/year;
- Everywhere in the humid and sub-humid zone a trend towards shortening fallows was observed;
- Soil conditions are highly variable over short distances along toposequences, especially in the basement complex area; and
- Important factors militating against intensified production in the moist savannah were nutrient deficiencies, *Striga* and weeds, including *Imperata*.

Some findings from the experimental research:

ON SOM AND BIOMASS:

- There was only a weak relationship between total N-content of the soil and crop N-uptake, and a much better one with N-content in the particulate organic matter fractions;
- The Rothamsted Carbon Model was able to simulate SOM accumulation under alley cropping and continuous cropping but substantial parameterisation was required; and
- In low-yielding fields in the DS and NGS, the effect on maize yield of a combination of fertiliser-N and biomass was the same as that of the same amount of N from fertiliser alone.

ON NUTRIENT RESPONSES:

- The geographic pattern of soil P-status and P-response was chequered, due to both inherent soil properties and previous land use (e.g. fertiliser use in cotton);
- There is wide variation among plant species and varieties within species in their ability to extract P from low-P soil, due to their association with AMF and possibly root exudates;
- Incorporation of RP into compost did not have a measurable effect on P-availability;

- Preliminary nutrient response tests in 'groundnut fields' in southern Cameroon showed only a response to K by cassava; and
- *Pueraria* could be a good indicator crop for low-K soil.

ON LEGUMES AND P-AVAILABILITY:

- Some legumes, including *Mucuna, Lablab* and cowpea were found to be able to unlock P from RP, benefitting subsequent maize; and
- Association with AMF will often result in better ability of legumes and cassava to extract P from P-deficient soil.

ON ALLEY CROPPING AND LEGUMINOUS COVER CROPS:

- Alley cropping and leguminous cover crops continued to show very variable results, depending on soil conditions, management and the species used; and
- In short fallows *Chromolaena* often performed equally or better than other fallow types, including alley cropping.

5.5.2 Uptake and impact

A crucial stage in the research-development process is the assessment of adoption or non-adoption by farmers of the experimented and demonstrated technology. This had been an area of relative opacity in the past, where more effort and more adequate methodology would have been needed. In coming years, the study of uptake and impact would receive increasing emphasis, but for most of the technologies tested and demonstrated during this episode time has been too short to really ascertain genuine technology adoption. The exceptions are *Mucuna* and alley cropping.

ADOPTION OF *MUCUNA*

There was a massive effort in the second half of the 1990s to disseminate the *Mucuna* technology to farmers in Benin, by the World Bank-supported national extension service and by SG-2000. At the peak of this thrust in 1996, seed was distributed to 14,000 farmers, many of which used it primarily to control speargrass, especially in the southern part of the country where this weed is important (Manyong et al., 1999). Profitability for farmers was high as long as SG-2000 purchased the seed, making the species artificially into a commercial crop, but when they stopped doing that in 1996, the adoption declined by 25% (Douthwaite et al, 2002). The rise and following decline of *Mucuna* adoption are illustrated by Figure 5.8 (Manyong et al., 1999). There have been no further studies on adoption since then, but informal information suggests that the technology has not survived (Vanlauwe, pers. comm.).

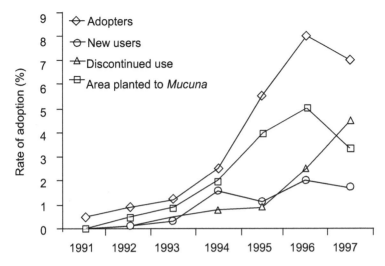

Figure 5.8 Dynamics of *Mucuna* fallow adoption in Southern Benin (1991–1997) (adapted from Manyong et al., 1999)

ADOPTION OF ALLEY CROPPING

A new study was undertaken in 1996 about adoption and adoptability of alley farming (Adesina et al., 1999). The study was deemed necessary because the authors were not convinced by the unfavourable outcomes of the earlier studies (Chapter 4) which, they felt, had not actually measured adoption. The new study was essentially based on questionnaire surveys in selected villages in Nigeria, Cameroon and Benin. The results are shown in Table 5.8. The percentages of interviewees who had been 'exposed' to the technology were quite high and so was the percentage of those of them who actually established alley farms. The

Table 5.8 Assessment of adoption and adoptability of alley cropping in Nigeria and Cameroon (number of farmers with the last 2 lines also indicating the percentage of farmers) (Adesina et al., 1999)

Households	Nigeria		Cameroon			Benin
	South-East	South-West	North-West	South-West	Centre	Mono
Interviewed	81	142	341	256	223	288
Exposed[a]	71	137	285	211	176	225
Established[b]	43 (54%)	96 (68%)	120 (35%)	86 (34%)	50 (22%)	72 (25%)
Retained[c]	25 (58%)	49 (51%)	115 (96%)	81 (94%)	42 (84%)	67 (93%)

Notes:
a Those interviewed, who had at least heard about the technology.
b Those who had planted hedgerows.
c Those who maintained the established hedgerows.

question therefore arises to what extent the choice of villages for the survey was influenced by previous on-farm research, even though according to Adesina et al. (1999), 'villages with and without extension activities [on agroforestry technologies] were selected for data collection'. Furthermore, a rather optimistic rendering is given of 'retention', which referred to those farmers who continued to use their hedgerows in some way, or modified the technology. This category also included farmers who had let their alley farms go into fallow. That could mean that the hedges, as far as they survived, had simply become part of the fallow vegetation[10].

In the case of Cameroon, where the 'retention' rate was reported to be very high, one would like to know to what extent the researchers influenced farmers to keep maintaining their hedgerows. Another crucial question in all cases would be whether any of the farmers had established or expanded alleys on their own initiative, but that question was not addressed. This new adoption study thus hardly refutes the conclusions of the earlier ones that alley cropping fields were not being adequately managed and that the technology was not being actively adopted (Chapter 4). In a summary paper on the history of alley cropping adoption, Douthwaite et al., (2002) also concluded that the high rates of adoption reported in earlier studies were largely explained by a mistaken interpretation of the term 'adoption', which should only be used when farmers planted or extended alleys, or at least permanently maintained the ones they had.

5.6 Emerging trends

Extensive and often detailed surveys of socio-economic and physical determinants of farmers' systems were conducted, as well as a range of diagnostic on-farm trials. The area-wide surveys were useful to characterise the Benchmark areas and highlighted the importance of cocoa and other perennials, intensive horticulture and livestock as income sources. This brought these commodities into the foreground as necessary targets for the Institute's research, if it was to transform into a genuine Research for Development mode. The question is justified, however, whether many of the conclusions from the detailed time-consuming surveys at village level about farmers' constraints and opportunities could not have been drawn in a much less complicated manner, by careful informal observation, as exemplified by the work of van Gils and others in southern Cameroon in the 1970s (e.g. van Gils, 1975). Kleene et al. (1989) also arrived at that conclusion when reviewing their own elaborate surveys in southern Mali.

In retrospect, the vastness and patchiness of the humid and sub-humid ecoregion would make it practically impossible to attain the ambitious goal of identifying or developing and promoting specific technology for each set of conditions occurring in the region. In fact, the Benchmark setup, which was intended to form EPHTA's organisational framework, was initiated only in the Forest Margins and the DS and NGS Benchmarks, where the Institute had already major on-going activities prior to the project. Even there it hardly went beyond the characterisation stage and died a more or less silent death by the end of this episode. Nevertheless, a strong defence of its rationale was formulated by its major protagonists in the early 2000s (Douthwaite et al. 2005) and some of EPHTA's concepts would continue to influence the Institute's programs in years to come.

An intriguing question arises from the limited interest particularly in the SGS, compared with the NGS. The choice to concentrate (scarce research resources) in the latter was probably motivated by a more dynamic production sector there, so the question is: what is holding back development of the vast and thinly populated SGS (called the 'middle belt' in Nigeria) and what is limiting its productive potential? This question had so far not been asked, at least not explicitly.

As far as the characterisation of soils is concerned, the attention of research in this and the preceding episode had shifted from the study of soil profiles and chemical and physical properties to biological and cropping factors and the effect of land use on the all-important topsoil (IITA, Project 1, 1999b):

> Before attempting to make any change to the practices used by farmers to maintain soil fertility status, it is necessary to diagnose how organic resources, such as crop residues or animal manure and inorganic inputs, are currently used. [...] it is also necessary to relate the information obtained to a diagnosis of the general nutrient status of the soils in the study domain.

A renewed awareness was arising, however, of the importance of the properties of the soil matrix and their implications for the soil's productive potential (IITA, Project 2, 1999c):

> Extrapolation of results from on-farm trial sites to the larger domain will only be possible if the fields are chosen such that they cover the important soil units present in the landscape. Detailed soil mapping of the villages prior to field selection and a pedological characterisation of experimental fields selected thereafter is indispensable.

The BNMS-I project took up the challenge and intended to develop a GIS-based decision support system for diffusion of the developed BNMS technologies, and a decision support system to calculate the nutrient balance for a wide range of cropping systems and management practices. That would have required mapping of the mandated area on the basis of characteristics with relevance for extrapolation of its results, soil properties being prominent among them, as well as an inventory and interpretation of a wide range of experimental results (BNMS-I, 1998). These very ambitious goals could not be achieved within BNMS-I's time span, because of the difficulties of accessing and interpreting the existing soil and soil fertility information and 'the lack of expertise in GIS in IITA' (BNMS-I, 2000). Nevertheless, these objectives point forward to the challenges soil and soil fertility research were going to tackle with the adoption of a Research for Development orientation in years to come.

They also point backward to the original concept of the Institute's soil characterisation research and raise the question of how current soil characterisation work can be linked with and benefit from that of the 1970s and 1980s. In most publications from this and the following episodes, soils were classified according to the FAO/UNESCO System (now called the World reference Base, WRB), while Soil Taxonomy was used in the early years and even sometimes in this episode, but there was no reference to the correlation between the two systems. This correlation is reasonably well established only at the highest level, with a tentative correlation for specific benchmark soils given by IITA soil scientists in the early years (Annexe II). Further correlation at lower classification levels is currently being attempted by ISRIC and other international soil research institutes, with reasonable success (Leenaars, ISRIC, pers. comm.). In order to exploit the Institute's large amount of earlier soil data, collaboration should be established with ISRIC.

The perceived potential of herbaceous legumes as intercrops or short fallow species continued to drive research on these species during this episode. Sensible requirements for their adoptability were formulated: (i) the legume would have to solve one or more specific problems associated with a target cropping system, such as low fertility and/or a major weed problem, (ii) there had to be a direct return from the legume in terms of food, fodder or other useful products, not just a 'postponed return', such as improved fertility or weed control and (iii) the problems involved in seed production would have to be solved.

By the early 2000s, however, the interest seems to have waned, due to the steeply declining use by farmers in Benin of the most successful species, *Mucuna*. Research to increase the palatability of its grain by detoxification, which started in the mid-1990s, apparently was no longer pursued. CIEPCA wound up in 2002, and the LEXSYS package experienced disappointing interest from the field. It was transferred to the School of Agricultural and Forest Sciences, University of Wales in 2002/2003 for unexplained reasons, where it remains in a comatose state until this day[11].

Two important findings led to a further rethinking of the alley cropping concept. One was the observation that an important impediment to the adoption of alley cropping was its sensitivity to insufficient or untimely pruning (e.g. Versteeg et al., 1998). The other was root competition by the hedgerows with the crops planted in the alleys, especially in shallow soils or acid subsoils. Proliferation of hedgerow roots beyond the experimental plots also put doubts on reported positive yield effects of alley cropping measured in small plots at research stations (Chapter 4, Hauser et al., 2006).

In Benin, farmers' frequent inability to keep the hedgerows under control led to a modification whereby hedges were planted in double rows and the spacing between them was increased, while maintaining the same overall shrub density (Versteeg et al., 1998). In Cameroon several different arrangements were tried in farmers' fields. There was large and seemingly unsystematic variation in dry matter production between treatments across villages and farmers and no significant trends emerged. For the farmers, the effect of the shrubs (*Calliandra*) on *Chromolaena* turned out to be their most important consideration, which made them opt for the standard alley arrangement as the best (Nolte et al., 2003).

In retrospect, when even for researchers, it was often difficult to judge whether there were significant benefits in terms of crop yield and soil fertility enhancement, it is not surprising that farmers were not keen on adoption.

The decline in adoption of *Mucuna* and the abandonment or neglect of many alley cropping fields, revealed by several impact studies, led to the virtual abandonment of these technologies by research and development organisations. The *Mucuna* story was well documented in several detailed studies, but no attempt was made to bring together all that was known about the more complex alley cropping system, its remaining potential for specific conditions and the causes of its failure (CGIAR Science Council, 2008). It appears as if the technology, in which a great amount of energy had been invested, simply dropped out of sight.

Practically all on-farm work in this episode was researcher-controlled. Earlier problems with such trials, e.g. that farmers would steer trials to poorer soils, the difficulty to find more or less equal representation for stratified samples, the eternal problem of late planting and poor maintenance, were also experienced in some on-farm work in this episode. However, there was also a trend towards more realistic on-farm experimentation, with farmers carrying out the tests in their actual production fields, instead of in fields specially chosen for the occasion, the use of single replicates per farmer and statistical methods to explain variability, rather than *a priori* stratification to capture such variability. Another new element was to offer farmers a 'basket of options' to choose from, rather than fixed packages, based on the observation that farmers would choose those components from the demonstrations which best fitted their needs.

Soon after the publication of the 2001–2011 Strategic Plan, however, the Institute moved away from the eco-regional approach towards more direct involvement with the farmer through involvement in various development-oriented projects and the adoption of a *Research for Development* (R-for-D) approach. This was endorsed later by the 2007 External Program Review and the CGIAR Science Committee (CGIAR, 2008), which recommended that the centre should:

> develop an overall strategy to contribute to R-for-D and strengthen the alignment between such a strategy and the Center's [core] projects and individual grant projects.

The Institute's research on soils and soil fertility management has since continued to be inspired by the goal of contributing directly to development and most of its research has moved on-farm.

The new ambition, to work directly with and at the service of farmers and other stakeholders as a way to attain real impact was yet to be realised. That ambition would be the main thrust of the next episodes.

Notes

1 Defined by Jagtap (1995) as the period when precipitation exceeds half the potential evapotranspiration.

2 The Bauchi Benchmark (Chapter 4) was given up as part of the agreement under EPHTA, which opted for a single Benchmark in the Northern Guinea savannah.
3 A major example was the Belgian-funded collaborative project between Leuven University (KU-Leuven) and IITA on 'Process-based studies of soil organic matter dynamics in relation to the sustainability of agricultural systems in the tropics', and its successor, the Balanced Nutrient Management System (BNMS) project, which started in 1997.
4 This is the list since 2000, in the earlier years the denominations were somewhat different.
5 *Parturient montes, nascetur parvulus mus*, so to speak.
6 Farmers' practice would be the reverse, with cowpeas in the second season.
7 Conclusion not consistent with data in table of IITA (1999a).
8 Interestingly, similar techniques were tried in the 1930s and 1950s in rubber plantations in Indonesia to get seedlings established in *Imperata*-infested fields (van der Meulen, 1976).
9 The cumulative dry matter yields over all cropping cycles were similar for the two sites.
10 As some recent observations suggested has also happened in Ayepe and Alabata in South-West Nigeria (Mutsaers, pers. comm.).
11 At the time of writing it could still be downloaded from http://www.bangor.ac.uk/senrgy/content/other/LEXSYS/LEXSYS.zip.

References

Adesina, A.A., O. Coulibaly, V.M. Manyong, P.C. Sanginga, D. Mbila, J. Chianu and D.G. Kamleu, 1999. *Policy shifts and adoption of alley farming in West and Central Africa.* Impact paper, IITA, Ibadan.

AFNETA, 1992. *The AFNETA Alley Farming Training Manual. Volume 1, Core Course in Alley Farming; Volume 2, Source Book For Alley Farming Research.* IITA, Ibadan.

Asadu, R. and F. Nweke, 1999. *Soils of Arable Crop Fields in Sub-Saharan Africa: Focus on Cassava-growing Areas.* COSCA Working Paper No. 18. IITA, Ibadan, 182 pp.

BNMS-I, 1998. Balanced nutrient management systems for maize-based systems in the moist savanna and humid forest zone of west Africa. *Annual Report 1997.* IITA, KULeuven, DGDC.

BNMS-I, 2000. Balanced nutrient management systems for maize-based systems in the moist savanna and humid forest zone of west Africa. *Annual Report 1999.* IITA, KULeuven, DGDC.

BNMS-II, 2003. Achieving development impact and environmental enhancement through adoption of balanced nutrient management systems by farmers in the west African savanna. *Annual Report 2002.* IITA, KULeuven, DGDC.

Carsky, R.J., S.A. Tarawali, M. Becker, D. Chikoye, G. Tian and N. Sanginga, 1998. *Mucuna – Herbaceous Cover Legume with Potential for Multiple Uses.* Resource and Crop Management Research Monograph No. 25, IITA, Ibadan.

CGIAR Science Council, 2008. *Report of the 6th External Program and Management Review of the International Institute of Tropical Agriculture (IITA).* CGIAR Science Council Secretariat, Rome, Italy.

Diels, J., B. Vanlauwe, M.K. van der Meersch, N. Sanginga and R. Merckx, 2004. Long-term soil organic carbon dynamics in a subhumid tropical climate: ^{13}C data in mixed C_3/C_4 cropping and modeling with ROTHC. *Soil Biology & Biochemistry*, 36, 1739–1750.

Douthwaite, B., V.M. Manyong, J.D.H. Keatinge and J. Chianu, 2002. The adoption of alley farming and *Mucuna:* lessons for research, development and extension. *Agroforestry Systems*, 56, 193–202.

Douthwaite, B., D. Baker, S. Weise, J. Gockowski, V.M. Manyong and J.D.H. Keatinge, 2005. Ecoregional research in Africa: learning lessons from IITA's Benchmark Area Approach. *Experimental Agriculture*, 41, 271–298.

Hauser, S., 2000. Effects of fertiliser and hot-water treatment upon establishment, survival and yield of plantain (*Musa* spp., AAB, French). *Field Crops Research*, 66, 213–223.

Hauser, S. 2006. Soil temperatures during burning of large amounts of wood, effects on soil pH and subsequent maize yields. Tropentag 2006, 11–13 October, Bonn, Germany. Available online at www.tropentag.de/2006/abstracts/full/54.pdf.

Hauser, S., 2008. Groundnut/cassava/maize intercrop yields over three cycles of planted tree fallow/crop rotations on Ultisol in Southern Cameroon. *Biological Agriculture and Horticulture*, 25, 379–399.

Hauser, S. and C. Nolte, 2002. Biomass production and N fixation of five *Mucuna pruriens* varieties and their effect on maize yields in the forest zone of Cameroon. *Journal of Plant Nutrition and Soil Science*, 165, 101–109.

Hauser, S., J.N. Ndi and N.R. Hulugalle, 2000a. Yields of maize/cassava intercrops grown with hedgerows of three multipurpose trees on an acid Ultisol of Cameroon. *Agroforestry Systems* 49, 111–122.

Hauser, S., J.N. Ndi and N.R. Hulugalle, 2000b. Performance of a maize/cassava intercrop in tilled and no-till *Senna spectabilis* alley cropping on an Ultisol in southern Cameroon. *Agroforestry Systems* 49, 177–188.

Hauser, S., J. Henrot and A. Hauser, 2002. Maize yields in a *Mucuna pruriens* var. *utilis* and *Pueraria phaseoloides* relay fallow system on an Ultisol in southern Cameroon. *Biological Agriculture and Horticulture*, 20, 243–256.

Hauser, S., C. Nolte and R.J. Carsky, 2006. What role can planted fallows play in the humid and sub-humid zone of West and Central Africa? *Nutrient Cycling in Agroecosystems*, 76, 297–318.

Hauser, S., L. Norgrove and J. Nkem, 2006. Groundnut/maize/cassava intercrop yield response to fallow age, cropping frequency and crop plant density on an Ultisol in southern Cameroon. *Biological Agriculture and Horticulture*, 24, 275–292.

Hauser, S., B. Bengono and O.E. Bitomo, 2008. Short- and long-term maize yield response to *Mucuna pruriens* and *Pueraria phaseoloides* relay fallow and biomass burning versus mulching in the forest zone of southern Cameroon. *Biological Agriculture and Horticulture*, 26, 1–17.

Houngnandan, P., P. Sanginga, P. Woomer, B. Vanlauwe and O. Van Cleemput, 2000. Response of *Mucuna pruriens* to symbiotic nitrogen fixation by rhizobia following inoculation in farmers' fields in the derived savanna of Benin. *Biology and Fertility of Soils*, 30, 558–565.

IITA, 1995–1997. *Annual Reports, 1995–1997*. IITA, Ibadan.

IITA, 1998. Project 10. Farming Systems Diversification. *Annual Report 1997*. IITA, Ibadan.

IITA, 1999a. *Strategic Plan 2001–2010, Supporting Document*. IITA, Ibadan.

IITA, 1999b. Project 1. Short fallow systems to arrest resource degradation due to land-use intensification. *Annual Report 1998*. IITA, Ibadan.

IITA, 1999c. Project 2. Agroecosystems Development Strategies and Policies. *Annual Report 1998*. IITA, Ibadan.

IPCC, 2001. *Climate Change 2001: The Scientific Basis. Third Assessment Report, Intergovernmental Panel on Climate Change.* Cambridge University Press, Cambridge.

Iwuafor, E.N.O., K. Aihou, B. Vanlauwe, J. Diels, N. Sanginga, O. Lyasse, J. Deckers and R. Merckx, 2002. On-farm evaluation of the contribution of sole and mixed applications of organic matter and urea to maize grain production in the savanna. In: B. Vanlauwe, J. Diels, N. Sanginga and R. Merckx (eds). *Integrated Plant Nutrient Management in Sub-Saharan Africa, From Concept to Practice.* CAB International, Wallingford.

Jagtap, S.S., 1995. Environmental characterization of the moist lowland savanna of Africa. In: B.T. Kang, I.O. Akobundu, V.M. Manyong, R.J. Carsky, N. Sanginga and E.A. Kueneman (eds) *Moist Savannas of Africa: Potentials and Constraints for Crop Production.* IITA, Ibadan.

Juo, A.S.R. and M.K. van der Meersch, 1983. Soil degradation. *IITA Annual Report 1982.* ITTA, Ibadan.

Kleene. P., B. Sanogo and G. Vierstra, 1989. *A partir de Fonsébougou: Présentation, objectifs et méthodologie du Volet-Fonsébougou (1977–1987).* KIT, Amsterdam.

McHugh, D. and J. Kikafunda-Twine, 1995. Ten years of Farming Systems Research in North West Highlands of Cameroon. *Resource and Crop Management Research Monograph No. 13.* IITA, Ibadan, Nigeria.

Manyong, V. M., K.O. Makinde and J.O. Olukosi, 1998. Delineation of resource-use domains and selection of research sites in the northern Guinea savannah ecoregional benchmark area, Nigeria. Paper presented at the launching of the northern Guinea savannah ecoregional benchmark area, Institute of Agricultural Research, Ahmadu Bello University, Zaria, 2 December, 1998.

Manyong, V.M, V.A. Houndékon, P.C. Sanginga, P. Vissoh and A.N. Honlonkou, 1999. *Mucuna Fallow Diffusion in Southern Benin.* IITA, Ibadan.

Manyong, V.M., K.O. Makinde, N. Sanginga, B. Vanlauwe and J. Diels, 2001. Fertiliser use and definition of farmer domains for impact-oriented research in the Northern Guinea savanna of Nigeria. *Nutrient Cycling in Agroecosystems,* 59, 129–141.

Menzies, N.W. and G.P. Gillman, 1997. Chemical characterization of soils of a tropical humid forest zone: a methodology. *Soil Science Society of America Journal,* 61, 1355–1363.

Mutsaers, H.J.W., P. Mbouemboue and Mouzong Boyomo, 1981. Traditional foodcrop growing in the Yaounde area (Cameroun). I. Synopsis of the system. *Agro-Ecosystems,* 6, 273–287.

Mutsaers, H.J.W., G.K. Weber, P. Walker and N.M. Fisher, 1997. *A Field Guide for On-Farm Experimentation.* IITA/ISNAR/CTA, Ibadan.

Nolte, C., T. Tiki-Manga, S. Badjel-Badjel, J. Gockowski, S. Hauser and S.F. Weise, 2003. Effects of *Calliandra* planting pattern on biomass production and nutrient accumulation in planted fallows of southern Cameroon. *Forest Ecology and Management,* 179, 535–545.

Slaats, J.J.P., 1995. *Chromolaena odorata* fallow in food cropping systems: An agronomic assessment in south-west Ivory Coast. Doctoral Thesis, Wageningen University, The Netherlands.

Smaling, E.M.A., J.J. Stoorvogel and B.N. Janssen, 1993. Calculating soil nutrient balances in Africa at different scales. *Fertilizer Research,* 35, 227–235.

TAC, 1996. *Report of the Fourth External Programme and Managment Review of the International Institute of Tropical Agriculture (IITA).* TAC Secretariat, FAO, Rome.

Tchienkoua, G.M, and N.W. Menzies, 1996. Physiography and soils in the humid part of south Cameroon. Internal document, IITA, Ibadan.

Tian, G., B.T. Kang, G.O. Kolawole, P. Idinoba and F.K. Salako, 2005. Long-term effects of fallow systems on crop production and soil fertility maintenance in West Africa. *Nutrient Cycling in Agroecosystems*, 71, 139–150.

van der Meulen, G.F., 1976. A real Green Revolution. Unpublished mimeograph, ACBT, The Hague (cited in Mutsaers, 2007).

van der Pol, F., 1992. *Soil Mining: An Unseen Contributor to Farm Income in Southern Mali.* Bulletin 325. Royal Tropical Institute, Amsterdam.

van Gils, L., 1975. Lexique des problèmes agricoles, des 'solutions' pratiquées par la population et des 'améliorations' à préconiser. Zone d'application et de vulgarisation de l'Ecole Nationale Supérieure Agronomique, située dans l'Arrondissement d'Obala. Mimeograph, ENSA, Nkolbisson, Cameroon.

Vanlauwe, B., C. Nwoke, J. Diels, P. Sanginga, R. Carsky, J. Deckers, and R. Merckx, 2000. Utilization of rock phosphate by crops on a representative toposequence in the Northern Guinea savanna zone of Nigeria: response by *Mucuna pruriens*, *Lablab purpureus*, and maize. *Soil Biology and Biochemistry*, 32, 2063–2077.

Vanlauwe, B., J. Diels, O. Lyasse, K. Aihou, E.N.O. Iwuafor, N. Sanginga, R. Merckx and J. Deckers, 2002. Fertility status of soils of the derived savanna and Northern Guinea savanna and response to major plant nutrients, as influenced by soil type and land use management. *Nutrient Cycling in Agroecosystems*, 62, 139–150.

Versteeg, M.N and V. Koudokpon, 1993. Participative farmer testing of four low external input technologies, to address soil fertility decline in Mono Province (Benin). *Agricultural Systems*, 42, 265–276.

Versteeg, M.N., F. Amadji, A. Eteka, A. Gogan, and Y. Koudokpon. 1998. Farmers' adoptability of *Mucuna* fallowing and agroforestry technologies in the coastal savanna of Benin. *Agricultural Systems*, 56, 269–287.

Woomer, P., O. Bajah, A.N. Atta-Krah and N. Sanginga, 1995. Analysis and interpretation of alley farming network data from tropical Africa. In: B.T. Kang, O. Osiname and A. Larbi (eds), *Alley Farming Research and Development*. Proceedings of the International Conference on Alley Farming, IITA, Ibadan.

6 Research for development

From the virtual to the real farm – 2002–2011

6.1 Scope, approaches and partnerships

A new Strategic Plan for 2001–2010 was developed by the end of the last episode (1995–2001), based on the then organisational model with three Divisions, one of which was the Resource and Crop Management Division (RCMD), an eco-regional focus and 16 projects. Immediately after the arrival of the new Director General in 2001, there was another reshuffle of the research setup, into six thematic projects, each with its own coordinator, in replacement of the earlier 16 projects, to make management more efficient (IITA, 2003):

- Project A. Preserving and Enhancing Germplasm and Agrobiodiversity
- Project B. Developing Biological Control Options
- Project C. Impact, Policy and Systems Analysis
- Project D. Starchy and Grain Staples in East and Southern Africa
- Project E. Diverse Agricultural Systems in the Humid Zone of West and Central Africa
- Project F. Improving and Intensifying Cereal-Legume Systems in the Moist and Dry Savannahs of West and Central Africa

Soil and soil fertility management research was divided over the projects D, E and F. The ecoregional dimension of previous episodes is still apparent in these projects, but the concepts underlying the research would soon be overhauled and the ecoregional approaches would be abandoned.

The Plan was followed initially during this episode, but the new Director General, who arrived in the course of 2001, set a significant reorientation in motion. His vision was laid down in a 'philosophical' paper entitled 'An approach to hunger and poverty reduction for Sub-Saharan Africa' (Hartmann, 2004), which, according to the 2007 EPMR (External Program and Management Review) Panel, 'started to alter the way IITA addressed its research mission'. It would eventually result in an approach to research which came to be known as *Research-for-Developmen* (R-for-D). It changed the institute's vision about its role, from being a supplier of improved production technology at the service of development, to being an active partner in the development process itself, together with other stakeholders. This could even mean involvement in market development, if market opportunities were a priority constraint, a domain that was traditionally considered outside the competence of an agricultural research institute. It would also imply a more 'holistic' approach to farm productivity and soil fertility management, whereby nutrient flows in the farm would be considered rather than the effects of applied nutrients in particular soils on particular crops. In the francophone world, the former approach had actually been pursued since the days of Farming Systems Research (FSR) in the 1970s and 1980s and was based on a concept of participatory learning developed in Senegal and Mali, entitled '*Conseil de Gestion aux Exploitations Agricoles*'[1] (Kleene et al., 1989, Faure, Kleene and Ouedraogo, 1998; Defoer and Budelman, 2000).

In respect of the actual research content, this episode saw a variety of inherited and new initiatives in soils and soil fertility management research and a shift from ecoregional to 'crop-based systems' research, from the development of 'international public good' or prototype technology to integrated production packages for chosen product chains, and from developing and testing improved technology to a wide range of activities from technology 'discovery' to delivery.

In 2006, there was another reshuffle of the institute's programs, in order to align research with the emerging vision about its task. This reshuffle resulted in five new programs, four of which were to integrate all research activities around crop-based systems:

- Banana and Plantain Systems;
- Root and Tuber Systems;
- Cereal and Legume Systems; and
- High Value Products/Horticulture and Tree Systems.

Some research continued in the earlier benchmark sites, but the benchmark concept, which had been typical for the eco-regional approach (Chapter 5), was no longer followed. The Institute also withdrew from the Humid Forest Station and transferred it to the Cameroonian research organisation IRAD (*Institut de Recherche Agricole pour le Développement*) in 2006, mostly because of the decline in core funding. For similar reasons, IITA handed back the IITA-managed research farm Namulonge, Uganda, to NARO (National Agricultural Research Organization). New target areas were chosen on the basis of (i) high population density, (ii) high priority challenges, (iii) productive potential and (iv) strength of partner organisations.

An important factor in the choice of research locations was the shift in funding modalities, from predominantly core, unrestricted funding up to the mid-1990s to mainly project-based funding[2]. This limited the liberty to choose its research locations, project donors usually having their own geographic preferences. It also affected the research agenda because of the partnerships with advanced research institutes that often came with it. Examples are the Belgian-funded BNMS-II ('Balanced Nutrient Management Systems') and CIALCA ('Consortium for Improving Agriculture-based Livelihoods in Central Africa')

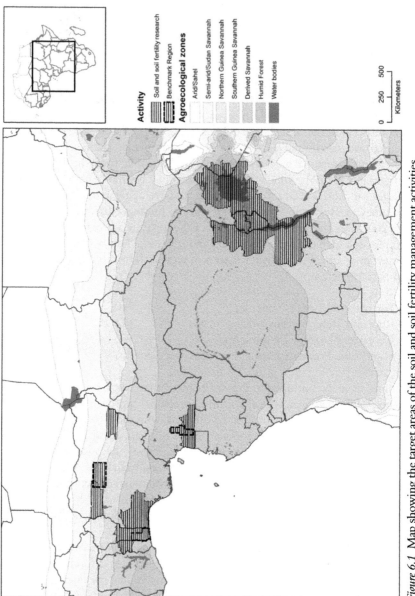

Activity

Soil and soil fertility research

Benchmark Region

Agroecological zones

Arid/Sahel

Semi-arid/Sudan Savannah

Northern Guinea Savannah

Southern Guinea Savannah

Derived Savannah

Humid Forest

Water bodies

0   250   500

Kilometers

Figure 6.1 Map showing the target areas of the soil and soil fertility management activities

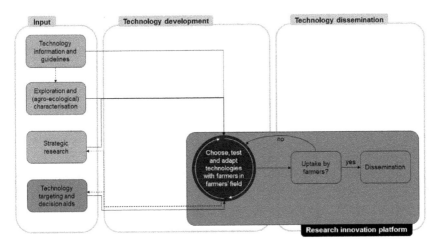

Figure 6.2 IITA's technology development and dissemination model, 2002–2011; towards multi-stakeholder research for development through on-farm technology development, testing and dissemination

Shading indicates relative emphasis on the various components with darker ones indicating a relatively higher emphasis

projects whose activities will be extensively discussed in this chapter. Another example was PROSAB ('Promoting Sustainable Agriculture in Borno State'), carried out with CIDA funding in an area of Nigeria where IITA had not traditionally been active: the north-eastern part of the country. The motivation was to contribute to the development of 'one of Nigeria's most resource-limited states, where 60% of the rural community live below the poverty line of 1 USD per day' (IITA, 2006), through the implementation of the new R-for-D.

Figure 6.2 pictures the overall strategy adhered to whereby the central circle represents the core approach of integrated technology development and testing as a joint activity of all stakeholders, carried out under real farm conditions.

As a consequence of the multiple funding sources, there was a proliferation of cooperative arrangements with regional and international organisations, which affected the coherence of the research activities and resulted in a rather fragmented picture. Some initiatives[3] aimed at both the development and dissemination of specific technologies, while others[4] aimed primarily at technology dissemination, through multi-stakeholder platforms and participatory applied research.

All projects professed to subscribe to the R-for-D 'philosophy' and methods. The Sub-Saharan Africa Challenge Program (SSA-CP) differed from the other programs in that its subject matter was primarily institutional rather than technical innovation, viz. 'evaluating whether Integrated Agricultural Research-for-Development (IAR4D) works' (ISPC, 2010), which is indeed a pertinent question.

6.2 Characterisation of soils, farms and farming systems

The rise of R-for-D projects entailed a need for new surveys, both as a reconnaissance and research targeting tool at project initiation and for more in-depth characterisation of farmers' systems and soils. Soil characterisation and mapping was no longer a stand-alone activity, but was meant as an integral part of the project design and implementation.

6.2.1 Surveys of farming systems

Research-for-Development, of course, demanded good knowledge of local farmers' conditions challenges and a shared vision about priority problems to tackle. Two of the programs initiated during this episode, PROSAB and CIALCA, therefore started off with extensive surveys and participatory problem analysis, while BNMS-II inherited information on the farming systems in the West African savannah from its predecessor, BNMS-I.

The mandated zone of the PROSAB project in Borno State, Nigeria, extended across three major sub-zones of the African Savannah: the Sudan (SS), Northern (NGS) and Southern Guinea Savannah (SGS). At the project's commencement, participatory surveys, community analysis and social mobilisation activities were undertaken in the three agro-ecological zones (AEZ) (Kamara et al., 2009). Information was collected on livelihood sources, crop production problems and farmers' coping strategies. Farmers identified a wide range of problems including *Striga* infestation in both cereals and cowpea (*Vigna unguiculata*), poor soil fertility and lack of fertiliser and improved varieties.

The overall goal of CIALCA, operating in the Great Lakes area of central Africa, was the improvement of Agriculture-based Livelihoods in Central Africa through three interlinked projects, led by three CGIAR centres (IITA, the Tropical Soil Biology and Fertility Institute of the International Center for Tropical Agriculture (TSBF-CIAT) and Bioversity International). The aim of the IITA-led project was to 'develop and disseminate, in partnerships with all stakeholders, technologies that improve the sustainability and profitability of banana-based cropping systems' (CIALCA, 2008). A number of 'mandate areas'[5] were therefore delineated, with action-sites for on-station research and satellite sites for wider demonstration of proven technology, resembling the Benchmark approach of the 1990s. The research process followed the Farming Systems and Livelihood Analysis sequence, consisting of Participatory Rural Appraisals (PRA) and selection of research sites with National Agricultural Research System (NARS) partners, followed by a baseline survey and finally detailed diagnostic studies to fine-tune information on priority research themes. Restitution and planning workshops were an intrinsic part of the methodology. The surveys were carried out jointly by the CIALCA partners. The themes chosen for research and development activities by the Consortium were the improvement of the banana-based systems through improved varieties and more effective resource use, including Integrated Soil Fertility Management

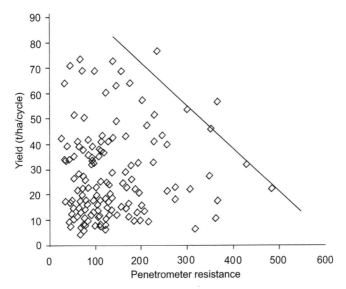

Figure 6.3 A plot of banana yield against soil compaction, illustrating the large yield variation in the Great Lakes area (adapted from CIALCA, 2009)

(ISFM), and by strengthening the role of grain legumes (beans (*Phaseolus vulgaris*), soybean (*Glycine max*) and groundnut (*Arachis hypogaea*).

An impression of the large variability of banana (*Musa* spp.) yield, presumably embracing all types, from table to beer bananas, can be gleaned from Figure 6.3 (here shown as a function of soil penetrometer resistance), suggesting an absolute maximum of about 75 t/ha/crop cycle. Highest yields were recorded for the western parts of the area with soils of volcanic, metamorphic and alluvial origin and rainfall often exceeding 1,400 mm/yr.

Besides banana diseases, farmers mentioned drought stress as a major problem especially in the central and eastern parts with annual rainfall below 1100 mm. Soil fertility problems were identified as important across the entire area, though more prominent on old Ferralsols and Acrisols in Central and southern Rwanda, Central Burundi and the Walungu axis in the Democratic Republic of Congo (DR Congo).

6.2.2 Surveys and characterisation of soils and soil fertility

In earlier years, IITA's research on soils and soil fertility followed the conventional pathway of the time:

- Characterise and group the soils in an area through soil survey and physical and chemical laboratory analysis;
- Carry out nutrient response trials in representative benchmark soils and relate the results to the chemical soil analysis; and
- Formulate nutrient recommendations for different soil groups and crops.

Effective implementation of this method requires a comprehensive soil database and an extensive and expensive network of testing sites. Since this turned out to be too large a task for a relatively small institute with a huge mandate area, a modified approach was tried for some time, borrowed from the US Benchmark Soils project: the Fertility Capability Classification (FCC) (Chapter 2). Although this system had merit for sub-Saharan Africa (SSA) conditions and was more in line with IITA's capabilities, it was never really adopted as the way to go. As a result, for many years a lot of research on soil fertility and crop nutrient responses went on without a discernible framework which could have helped to extrapolate the findings to similar conditions elsewhere.

During the present episode, some projects still followed the conventional characterisation approach of soil surveys and soil analyses as a basis for the choice of appropriate technology, now using the 'World Reference Base for Soil Resources' (WRB) taxonomy (Jones et al., 2013), but there was also a shift to alternative approaches, which were less dependent on regional soil databases and local soil surveys, in particular foliar nutrient diagnostic tools ('Diagnosis and Recommendation Integrated System' – DRIS and 'Compositional Nutrient Diagnosis' – CND tools) and nutrient response models ('Quantitative Evaluation of the Fertility of Tropical Soils' – QUEFTS). They were potential tools in the new thrust towards a more portable system of characterisation of soil resources.

A detailed soil survey was carried out in two research villages in Central Togo in support of the research on balanced nutrient management by the BNMS-II project (BNMS-II, 2007). Three broad map units were distinguished, according to their toposequence position: S1: hill crest an upper slope; S2: middle slope and S3: lower slope and valley bottom. Characteristic soil profiles were classified using the WRB system, physical and chemical analyses of the profiles were carried out and current land use and farmers' own assessment of land quality were recorded. The total area surveyed was 185 km². The idea was that once the survey, the analyses and the mapping were complete, 'morphological soil characteristics observed with the auger can be linked to a soil type for which a full soil profile characterisation has been done during the soil survey' (BNMS-II, 2007). This is of course the traditional approach, similar to that of the Benchmark Soils project of the 1970s and 1980s, except that the latter used the USDA (United States Department of Agriculture) Taxonomy.

The soil data and maps were used for the choice of representative experimental fields, in order to extrapolate findings to a wider area. The BNMS-II Final Report also states that the maps should be used in future by national institutions and even farmers, but there is no further record on whether that optimistic scenario materialised. That raises questions about the usefulness of such, probably quite costly, surveys, apart from the training aspects for young scientists which was certainly one of the project's aims.

As for the PROSAB project in Borno State, north-east Nigeria, extending across the Derived Savannah (DS), SGS and NGS AEZs, soil survey data from the 1970s showed medium- to heavy-textured soils of volcanic origin in the SGS and part of the NGS and lighter soils from sandstone in the NGS and SS (Kwari et al.,

2009, 2011). Participatory surveys highlighted the fertility gradient at the micro-level, from the continuously cropped compound fields close to the settlements to the bush fields where cropping alternated with fallows of at least two years. The surveys produced the usual set of priority problems (Kamara et al., 2009): lack of capital and credit, poor soil fertility, lack of fertiliser, seed and equipment, storage problems, postharvest pests, *Striga* infestation and insect pest attacks on cowpea.

'Participatory demonstration trials' with maize (*Zea mays*) and soybean were carried out across the three zones in 2004 in about 200 farmers' fields, with 100 kg N, 22 kg P and 21 kg K/ha applied to the maize and 15 kg/ha P as single super phosphate (SSP) to the soybean. Standard management practices were probably recommended, but it is not clear to what extent they were actually applied by the participating farmers. Soil analyses were carried out in 137 of the fields to establish the relationship between soil nutrient status and crop yield. Mean values for measured soil parameters and maize and soybean yields from these 137 fields, disaggregated by field type (compound and bush), are shown in Table 6.1.

Table 6.1 Mean values for chemical soil parameters and crop yields in trial fields in North-East Nigeria (Kwari et al., 2009, 2011)

Soil parameters	Southern Guinea Savannah		Northern Guinea Savannah		Sudan Savannah	
	Compound	Bush	Compound	Bush	Compound	Bush
Maize yield (2004) (kg/ha)	2,442	2,754	2,514	2,550	2,241	1,765
Soybean yield (2004) (kg/ha)	1,799	2,527	1,701	1,906	1,816	1,553
Sand (%)	27.2	28.7	46.4	33.0	46.6	45.6
Silt (%)	38.0	38.0	26.4	38.1	38.7	37.5
Clay (%)	34.5	33.4	27.2	28.9	14.7	16.9
pH	6.40	6.33	6.48	6.42	6.65	7.41
Organic C (%)	1.21	1.09	0.92	1.26	0.89	0.82
Total N (%)	1.6	1.5	1.5	1.9	1.5	1.5
Available P (mg/kg)[a]	4.24	4.65	4.70	3.68	2.51	1.49
Exchangeable K (cmol$_c$/kg)	0.60	0.58	0.54	0.68	0.61	0.44
Available S (mg/kg)	10.49	9.90	8.41	9.17	8.44	8.86
Available Zn (mg/kg)	0.52	0.40	0.27	0.27	0.25	0.23
Available Cu (mg/kg)	0.07	0.10	0.09	0.11	0.07	0.07

Note:
a Probably Mehlich-3 test, which gives results comparable with Bray-I.

Considering the fairly heavy fertiliser rate, the average yields of maize were low (1.8–2.7 t/ha) and so was the maize-to-soybean yield ratio, which would be expected to be close to 3 under unstressed conditions. A yield gap analysis carried out for the individual fields would have been most useful here.

Multiple regression of the yields on measured soil parameters across ecologies and field types produced a confusing pattern. Although scattered significance was recorded there was no clear pattern emerging for both major and micro-nutrients. Considering the large number of trial sites, more advanced multivariate analysis of the data could perhaps have generated more clarity here. At most, the results, or rather their absence, reinforce the frequent observation that the micronutrient status of the savannah soils was, as yet, poorly understood (CGIAR, 2008).

More generally, the problem with this type of on-farm trial is their hybrid nature: they are neither really farmer-managed, nor are they researcher-controlled. As a result, the outcomes become highly artificial as they neither reflect the variability due to farmer management practices, nor do they eliminate that variability as in the case of full researcher control. Accounting for farmer variability was now becoming a major issue and would be addressed in several projects in this (Section 6.3) and the following episode.

Detailed diagnostic studies which were also part of the CIALCA start-up activities included foliar analysis of bananas across the Consortium's mandate areas, rather than only soil analyses to identify possible nutrient deficiencies (CIALCA, 2008). The raw nutrient concentration data pointed to P and Mg deficiencies in strongly weathered soils, such as the Walungu axis south of Bukavu, notorious for its poor crop yield and K-deficiency in soils derived from acidic rocks such as in Kibuye, Ruhango (both Rwanda) and Gitega (Burundi)[6]. Areas where external nutrient inputs (e.g. external mulch) were frequently applied (e.g. Kibungo) tended to have less nutrient deficiency problems.

Nutrient omission pot trials were carried out with soil from eight locations along the Walungu axis in Sud-Kivu to identify the causes of their pronounced unproductivity. Maize suffered from severe P-deficiency in all treatments, even where the equivalent of 120 kg/ha P was applied. In another pot trial, the P-deficiency symptoms were relieved by a ten-fold P application rate. Second to P as the most deficient nutrient, N was identified as a critical nutrient, but also important deficiencies in K and indications of Mg deficiency were found, the latter two particularly in soils that farmers themselves considered as poor (Pypers et al., unpublished manuscript). Micro-nutrient deficiencies were suspected (CIALCA, 2008), and low concentrations of Zn were sometimes observed in western Rwanda, but these were never associated with low yields, so no proof was found for a micronutrient problem at any significant scale in the CIALCA region (van Asten, pers. comm.). Field trials were carried out simultaneously with the pot trials to find the best combinations of mineral and organic fertilisers to correct the soils' fertility status. They will be discussed in Section 6.3.

The P-status of savannah soils, always a matter of great interest, came even more to the fore during this episode, because of the mounting emphasis on soybean and cowpeas as components for productive cropping systems. Adequate P-nutrition is essential for satisfactory development of the legumes and therefore for their contribution to the N-economy of the system. Immobilisation of P is common in savannah soils and its occurrence and the means of overcoming it were important research topics. One of the objectives of the ACIAR project, for example, was 'to select soils that are low in available P and which also support maize/grain legume cropping systems in representative agroecological zones and to characterise their soil P status and P chemistry' (ACIAR/IITA, 2005), thus sounding like an echo of the benchmark projects of earlier times. The immediate purpose was to use those soils to screen cowpea and soybean varieties for their ability to access tightly bound P. The project documentation does not provide further information on the soils chosen for this work.

The relation between laboratory-measured soil-P content and crop growth was known to be often problematic. The degree of immobilisation varies strongly with soil type and there are great differences in the ability of plant species and varieties within species to access soil-P. One step towards a better understanding of P-accessibility and crop response to applied P, undertaken during this episode, was fractioning of soil-P into fractions with different levels of accessibility. Nwoke et al. (2003, 2004) conducted sequential soil P fractionation on samples from the West African DS and NGS to assess the influence of soil characteristics and management on soil-P pools. They observed generally low resin-extractable P, indicating low P availability, generally higher levels of labile P in DS soils than NGS soils, with site-dependent effects of organic and mineral fertiliser inputs on labile soil fractions, and little impact of P inputs on stable P fractions. P sorption was considered generally low in most soils (but with important exceptions in the NGS), and the authors concluded that small quantities of P inputs, calculated from P uptake rather than P sorption, suffice.

6.3 Technology development

Although in this period, the work on soils and soil fertility management was carved up in a fairly large number of mostly bilaterally funded projects scattered across SSA, this section presents an analysis from a subject matter, rather than a project perspective. It should thereby be kept in mind that most activities were embedded in a multi-stakeholder environment centred on farmers' production systems, rather than on individual crops or production factors in isolation.

A number of themes can be observed, some carried over from previous episodes, some being new initiatives. Carry-over themes were the work on:

- N-use efficiency and the interaction or synergism between fertiliser-N and organic matter;
- The (potential) role of legumes as components of cropping systems, whereby emphasis shifted from 'auxiliary' to multi-purpose legumes; and
- P-nutrition and the ability of legumes to access 'sparingly available' P-sources.

New themes were concerned with:

- Integrated plant nutrition and fertility management in cereal- and cassava (*Manihot esculenta*)-based systems;
- Perennial crops-based agriculture in the Great Lakes area of Central Africa; and
- Sustainable tree crop production in the humid area, in particular with cocoa (*Theobroma cacao*) in what came to be called 'cocoa agroforests'.

Most of the work was now carried out on-farm, with varying degrees of farmer management. Technologies for improved soil fertility management were embedded in farmers' cropping systems, rather than tested in isolation in separate fields. Nevertheless, a distinction can still be made between trials primarily meant for technology development and trials where on-farm validation or dissemination was the primary objective (sometimes called 'demonstrations' in project reports). In the PROSAB project the former were called 'mother trials', which were essentially researcher-managed trials in farmers' fields, the latter 'daughter trials', with a small number of treatments and a strong demonstration objective. As before, this distinction is reflected in the structure of this chapter, with separate sections on 'Research Activities and Outputs' and 'Technology Validation and Dissemination'.

The shift in research objectives, from the development of 'International Public Goods' (IPGs) to applied research in direct support of local development meant the tacit abandonment of the earlier aims of building databases and prototype technologies with relevance for large areas. Although some basic work continued with wider relevance for the farming systems in the humid and sub-humid zone, the emphasis was now on assembling 'baskets of technologies' for particular intervention areas. The lack of a reliable survey- and experiment-based framework to determine location-specific nutrient requirements led the researchers to experiment with alternative systems to assess such requirements, such as the DRIS and CND tools and the QUEFTS modelling framework.

6.3.1 Soil and soil fertility management

Research on soil and soil fertility management was now mostly carried out on the real farm, with only limited back-up research on the station, a trend which had taken shape during the previous episode (Chapter 5). This clearly posed the methodological dilemma which had also faced earlier generations of on-farm workers: how much of the management of research trials is left in the hands of the farmers themselves and, if the answer is: 'most of it', what analytical methods are available to draw valid conclusions from trials. It also makes the distinction between research trials, treated in this section, and on-farm trials, treated in the next, somewhat artificial. We will nevertheless continue to make that distinction, whereby the difference consists mainly in the degree of researcher control, which is high to total in the first case and low to absent in the second.

Soil erosion control

The CIALCA project in its second year started a research program on erosion control in South Kivu Province of east DR Congo where annual crops were grown on unprotected slopes with gradients of up to 40%. Since it was apparently felt that existing knowledge was not sufficient to go directly to the farm, a complex on-station trial was set up to test several options. The trial was to run for several years in order to produce meaningful results. Also, several herbaceous and woody species were screened jointly by researchers and farmers for their suitability for bund planting.

The station trial was started in 2007 with a soybean-maize rotation to test the effects of three factors: (i) physical embankments versus no embankments, (ii) minimum or conventional tillage and (iii) *Calliandra calothyrsus* hedgerows planted along the contours or no hedgerows (CIALCA, 2008). Constructing physical embankments and hedgerows, however, resulted in a loss of planted area of respectively 27 and 20%. Results in the first year were affected by soil movements for bund construction, reducing soybean yield per unit planted area. In the other treatments, loss of planted area was not compensated by higher soybean yield per unit area.

Without any form of soil protection and under traditional tillage, about 2.4 cm of the topsoil had been washed away after seven seasons. Even more soil (3 cm) was lost under zero-tillage, due to a single heavy downpour undoing all the positive effects on soil conservation. The sole use of physical embankments was more effective than *Calliandra* hedgerows to control erosion, reducing soil loss by 70 vs 24%. Despite substantial improvements in soil conservation by embankments and hedgerows, crop yields in following years remained highest in the treatment without these erosion control measures and under traditional tillage. The lower yield with physical embankments could be fully explained by the loss in cropped area, seasonal application of fertiliser and manure presumably having restored soil fertility levels. Yield losses in treatments with *Calliandra* hedges were larger than expected based on the area loss, suggesting that the soybean and the hedgerows competed for light, water and/or nutrient resources. From these results, it was evident that promoting erosion control measures to farmers will remain an arduous task, as it does not entail benefits for crop production in the short or medium term. Particularly establishing physical embankments requires enormous amounts of labour. In fact, this is a lesson learned long ago in Rwanda, where embankments would only be constructed when imposed by government. Further efforts focused on the added benefits from the hedgerows, for example by providing fodder for livestock or stakes for climbing beans (CIALCA, 2011).

The choice of species for contour planting was very important, because farmers would be expected to use them for various purposes other than just slope control. Participatory evaluation of seven herbaceous and three woody species was therefore done with farmer groups in six sites. The most important farmer-chosen criteria were the production of biomass as a green manure, effective rooting and suitability as a forage, along with several minor criteria.

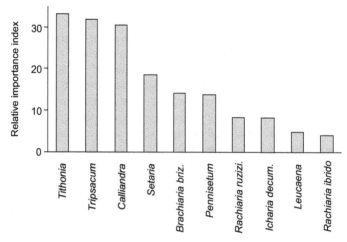

Figure 6.4 Ranking of species for bund planting based on farmer criteria (adapted from CIALCA, 2009)

A weighted index was calculated to rank the different species, with the results shown in Figure 6.4. *Tithonia diversifolia*, the species with the highest score was well known to farmers as a good green manure, though little used. *Calliandra* was preferred mainly for forage and to produce stakes for climbing beans.

No further work was done on physical embankments or bund planting, mainly because the station tests had shown no immediate yield advantage from slope protection. Investing in technologies for slope protection with a longer-term effect would therefore be hard to sell to farmers with an uncertain hold on the land.

Soil fertility management

REMEDIES FOR VERY UNFERTILE LANDS

The unfertile lands along the Walungu axis in South Kivu posed a challenge to the CIALCA project to identify its causes and find remedies (Section 2.2). Exploratory trials were conducted with several farmer groups to further quantify the problem. The treatments included farmyard manure (FYM) (5 t dry matter (DM)/ha), NPK, mavuno fertiliser (NPK enriched with micronutrients), lime (4 t/ha) and *Tithonia* leaf residues (5 t DM/ha). NPK and mavuno fertilisers were applied at 10 kg P ha^{-1}, and maize was supplemented with 40 kg N ha^{-1} as urea. Fertiliser rates were kept at relatively low rates; due to the high cost of fertiliser in the area ($1.5–$2.0 per kg), moderate or high rates would unlikely be affordable. Although all these treatments increased maize and bean yields, the yields fell far short of their potential, especially for maize (Figure 6.5).

Based on the results from the pot trials (Section 6.2.2), P and Mg deficiency was suspected as main cause for the overall low yields, while micronutrient

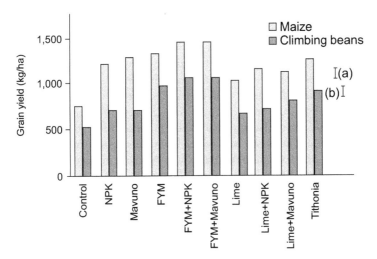

Figure 6.5 Effect of amendments on maize and bean yields in unfertile Walungu soil with '(a)' referring to the standard error of the difference (SED) to compare treatments for maize and '(b)' to SED to compare treatments for climbing beans (adapted from CIALCA, 2008)

deficiencies were also thought to play a role. It turned out, however, that low yield potential of the maize variety used in the trial (*Katumani*) was an important factor, possibly because of its genetic degeneration. With a few hybrids and open pollinated varieties from Kenya, yields up to 6 t/ha were obtained with similar manure and fertiliser inputs (Pypers, pers. comm.).

ANALYSIS OF NUTRIENT DEFICIENCIES AND NUTRIENT RESPONSES

The depletion of the nutrient stocks of the soils of SSA due to increased yields or increased cropping intensity remained a concern and new approaches were sought to assess the nutrient status of a soil and predict likely crop responses to applied nutrients.

Tissue analysis had occasionally been used in earlier years to diagnose possible nutrient deficiencies, but there had been no systematic application of the technique. The BNMS-II project introduced the DRIS tool based on tissue analysis. It uses relative values of nutrient content rather than absolute values. For each nutrient, an index is calculated based on the ratios of the content of that nutrient with those of all other nutrients under consideration and their 'optimum' values or norms, obtained from a database of high yielding crops. A negative DRIS index for a particular nutrient indicates an imbalance or deficiency for that nutrient.

The appropriateness of DRIS analysis for African moist savannah conditions was tested through a 'nutrient omission' trial with maize, carried out in 2004 and 2006 in altogether 53 farmer fields in Nigeria and Togo (Nziguheba et al., 2009). The full nutrient package was 120 kg N, 40 kg P, 80 kg K, 26 kg S, 5 kg Zn and 1 kg B/ha, and in each of the other treatments one of these nutrients except N was omitted. In 2004 maize yields in Togo were at a reasonable level, ranging from 1,530 kg/ha for 'Farmer Practice' to 3,300 kg/ha for the full fertiliser treatment (Table 6.2). The effects of omitting P and S were significant, but the difference between full and half P-rate was not. In Nigeria, mean yields were very low, ranging from about 650–750 kg/ha, with no significant differences between treatments. One possible explanation given for the low yields was that the fertiliser had been placed in bands too close to the seedlings, resulting in moisture stress and increased acidity, the latter causing Mn-toxicity (BNMS-II, 2005). In 2006, the tests were therefore only repeated in Togo in 2006.

Maize yields and DRIS indexes for both years are shown in Tables 6.2 and 6.3. A negative mean DRIS index indicates a predominance of negative values but may hide positive values in some fields. On average, omission of P and S resulted in significantly lower yield, except for S in Affem, 2004, where the differences fell short of significance. DRIS analysis also pointed to P- and S-deficiency, except in Affem in 2006, even though omission of S resulted in the largest drop in yield (Affem, 2006). The average DRIS index for that treatment pointed to Ca- and Mg-, not S-imbalance. At the individual field level, however, S-indices were also negative in 62% of the fields in Affem (and in all fields in Sessaro). Omission of P resulted in negative P-indices in *all* fields. The occurrence of negative Ca and Mg indices in many cases was striking, suggesting hitherto under-reported deficiencies of these elements, especially Ca. Zn indices were negative in all fields and in a

Table 6.2 Maize yields in an on-farm 'nutrient omission' trial, Togo, 2004 and 2006 (Nziguheba et al., 2009)

Treatment	Omitted nutrient	Mean maize grain yield (kg/ha)			
		Affem		*Sessaro*	
		2004 (n=8)	2006 (n=10)	2004 (n=10)	2006 (n=10)
Farmers	-	2,121	1,907	923	1,351
Full package[a]	-	3,329	3,865	3,311	4,149
P-0	P	2,439	2,460	2,323	2,789
P-20	½ P	3,137	3,227	3,089	3,949
K-0	K	3,148	3,503	3,058	4,318
S-0	S	2,900	2,532	2,521	3,387
Zn-0	Zn	3,195	3,066	3,199	3,915
B-0	B	3,080	3,719	3,071	4,139
LSD[b]		609	717	475	647

Notes:
a Full nutrient treatment, with N: 120, P: 40, K: 80, S: 26, Zn: 5, B:1 kg/ha.
b 'LSD' means 'least significant difference'; confidence level not given.

Table 6.3 Nutrients with mean negative DRIS indices in an on-farm 'nutrient omission' trial, Togo, 2004 and 2006 (Nziguheba et al., 2009)

Treatment	Omitted nutrient	Negative DRIS indices for:			
		Affem		Sessaro	
		2004 (n=8)	2006 (n=10)	2004 (n=10)	2006 (n=10)
Farmers	None	Mg, Ca	Ca, Mg	S, Ca, N, Mg	–
Full package	None	Mg	Ca	–	–
P-0	P	P	P, Ca	P	P
P-20	½ P	P	Ca, P, Mg	S	–
K-0	K	–	P	S	–
S-0	S	S	Mg, Ca	S	S
Zn-0	Zn	Mg	Ca, Mg	S	–
B-0	B	Mg	Ca	Mg, S	–

majority they were even the most negative of all. Although it confirms widespread Zn deficiency in the savannah, the authors suspected that the low values were at least partly due to inappropriate norms for nutrient ratios involving Zn.

A large 'historical' data set for maize from Nigeria and Benin was also subjected to DRIS analysis by Nziguheba et al. (2009). In most cases, N and P indices were negative where these elements were not supplied to the crop. In many cases, S indices were positive, especially where S-containing fertiliser (SSP) had been applied over many years. Negative Ca and to a lesser extent Mg indices were found in several cases, confirming the need to pay more attention to these elements.

Finally, the DRIS indices, lumped together for all these trials showed a very wide scatter, as illustrated for Ca and P in Figure 6.6. This is a result of the complex relationship between the indices and nutrient responses. The P-index in this data set, for example, could even suggest a *decline* of maize yield with increasing P-index, contrary to Ca. This is due to the fact that the indices are about balances between nutrients: a field may have good balance between analysed nutrients, while another nutrient or a different management factor may cause low yield. The anomalous trends for P in Figure 6.5 were caused, at least partly, by the data from Samaru with very low yields due to other factors, perhaps including Mn-deficiency (Nziguheba et al., 2009). The authors concluded that DRIS analysis had been shown to be capable of identifying nutrient limitations for maize in the moist savannah areas. Although this conclusion seems justified, more work would be needed to establish or validate norms for a wide range of nutrient ratios and crops under African savannah conditions and develop a practical DRIS-based tool for use by on-farm researchers and developers.

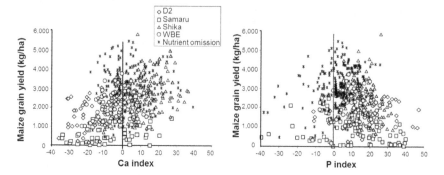

Figure 6.6 Scatter plot of DRIS (Diagnosis and recommendation integrated system) indices vs. maize yield in a large number of maize trials in West Africa (adapted from Nziguheba et al. (2009), with that paper also containing details)

SOIL FERTILITY MANAGEMENT FOR CASSAVA; BOUNDARY LINE AND YIELD GAP ANALYSIS

Mineral nutrition of cassava had received relatively little attention in earlier years, in spite of the great and increasing importance of the crop all over Africa. It had been widely believed that cassava was usually grown in infertile soil and caused further soil degradation, a belief that was finally put to rest by COSCA (Asadu and Nweke, 1999).

A large on-farm study was carried out in 2004 and 2005 in mid-altitude Uganda and Kenya to study farmers' cassava production systems and investigate the crop's response to 'improved practices', in particular varieties and fertiliser (Fermont et al., 2009a,b). We will dwell in great detail on this work for three reasons: (i) it explores fertiliser response of improved cassava varieties over a range of ecological conditions in farmers' fields, (ii) it demonstrates the challenges of drawing quantitative conclusions on the effect of farmer- and site-related factors on crop production and nutrient responses using standard analytical tools and (iii) it introduces a more powerful analytical approach, not used earlier in on-farm trial analysis at IITA, viz. boundary line analysis.

As a first step, farmers' cassava production systems were studied through structured farm surveys, recording biophysical and socioeconomic conditions and estimated current cassava yields from farmer recall (Table 6.4). Analysis of the yield data showed, unsurprisingly, that the highest yields were obtained by the better-off farmers and only weak relationships were found with individual factors such as labour availability and number and timing of weeding rounds.

The effects of variety, fertiliser and improved practices on farmers' cassava yields were then investigated in two large sets of multi-locational trials conducted in two successive years in farmers' fields, representing a range of environments under partial farmer management. Treatments and cassava yields are shown in Table 6.4[7] ('treatment' 1 is the farmer recall data). The cassava in the trial fields was grown without intercrop, possibly to reduce experimental variation. Apart

Table 6.4 On-farm cassava yields (t/ha) with improved varieties and fertiliser in different locations and on research stations in Kenya and Uganda (Fermont et al, 2009a,b). Refer to the text for explanation of the design

Treatment	Mean fresh cassava root yield (t/ha)											
	Farmer recall[a]	2004 Trial treatments						2005 Trial treatments				
	1	*n*	*2*	*3*	*4*	*5*	*n*	*6*	*7*	*8*	*9*	*10*
Fertiliser			–	–	NPK	NPK		–	PK	NK	NP	NPK
Kenya			TMS 30572	MM 96/5280	MM 96/5280	TMS 30572		MM96/5280				
Kwang'Amor	7.9	10	9.9	11.6	16.2	13.9	7	9.7	9.7	16.7	16.9	24.9
Mungasi	6.4	7	10.8	14.0	16.8	15.3	8	14.8	17.3	15.9	18.0	23.6
Nambale	–	9	7.4	9.4	14.8	13.0	9	18.7	22.7	25.5	24.8	29.0
Ugunja	6.1	9	5.2	6.4	11.2	6.2	10	14.3	21.2	17.9	17.7	27.3
KARI Station	–	8	10.8	12.4	17.7	16.3	4	17.4	17.2	20.0	19.6	23.4
Weighted mean	–	–	8.7	10.6	15.2	12.8	–	14.8	18.2	19.3	19.5	26.1
Uganda			TMS 30572	TMSI 92/0067	TMSI 92/0067	TMS 30572		TMSI92/0067				
Minani	–	6	13.0	19.1	26.7	16.5	8	14.4	17.2	17.7	21.6	24.2
Kisiro	8.3	8	11.9	15.7	17.1	13.7	8	14.7	21.6	20.5	24.1	27.5
Kikooba[b]	11.2	–	–	–	–	–	–	–	–	–	–	–
Chelekura[b]	11.7	–	–	–	–	–	–	–	–	–	–	–
NaCRRI Station	–	4	15.5	17.0	16.8	16.5	4	25.7	28.4	25.7	25.9	26.2
Weighted mean	–	–	13.1	17.1	20.2	15.3	–	16.8	21.2	20.4	23.5	25.9
Overall mean	8.6		10.0	12.5	16.8	13.5		15.6	19.2	19.7	20.9	26.0

Notes:
a Recall data only from subset of sites.
b No trials conducted, only recall data.

from cassava yields, rainfall was recorded and soil analysis was carried out for each field, while weediness and pests and diseases were scored regularly. These data were used to account for the large variation among farmers of cassava yield and fertiliser response. Replicated trials with the same treatments were carried out simultaneously in research stations in both countries.

For the purpose of the analysis, the trial results were disaggregated by the authors into two (overlapping) sets of treatments, representing a 'stepwise' and a 'nutrient omission' series, as follows (Table 6.4):

A. The stepwise trial in 2004, consisting of four treatments to study the cumulative effect of three factors, as follows (Fermont et al., 2009a):

1   Farmers' practice, reference yields obtained through the farm survey, therefore not really part of the trial (farmer recall in the table);
2   'Improved crop establishment', the trial's control treatment, with an improved variety common in the area (TMS 30572), recommended spacing, early planting and no fertiliser;
3   As 2 + recommended genotype (MM96/528 or TMSI92/0067); and
4   As 3 + NPK fertiliser.

B. A nutrient omission trial; in 2004 it consisted of treatments from the stepwise trial, viz. those without fertiliser and those with the full NPK package (100:22:83), while in 2005 three extra treatments were added where N, P or K was omitted, making a total of five treatments (numbered 6–10 in Table 6.4).

In the 'stepwise trial' of 2004, the difference between average farmer yield (treatment 1) and the yield with 'improved crop establishment practice' (treatment 2), besides being non-significant, was biased by the presence of the two research stations and the high-yielding site Minani in the 'improved establishment' treatment only. So it is safe to conclude that improved establishment contributed little, if anything. What remains is a stepwise trial with three treatments only. Replacing the older TMS 30572 by newer varieties (treatment 3 vs 2) gave an appreciable yield increase averaging 2.5 t/ha, while the additional effect of NPK averaged 4.3 t/ha.

In 2005, overall yields in the trials were much higher than in 2004 and so was the effect of the full NPK package compared to no fertiliser: 4.3 and 10.4 t/ha respectively. This was attributed to better rainfall distribution, the varieties used, soil fertility of the trial fields and fertiliser application method. In 2005, the nutrient omission treatments in most locations showed significantly lower yield relative to NPK when N or P were omitted. The effect of omitting K was also significant in most Kenyan sites but not in Uganda.

More interesting than average yield is the explanation of the differences among farmers. First we look at farmers' yield variability in the absence of fertiliser, followed by the analysis of the differences in nutrient effect.

Multiple regression and boundary line analysis were used to explore the influence of farmer- and field-related variables on cassava yield with improved varieties and crop establishment, but in the absence of fertiliser (Fermont et al., 2009a). The analysis consisted of three steps:

1   Multiple regression of cassava yields for treatments without fertiliser (treatments 3 and 6) on 'independent variables' measured in the farmer's field;

2   Boundary line analysis of the same data to identify relationships between yield and independent variables for the best performing fields, as an upper level for those relationships; and

3   Prediction of cassava yield in each trial field assuming a Liebig-type relationship between the effects of all the variables and the yield in that field.

Standard multiple regression accounted for only 38% of yield variability in Kenya but a high 82% in Uganda, with quite different sets of explaining variables (Table 6.5). Of these variables, 'days to harvest' is of a different nature than the other variables in that it does not affect crop growth, but is rather a conscious farmer decision to terminate it, probably for nutritional or economic reasons. If we exclude this variable, the variation accounted for in Uganda decreases to 46%, not much better than for Kenya. Thus, the overall explanatory power of the regression analysis was quite low, with weed control standing out. The authors therefore looked at an alternative analytical method for more clarity.

Boundary line analysis was used by Fermont et al. (2009a) to sharpen the analysis of the seemingly weak relationships between measured parameters and cassava yield under farmer conditions. From the plots of all farmers' cassava yields against soil pH and clay content, for example (Figure 6.7), it was postulated that

Table 6.5 Contribution by independent variables to explained variability of cassava yield; data by Fermont et al., 2009a

Decomposition of variation		Kenya	Uganda	Overall
		Proportion of total variation explained by full regression equation (%)		
		38	82	58
		Proportion of total explained by (%)		
Rainfall – related	9–12 MAP	48.8	–	30.2
	Total	–	0.1	–
Soil texture – related	Clay	–	–	6.2
	Silt	–	–	14.9
Soil chemistry – related	pH	16.1	–	3.8
	Available P	–	–	10.1
	Exchangeable Ca	–	4.7	–
	Exchangeable Mg	–	–	11.2
Crop management – related	Weed control	32.2	51.2	23.6
	Days to harvest	3.2	44.0	–

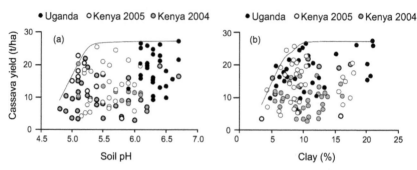

Figure 6.7 Boundary lines for cassava yields under improved crop establishment and improved genotypes for (a) soil pH; (b) clay content (adapted from Fermont et al., 2009a)

the boundary curves of the data clouds represent the maximum possible effect of those variables. Logistic functions with a common maximum were then fitted to the boundary data points for those variables which showed a clear structure, viz. soil fertility parameters, texture, pests and diseases and weed control. The data clouds which lacked structure were left out. The common maximum of the curves was estimated from the joint data sets at 27.3 t/ha, representing the *attainable yield* if all measured variables were at their optimum level.

For each individual field boundary line yields were then read from the graphs, corresponding with the field's value of each variable, that being the best possible yield if all other variables were at their optimum level. The lowest of those was the predicted yield for that field, and the corresponding variable was assumed to be the most limiting factor, assuming responses to obey von Liebig's law of the minimum. This was described as the 'multivariate yield model based on the boundary lines'. Figure 6.8 shows the predicted yields for all the fields using the same model for Kenya and Uganda, except weed management and rainfall, for which separate boundary lines had to be used for reasonable prediction. At the 1:1 line predicted and measured values are equal and, if the model captures most of the variation, all data would cluster along that line. In a majority of the fields, however, yield was considerably lower than predicted, due to unidentified causes and constraint interactions operating in the individual fields.

Two kinds of yield gap were now defined for each field. The first is the gap between the 'predicted' yield according to Figure 6.8 and the theoretically attainable yield, which could be obtained in case all the variables were at their optimum (27.3 t/ha in this case). The second is the unexplained yield gap in a particular field between the actual and the predicted yield.

Including rainfall in the definition of the first yield gap is dubious, because there is no way farmers can correct rainfall shortage, except perhaps to some extent by adjusting planting dates. In order to bridge the remainder of the first gap, farmers in most cases have to address several constraints simultaneously, including constraints related to soil fertility. The second, unexplained yield gap may be due to factors such as micronutrient deficiencies or other unidentified

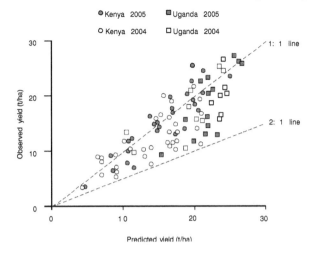

Figure 6.8 Predicted versus observed cassava yields using the multivariate boundary line model (adapted from Fermont et al., 2009a)

physical causes, but it will often be related also to farmers' management skills, the shortage of which needs a different kind of remedy than soil science can offer.

The response to fertiliser turned out to be extremely variable and its relationship with individual soil properties was weak (Figure 6.9, Fermont et al., 2009b), except that fertiliser response was substantially larger in 2005 when rainfall distribution was

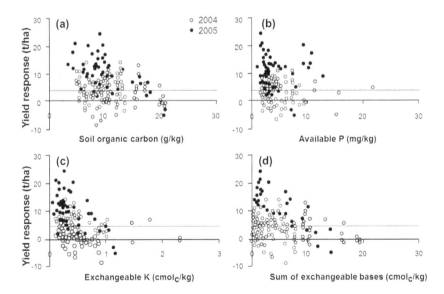

Figure 6.9 Yield response to NPK fertilizer in 8 sites in Kenya and Uganda in relation to (a) Soil organic carbon; (b) available P; (c) exchangeable K; and (d) sum of exchangeable bases (adapted from Fermont et al., 2009b)

more favourable. Somewhat despairingly the authors assumed nutrient limitations other than NPK and Arbuscular Mycorrhizal Fungi (AMF) to be at play, as well as 'interactions between multiple constraints'. A larger dataset and more powerful analysis would be needed to really dig into explaining fertiliser response as a function of soil properties, weather and management. Even then straightforward relationships are unlikely to be found explaining nutrient response under farmer conditions: it is a farmer's skill to simultaneously manipulate many factors to his advantage that leads to the best possible yield and obtain a good nutrient response.

SOIL FERTILITY MANAGEMENT IN HIGHLAND BANANAS; APPLICATION OF CND, DRIS, BOUNDARY LINE ANALYSIS AND QUEFTS

CIALCA, after its initial diagnostic surveys (Section 6.2.1), started an intensive on-farm program to help improve crop yield in highland banana-based systems through better resource use and crop management. Because of the widely varying soil and rainfall conditions and management practices in the Great Lakes area, reliable tools were needed to predict the likely effect of improved technology on banana yield. The program thereby opted for a combination of four tools based on tissue and soil analysis: CND, DRIS, boundary line analysis and QUEFTS modelling. The experience in the institute with these tools was still limited and an important part of the work revolved around testing and adapting them to the conditions and crops of the area. All these methods ultimately aimed at the development of a system to rationalise the use of fertiliser.

As a first step DRIS and CND indices were compared for their suitability to assess the nutrient status of bananas (Wairegi and van Asten, 2011). Both systems use ratios between nutrient contents rather than absolute values, as in older tissue-based methods. DRIS uses individual ratios between a nutrient and all other measured nutrients, while CND uses log transforms of the ratios between a nutrient and the geometric means of all nutrients under consideration plus a 'filling value'. In both cases, the indices for a particular field are a function of the transformed ratios calculated for that field and those calculated from data from high yielding populations, called 'norms'. There was a strong linear relationship between the DRIS and CND indices when applied to the nutrient contents of leaf tissue and bunch weights from 300 fields in different highland regions in Uganda (Figure 6.10) and both methods resulted in practically the same ranking of nutrients in terms of adequacy or deficiency. Since CND indices are more suitable for Principal Component Analysis than DRIS indices and easier to calculate, the former was considered most suitable for future development of fertiliser recommendations for East African highland bananas.

Compositional Nutrient Diagnosis analysis was then applied in a study of 179 banana plots in the same four regions in Uganda, 95 of which were demonstration plots, the remainder being neighbouring farmer-managed control plots (Wairegi and van Asten, 2010). The CND norms used were those determined in the above methodological study[8]. The demonstration plots received N, P and K, averaging 71, 8 and 32 kg/ha while the farmers were asked not to apply fertiliser and were

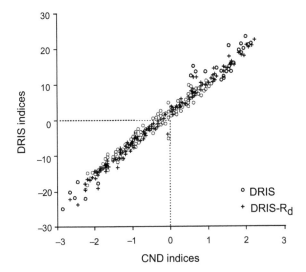

Figure 6.10 Relationship between CND and DRIS indices for highland bananas in Uganda (adapted from Wairegi and van Asten, 2011)

'CND' means 'Compositional nutrient diagnosis' and 'DRIS' means 'Diagnosis and recommendation integrated system'. 'DRIS-Rd' refers to a 'Rest value in addition to d measured nutrient contents', or basically an alternative way to calculate DRIS norms (see above paper for the detailed explanation)

stimulated (by a financial incentive) to apply external mulch in the demonstration plots. Average yields for the demo and control plots in the four regions are shown in Table 6.6, with significant differences in three of the four areas, explained mainly by the (higher rates of) fertiliser and mulch applied in the demonstration plots. There were no significant differences between the average CND indices.

From the indices for the control plots (Figure 6.11), it was concluded that the most deficient nutrients were P and K in the central area, N in the south, K in the south-west and Mg in the eastern part of the highland banana growing areas in Uganda. This was consistent with the absence in the east of a significant yield increment in the demonstration plots (Table 6.6), which did not include Mg

Table 6.6 Average banana yields in control and demonstration plots in four regions in Uganda (Wairegi and van Asten, 2010)

Area	Average bunch yield [t/ha/year]1	
	Control	*Demonstration plots*
Central	12.4	20.4***
South	9.7	19.7***
South-west	20.0	24.8*
East	25.5	32.6

Note:
*, *** denote significant difference with control, at 5% and 0.1% probability level, respectively.

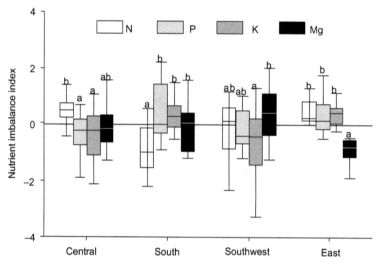

Figure 6.11 CND indices for bananas in four regions in Uganda; boxes and bars cover the 25–75% and 5–95% range, solid horizontal lines are median values different letters indicate significant (P<0.05) differences within regions (adapted from Wairegi and van Asten, 2011)

application, and with the largest yield increment in the south, where N deficiency dominated. Adjusting fertiliser rates was therefore recommended, whereby K-application should be increased in the south-west and Mg should be added in the east. Furthermore it was thought that single- or double-nutrient formulations would often be more profitable than the more expensive NPK fertiliser.

In another on-station study in Central and south-west Uganda (Nyombi et al., 2010), three banana cycles were grown under seven fertiliser regimes (Table 6.7). All plots also received 60 Mg, 6 Zn, 0.5 Mo and 1 B kg/ha/yr. The main purpose was to 'understand banana systems' and test the QUEFTS model for the elaboration of site-specific fertiliser recommendations, in replacement of the blanket recommendation for bananas in force in the country of 100N:30P:100K:25Mg kg/ha/yr.

The yield data rearranged in ascending order are shown in Table 6.7. In both areas, the lowest yields were recorded when K was omitted, clearly showing the importance of K for highland bananas. The effect was strongest in Ntungamo, where the yield at 250 kg K/ha/yr was also significantly lower than at 600 kg, consistent with the soil's average exchangeable K-content below the threshold of 15 cmol/kg. Furthermore, omitting N or P resulted in lower yield in Ntungamo, but more so for P, while reducing the N rate to 150 kg had no significant effect. Hence, the order of nutrient deficiencies in Ntungamo was K >> P > N.

In Kawanda, the results for N and P were less clear-cut, with the highest yields recorded when either N or P was omitted, a result which is difficult to explain. The much lower yields in the best producing plots compared with Ntungamo suggest an unidentified factor limiting yield, apart from K. The Kawanda field

Table 6.7 Banana bunch yield of highland bananas at different fertiliser regimes ranked in ascending order; totals for two cycles (Ntungamo, South-West Uganda) or three cycles (Kawanda, South-Central Uganda). Data from Nyombi et al. (2010).

Kawanda					Ntungamo			
Fertiliser application (kg/ha/year)			Total banana bunch yield (kg/ha/year)[a]		Fertiliser application (kg/ha/year)			Total bunch yield (kg/ha/year)[a]
N	P	K	2 cycles	3 cycles	N	P	K	2 cycles
0	0	0	23.6(a)	38.9(a)	400	50	0	15.6(a)
400	50	0	25.4(a)	44.3(b)	0	0	0	15.8(a)
150	50	600	25.3(a)	47.7(b)	400	0	600	35.3(b)
400	50	600	27.8(ab)	50.3(bc)	0	50	600	41.4(c)
400	50	250	28.4(ab)	50.6(bc)	400	50	250	44.3(cd)
0	50	600	33.0(b)	55.9(c)	150	50	600	48.5(de)
400	0	600	29.6(ab)	56.1(c)	400	50	600	52.3(e)
Mean:			27.6	49.1	–			36.2
Pooled SED:[b]				2.34	–			2.58

Notes:
a Yields followed by a different letter are significantly different (Newman-Keuls test, P<5%).
b 'SED' = 'standard error of the difference'.

had been used previously for maize and beans, while the Ntungamo field had been under 10-year fallow. The authors suggest that drought stress in Kawanda due to compacted soil could be part of the explanation.

Nutrient availability, uptake and conversion data measured in the Ntungamo trial were used to calibrate the QUEFTS model. This is a static model which relates crop yield to native and applied nutrients under non-limiting moisture conditions (Janssen and Guiking 1990; Janssen et al., 1990). Calibrated with parameters from the Ntungamo trial the model could reproduce the yields from that same trial reasonably well, in what was essentially a curve-fitting exercise. There was no mention of simulating the Kawanda trial results but the model over-predicted the yields in an independent data set from the same region[9], which was tentatively attributed to moisture stress or limiting nutrients other than NPK. The sensitivity to drought was confirmed in later field studies by van Asten et al. (2011a) and pot trial studies by Kissel et al. (2015). In addition, work by Ndabamenye et al. (2012) showed that the plant densities used in the nutrient omission trials (1,111 mats/ha) were well below farmer practice (1,400–2,700 mats/ha) with the exception of the driest regions (around 1,000 mm/yr rainfall) where plant densities were just above 1,000 mats/ha. In any case, it is clear from Table 6.7 that even in Ntungamo the agronomic nutrient use efficiency was low and certainly not economically acceptable. The authors hypothesised that applying mulch to improve moisture availability could help and that other limiting factors had to be considered to explain the low yield and nutrient response. This being 'the first attempt to calibrate QUEFTS for highland bananas', further work would be needed to assess the model's suitability as a tool in developing recommendations for bananas.

Data from this same trial, supplemented with pot trials and measurements in farmers' fields, were used to further study the response mechanism of highland bananas to K-, N- and moisture stress (Taulya, 2013). Farmers' fields were grouped into different N-K classes on the basis of CND analyses, corresponding with DM and bunch yield at different N-K combinations in the trial. It was found that highland bananas favour dry matter allocation to subterranean structures in case of K- and N-deficiency and that yield losses due to drought could be mitigated by adequate K-fertilisation. Contrary to these findings, van Asten et al. (2011a) had reported absence of K-moisture interaction in the same area. This difference was explained by the fact that the rainfall data used in the latter study did not cover early growth, contrary to the present study. It was concluded that 'K-mediated osmotic adjustment mitigates drought-induced DM production and yield losses *if the drought occurs early* (i.e. in the first 6 months from emergence) rather than late in the growth cycle'.

Apart from the immediate importance of such findings for crop management, it was argued that accounting for the plant's adaptation mechanisms to nutrient and moisture stress was needed for the simulation of banana growth. Dynamic crop modelling was now being considered as a potential tool in banana research and two PhD research projects included the development of a model for potential (Nyombi, 2010) and water-limited banana production (Taulya, 2015), based on the LINTUL ('Light Interception and Utilisation') model from Wageningen University.

A researcher-managed field study with highland bananas was carried out in 2007–2009 in three regions in Rwanda with contrasting ecologies and soil conditions with two objectives: (i) identifying the role of nutrients, soil moisture and planting density as yield-limiting factors, in distinct zones under low input conditions and (ii) advising on fertiliser recommendations.

Three banana varieties were grown at five planting densities without external mulch or fertiliser application (Ndabamenye et al., 2013). Mean bunch yields in the three sites (Table 6.8) showed significantly higher overall yield and a consistent yield increase with density for the two wetter sites. In the driest zone, where yields were lowest (Kibungo, 900–1000 mm) no effect of density was found. An analysis of the effect of annual rainfall on banana yield in Uganda by van Asten et al. (2011a) suggest a yield plateau at annual rainfall above 1500 mm (their Figure 5)[10], so Ruhengeri rainfall may on average be close to the maximum requirements of the crop.

Pearson correlation coefficients between banana yield and the soil content of some nutrients, which varied by zone, were significant (Table 6.9). It should be noted that the variations in soil nutrient contents within sites were limited because they were all measured in the same field in a randomised block trial.

Boundary line analysis only showed a saturation pattern for K-content, levelling off between 0.15 and 0.20 cmol/kg, which has been widely found as a threshold for K-effect for many crops in many different soils (Chapter 2)[11].

Table 6.8 Effect of planting density on banana bunch yield in a station trial in three zones in Rwanda (mean seasonal rainfall in experimental period between brackets) (Ndabamenye et al., 2013)

Site	Bunch yield (t/ha/year)[a]					Mean
	Density (stools/ha):					
	1,428	2,500	3,333	4,000	5,000	
Kibungo (931)	8.7(g)	6.1(g)	8.6(fg)	9.2(efg)	7.0(g)	8.3
Rubona (1039)[b]	9.5(efg)	11.4(ef)	13.1(bcde)	21.5(ab)	19.2(bc)	14.9
Ruhengeri (1366)	7.0(g)	11.9(ef)	16.2(cd)	21.8(ab)	25.0(a)	16.3
SE[c]	8.7					

Notes:
a Yields followed by same letter are not significantly different (P<0.05).
b Average for the first two years was 1205 mm, but only 707 mm in the third year.
c 'SE' = 'standard error'.

For pH and Mg, the boundary lines were flat (not given for P), suggesting non-limiting soil levels[12]. Yield gap analysis using the boundary lines for the factors which correlated significantly with yield showed large deviations from the predicted bunch yields (Figure 6.12), which means that the causes of yield variation were poorly understood.

In all zones CND analysis, using Wareigi norms, did not show significant differences for cultivars or densities. Overall nutrient imbalance, expressed as the sum of squared CND indices, however, was highest for the site with the highest yield (Ruhengeri), which could mean that even its 'attainable yield' was (far) below its potential because of unidentified nutrient shortages or other factors. The trial did not provide clear guidelines for fertiliser recommendations.

P-NUTRITION IN GRAIN LEGUMES AND CEREALS

Response to P by grain legumes is common in the moist savannah and much research was carried out in the previous and the present episode on the (potential) role of P in cropping systems embracing cereals and legumes. Compared to earlier years, there was more emphasis on the dynamics of P-nutrition in a cropping systems context and less on the regional variation in P-responses. The ability of some or many legume species to extract P from soils with low P-content and from rock phosphate (RP), and an increased availability of P to a succeeding cereal crop had been reported earlier (Chapter 5), but needed to be verified. The phenomenon was therefore further explored and process studies were carried out in different places to elucidate the mechanisms.

Pot studies in Benin (BNMS-II, 2008) showed that legume species varied in the amount of citric acid exuded by the roots, with soybean having the highest exudation, followed by cowpea and pigeon pea (*Cajanus cajan*), but the amount did not correlate significantly with P-content of soybean shoots. Nevertheless,

Table 6.9 Mean banana bunch yields and correlation with soil nutrient contents in three zones in Rwanda; station trial (Ndabamenye et al., 2013)

	Mean yield (t/ha)	Item	pH	Available P (mg/kg)	Exchangeable K (cmol/kg)	Exchangeable Ca (cmol/kg)	Exchangeable Mg (cmol/kg)
Kibungo	8.3	Value	5.5–5.9	4.00–5.14	0.27–0.44	3.10–3.69	1.02–1.35
		Correlation1	0.34*	NS	0.53*	NS	NS
Rubona	14.9	Value	5.6–5.9	7.10–11.77	0.02–0.18	2.02–3.15	0.48–1.00
		Correlation1	NS	0.46*	0.58*	NS	0.40*
Ruhengeri	16.3	Value	6.1–6.2	34.0–41.5	0.21–0.26	2.90–3.19	0.54–0.57
		Correlation1	- 0.30*	NS	NS	NS	NS

Note: * denotes significant correlation at 5% probability level; NS=non-significant correlation.

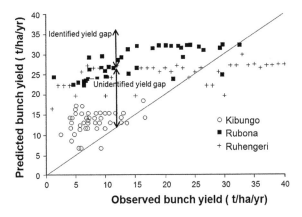

Figure 6.12 Observed banana bunch yield and yield predicted from boundary line analysis, Rwanda (adapted from Ndabamenye, 2013)

it was concluded that 'the secretion of citric acid appears important for P acquisition in P-limiting environments'.

In the same study, shoot dry matter and P accumulation correlated significantly $(r = 0.7)$ with root length in cowpea but not in soybean, maize or sorghum (*Sorghum bicolor*). This taken as proof that 'the cowpea genotype studied [depended], to a large extent, on the morphology of its roots for soil P acquisition', a rather strong conclusion. In another pot experiment, no relation was found between root length and P-accumulation by soybean accessions (BNMS-II, 2007).

Pot studies with soils collected from 40 farmers' soybean fields in the NGS of Nigeria showed a strong (logarithmic) relationship between soil Bray-1 P-content and soybean shoot yield without applied P, as percentage of yield with fertiliser-P (Figure 6.13). A yield level of 55% of potential yield, estimated from

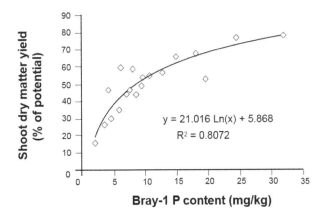

Figure 6.13 Relationship between soil Bray-1 P content and percentage of potential soybean shoot yield (adapted from BNMS-II, 2006)

the fitted curve, was obtained at about 10 ppm P and 80% at about 32 ppm P (BNMS-II, 2006). Of all the soils tested, 80% had a P-content below 10 ppm and would be expected to respond to P-application.

An on-station study on P-nutrition of groundnuts was carried out by PROSAB in soils in the Sudan and NGS of north-east Nigeria (Kamara et al., 2011) with an available P-content (Mehlich-3) below 3, far below the threshold for most crops. Table 6.10 shows pod and fodder yields as well as total annual precipitation in both years. The overall P-effect was significant in both years and both sites, but appeared to level off around 40 kg/ha. Interestingly, fodder yield was depressed by around 20% in both sites at a P application rate of 40 kg, compared with 20 kg/ha, presumably because of increased DM allocation to the grain.

Detailed greenhouse studies were carried out to elucidate the mechanism of the considerable enhancement of maize growth after *Mucuna pruriens* (velvet bean) supplied with phosphate rock (PR), as reported earlier (Chapter 3). These studies did confirm the beneficial effect of *Mucuna* on the following maize, but enhancement of P-availability alone could not explain the magnitude of the response (Pypers et al., 2007). In soils supplied with PR, the labile P quantity and P concentration in soil solution after plant growth were increased only if *Mucuna* was grown, confirming enhanced PR solubilisation in the legume-growing soils (Pypers et al., 2007). Grain yield and P uptake of a subsequent maize crop were twice as large when grown after *Mucuna* as following a first maize crop. This residual effect of velvet bean was even significant in treatments without PR application and could not be attributed to differences in P removal by the preceding crop, or differences in P availability in the soil. Furthermore, when legume residue was incorporated in soils where maize had been grown,

Table 6.10 Groundnut pod and fodder yield in two sites and two years in north-east Nigeria in response to P application (Kamara et al., 2011)

	Damboa (Sudan Savannah)		Wandali (Northern Guinea Savannah)	
	2005	2006	2005	2006
Annual rainfall (mm)	716	894	1,039	1,276
P-rate (kg P/ha)	Pod yield (kg/ha)			
0	1,062	1,568	1,918	1,717
20	1,114	2,340	2,057	1,852
40	1,460	2,474	2,140	2,093
	Fodder yield (kg/ha)			
Mean	3,260	2,852	2,773	3,365
LSD P-rate (5%)	197		189	

Note:
'LSD' = 'least significant difference'.

yields or P uptake of the subsequent maize crop were not affected, while P availability was significantly improved. Positive rotational effects of velvet bean were substantially larger than could be attributed to improvements in soil P availability. Maize yield significantly increased with increasing internal P concentration, but the maize following *Mucuna* followed a different relationship, with higher yields at identical internal P concentrations, compared with maize following maize (Figure 6.14). It was concluded that the elimination of other growth-limiting factors – hypothesised to be microbial in nature – by *Mucuna* caused the rotational benefits, more than improvements in P availability.

Inspired by earlier results in the West African savannah, the PROSAB project carried out on-farm verification trials on the effects of soil P-status and fertiliser-P application on the yield of legumes and cereals grown in rotation, as well as on the synergism between them. The researcher-managed trials tested the effect of three levels of fertiliser-P on two cycles of a soybean-maize rotation in the NGS (Miringa) and the Sudan Savannah (Azir) of north-east Nigeria (Kamara et al., 2007, 2008). The P-status of the soils in both locations was very low, viz. 1.7 ppm in Miringa and 2.4 ppm in Azir (Mehlich-3 test[13]). In each case 40 kg/ha of K was applied to the soybean and the subsequent maize crop received 50 kg/ha of N. There was no control treatment comparing the soybean-maize rotation with maize followed by maize.

Yields of soybean and subsequent maize are shown in Table 6.11. In 2004, the effect of P on soybean yield, averaged over the two sites, was just significant at a rate of 40 kg/ha. In 2005, it was highly significant at 20 kg and just significant up to 40 kg, although the large difference in that year looks anomalous and is not further explained. The effect of the P applied to soybean on the yield of the

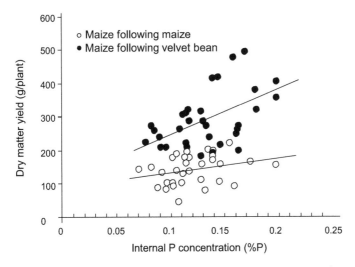

Figure 6.14 Maize dry matter yield and internal P-concentration in several treatments involving a maize-maize and velvet bean (*Mucuna*)-maize sequence (adapted from Pypers et al., 2007)

Table 6.11 Yield of soybeans and maize in rotation with three levels of P applied to the soybeans (data from Kamara et al., 2007, 2008).

	2004/2005 season		2005/2006 season	
	Miringa	*Azir*	*Miringa*	*Azir*
P applied to soybean (kg/ha)	*Soybean yield (kg/ha)*			
0		1,290		1,190
20		1,280		2,480
40		1,420		2,610
Mean	1,250	1,410	1,880	2,310
SED Year × P-level			100	
P applied to soybean (kg/ha)	*Maize yield (kg/ha)*			
0	1,991	2,023	2,353	2,606
20	2,432	1,999	2,470	2,926
40	2,558	2,863	2,580	3,185
Mean	2,327	2,295	2,468	2,906
SED P-level	109	158	174	213

Note:
'SED' = 'standard error of the difference'.

following maize crop was mostly significant up to 40 kg P/ha, with an average yield increment of 214 kg grain from 0–20 kg, and 340 kg grain from 20–40 kg P per ha. The authors commented that the maize yields and the magnitude of the yield increase due to P applied to the soybean were not very large, which 'may be due to some unknown limitations on the preceding soybean or on the maize crop itself thereby constraining their growth and yield'. The maize/soybean yield ratio was low, especially in 2005/2006, which also suggests 'unknown limitations' on the maize yield. In order to judge whether this was so, an estimate of the yield gap would be needed, for example from a simulation model.

In another series of researcher-managed on-farm trials in the savannah zones of western Nigeria and Togo, 13 soybean varieties were screened for their response to P and the effect on a following fertilised or unfertilised maize crop (Abaidoo et al., 2007). Treatments and yields for the combinations with the two best producing soybean varieties and averages for all varieties are shown in Table 6.12. Yields of the soybean were generally low, especially in Shika and Davié. Their response to triple super phosphate (TSP) was mostly significant and largest in Shika, while the effect of RP was only significant (and quite large) in Shika. Most varieties had a negative N-balance (difference between N exported with the grain and N derived from the plant's N-fixation), which according to the authors could have been due to their conservative estimation method of

Table 6.12 Yields in a soybean-maize rotation after different P-applications to the soybeans; mean soybean yields for 2001 and 2002, maize yields were given for 2003 only (data rearranged after Abaidoo et al., 2007)

	P applied (kg/ha)								
To soybean:	0 P			90 RP		30 TSP		Mean	
To maize:	0 P		15 TSP	0 P	0 P	0 P	0 P		
Treatment	Soybean	Maize		Soybean	Maize	Soybean	Maize	Soybean	Maize[a]
	Grain yields (kg/ha)								
Shika (Northern Guinea Savannah)									
TGM1566	688	967	2,405	752	2,394	1,235	1,674	892	1,860
TGx1456-2E	347	567	2,053	875	2,588	1,353	1,760	858	1,638
Mean 13 cultivars	433	885	2,434	668	2,091	1,216	1,636	772	1,538
Maize-maize[b]		449	1,544		1,632		1,422		1,168
Fashola (Derived Savannah)									
TGM1566	808	629	791	986	946	1,269	1,217	1,021	931
TGx1456-2E	1,004	631	1,134	1,009	1,186	1,288	1,104	1,100	974
Mean 13 cultivars	771	728	1,118	784	946	1,082	1,105	879	926
Maize-maize		729	739		1,063		1,136		976
Davié (Coastal Savannah)									
TGM1566	428	3,196	3,401	471	3,155	579	2,957	493	3,102
TGx1456-2E	548	3,125	2,369	695	3,327	804	3,230	682	3,347
Mean 13 cultivars	407	3,119	2,819	446	2,833	566	3,189	473	3,047
Maize-maize		3,366	2,646		3,267		2,637		3,090

SEM[c] P × Genotype:

	Shika	Fashola	Davié
Soybean	72	154	71
Maize	266	258	382

Notes:

a For the 0 P treatments only the plots where no P was applied to the maize was included.

b Same P-rate applied to first maize crop as to soybean in the other treatments.

c 'SEM' = 'standard error of the mean'.

biologically fixed N. The treatment effects on maize were very different in the three sites.

At Fashola in the DS, there was no effect of soybean on the following maize and a moderate effect of RP or TSP applied to the preceding crop (either maize or soybean). The effect was of the same magnitude as that of a smaller amount of P applied directly to the maize. Hence there was no rotational effect, only a P-effect. At Davié in the coastal savannah, maize yields were much higher than in the other sites, while there was no effect of either soybean or P on the maize. At Shika the beneficial effect of soybean on the following maize was quite strong at all levels of P applied to soybean. The effect of RP was considerably stronger than TSP, but not larger than a smaller amount of P applied directly to the maize. Considering the widely different yields of both crops in the three sites, the varying maize/soybean yield ratios and the unexplained differences in response to rotation and P-fertilisation, there must have been unidentified factors affecting the responses in these trials.

A word about maize/soybean yield ratios. Under conditions favourable for both crops the multiyear ratio has consistently been about 3 in the USA, because of the difference in carbohydrate pathways and the much higher carbohydrate requirement for the production of soybean grain, among other things. In comparison, much lower ratios were observed in exploratory (fertilised) trials across the PROSAB area and in the NGS and DS in West Africa, which again points to a shortfall in maize yield relative to what should be attainable and indicates that savannah conditions may often be more favourable for soybean than for maize. Crop modelling could help elucidate these issues.

INTERACTION BETWEEN FERTILISER AND APPLIED ORGANIC MATTER

In the previous episode, it was found that under certain conditions and in some years there was a positive interaction between fertiliser and organic matter (OM) applied as mulch or animal manure, in particular when rainfall was inadequate (Iwuafor et al., 2000; Vanlauwe et al., 2001). It was therefore hypothesised that the applied OM improved crop-water relationships and thereby the efficiency of nutrient uptake. Further studies were carried out in this episode to quantify and explain the interaction.

In 2002, an on-station trial was conducted in the DS (Davié, Togo) and the NGS (Kasuwan Magani, Nigeria) to quantify the combined effect of manure and fertiliser on maize yield (BNMS-II, 2003). The trial consisted of 15 combinations of urea, TSP and poultry or pig manure. A three-dimensional graph of the maize yields for the treatments with fertiliser alone is shown in Figure 6.15. Even at the highest fertiliser rates, which went up to 120/160 kg N/ha and 40 kg P/ha, the yields were quite low, especially in the NGS site, suggesting other limiting factors.

In order to investigate the differential effect of N and P supplied as fertiliser or as manure, saturation-type response functions, proposed by Balmukand (Black, 1993), were fitted to N and P uptake and maize yield. For maize yield the function was:

$$\frac{1}{Y} = \frac{1}{Y_{max}} + \frac{1}{f(N_s)} + \frac{1}{f(P_s)}$$

where Y is maize grain yield, $f(N_s)$ and $f(P_s)$ are simple functions of the N and P-supply, and Y_{max} is the maximum yield under non-limiting N and P. For the uptake of N and P similar functions were postulated. N and P supplies (N_s and P_s) consist of contributions from fertiliser, manure and soil stocks:

$$N_s = N_{fert} + eff_{manure-N} * N_{manure} + N_{soil}$$
$$P_s = P_{fert} + eff_{manure-P} * P_{manure} + P_{soil}$$

where *eff* is the efficiency of manure-supplied N or P, relative to urea and TSP.

Maize yields and N and P uptake in the entire dataset were fitted to these three functions, resulting in estimated maximum yields of 3930 and 3430 kg/ha in Davié and Kasuwan Magani respectively, and a relative efficiency of manure-N relative to urea in most cases far exceeding unity. The fact that the estimated maximum yield was well below the potential for hybrid maize in these areas was taken to indicate that factors other than N and P were limiting. Furthermore, as urea-N would normally be expected to be more readily available than manure-N, the fact that the (fitted) N-efficiency of manure exceeded unity was thought to indicate that the manure did not only contribute N and P but alleviated other constraints as well.

This interesting approach to the study of the interaction of mineral fertiliser and manure apparently was not further pursued.

In an on-station trial in Sékou (southern Benin), the effect of fertiliser-N, combined with *Senna siamea* mulch on maize yield, soil water availability and root development was tested (BNMS-II, 2005), in an attempt to validate and explain the earlier findings about synergism between mineral fertiliser and biomass (Chapter 5). There were two planting dates: April 19 and June 20. For the early planting date, grain and stover yield at a total N rate of 90 kg N/ha was lower for the urea-OM combination than for pure urea, but similar at the later planting date, when the overall yield level was much lower (Figure 6.16). It was presumed that the difference could be explained by more moisture stress for the later planting (rainfall data were not given), confirming earlier findings (Chapter 5). Root weight density was indeed higher for the combination-N treatment, especially in the late planting but less so in the early planting.

These results were taken as 'indications that improved water availability and improved root development are the possible mechanisms, which lead to positive interactions between OM and urea'.

A trial was conducted under CIALCA in Bas-Congo, DR Congo, to explore whether the interaction between inorganic fertiliser and biomass, which was sometimes observed in maize, occurred also in cassava (Pypers et al., 2012). The trial consisted of fertiliser levels of up to 1417 kg of 17:17:17, with the lowest rate of 283 kg combined with 2.5 t/ha dry matter of *Tithonia* or *Chromolaena odorata* as green manure (CIALCA, 2008). The trial was conducted in two locations with different soil fertility. No interaction was found between fertiliser and green manure in

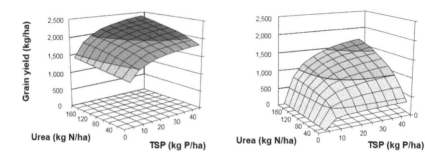

Figure 6.15 Maize yields at different levels of N and P applied as fertiliser only in Davié (left) and Kasuwan Magani (right) (adapted from BNMS-II, 2003)

either site, which was explained from the fact that improved synchronisation of N release and uptake was less important in cassava than in maize.

6.3.2 Cropping systems

The distinction between 'soil and soil fertility management' and 'cropping systems' was gradually becoming blurred, as more and more research tried to integrate technological components into the farmers' system. In this section, some 'terminal' research on herbaceous legumes as planted fallow and alley cropping is reported, as well as new research on crop-livestock interaction in the NGS and coffee (*Coffea* spp.)-banana intercropping in Uganda.

Integration of herbaceous legumes

Initially some interest persisted during this episode in herbaceous legumes as short-term fallow crops, stimulated by the reported large scale assisted, as well as the much lower autonomous, adoption of *Mucuna* by farmers, especially in Benin Republic (Manyong et al., 1999). The findings by BNMS-I, about the synergism between biomass and mineral N, provided additional arguments, even though farmers were much more attracted by its ability to control *Imperata cylindrica* than by its effect on fertility (Chapter 5). In the end, no further research was set up with *Mucuna*, however, possibly because of the rapid abandonment by farmers of *Mucuna* even in Benin (Chapter 5)[14]. Some further work was, however, undertaken with pigeon pea, which was (still) seen as having potential, because it did not suffer from some of the flaws of *Mucuna*, viz. its non-persistence in the dry season, its fire sensitivity and its practically non-edible grain.

In West and Central Africa, stands of indeterminate pigeon peas are often found in farmers' fields and backyards, without them playing a major role in the cropping systems. The BNMS-II project conducted a thematic survey in Benin in 2004 to assess the species' occurrence, its current role in the farming systems and its potential for further development. It was found that, in addition to their

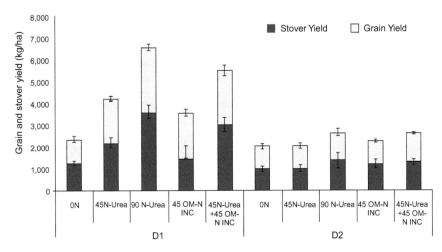

Figure 6.16 Maize grain and stover yields in an on-station trial in Sekou, Benin, with fertiliser and Senna pruning applications (referred to as 'OM' – 'organic matter') (adapted from BNMS-II, 2005)

'INC' refers to 'incorporated' since the prunings were incorporated in the topsoil. 'D1' and 'D2' refer to different planting dates ('D1' refers to April 19, 'D2' to June 8, 2004)

use as food and fodder, the leaves and other plant parts were often used to treat diseases such as malaria. It was also reported that 84% of the farmers 'desire [...] that the pigeon pea systems be properly developed in their villages' (BNMS-II, 2005). The validity of that conclusion seems doubtful, as farmers are known often to respond positively to suggestions from interviewers.

In spite of earlier disappointing experiences (e.g. Versteeg and Koudokpon, 1993), pigeon peas were further tested in the DS of Benin under BNMS-II, its perceived advantage being that it persisted during the dry season and produced edible grain. It was hoped that accessions could be found with more stable performance. As a first step, five determinate and five indeterminate varieties from ICRISAT plus a local variety were tested in researcher-managed trails with nine farmers, all intercropped into first season maize or as a second season crop after maize. The experiences were illustrative for the problems with researcher-managed on-farm trials, in particular complex ones with many treatments (16 in this case): 'poor pigeon pea establishment, poor performance of the pigeon pea varieties, damage by goats and pigs on some fields close to the villages, lack of interest from farmers, departure of the technician supervising the trials, and severe damage by insect pests' (Diels et al., 2004). The local variety outperformed all the others in terms of biomass production (scored, not measured) and the grain yield of all accessions was negligible, with the determinate varieties, planted as a second season crop after maize, being decimated by insects, much as cowpeas usually are in this ecology. The conclusion was that it would have been better to first screen the varieties on-station, which was done in the following year. Eight out of nine accession produced grain yields between 1,000 and 1,500 kg/ha, while one accession (ICP 8633) produced 2,674 kg/ha, unheard-of yields for pigeon peas, resulting from very careful (researcher-) management and three insecticide treatments. Biomass production varied strongly, ICP 8633 also having most litter and standing biomass at grain harvest. Further screening was thought to be needed, but nothing more was heard of it, partly because it was later understood that pigeon pea was considered a 'poor man's crop' in the target areas.

Integration of perennials in alley cropping

Active research on alley cropping essentially ceased in the late 1990s. Some of the existing long term station trials continued to be used for data collection on various soil fertility aspects (e.g. Nziguheba et al., 2009), but no new trials were set up.

Cocoa-based systems

In spite of the key role tree crops play in the agrarian economies in SSA, little concerted international attention had been paid to them, practically all development projects and research supporting them having addressed food crop production. The Sustainable Tree Crops Program (STCP) was set up in 2000 as an innovation platform aiming at technical, managerial and organisational improvement of tree crop production, essentially focused on the cocoa sector in the West and Central African cocoa belt. The project's key tools for the productivity improvement were the provision of seedlings for the rejuvenation of old cocoa plantations and Farmer Field Schools (FFS) for the dissemination of improved production technology. Some limited field research was carried out during this episode to generate or evaluate new messages which would feed into the FFS and other extension mechanisms.

A study in Côte d'Ivoire compared the profitability of improved production practices with current farmer practices (Assiri et al., 2012). The soils were generally found deficient in P and K, relative to the standard requirements for cocoa (Ahenkorah, 1981). Results of the study, carried out with several hundred FFS farmers over a period of four years, are shown in Table 6.13.

Table 6.13 Effect of improved practices and fertiliser on mean cocoa yield in the cocoa belt of Côte d'Ivoire (Assiri et al., 2012)

Season	Coca yield (kg/ha)		
	Farmer practices	*Improved maintenance + integrated pest management*	*Same + recommended fertiliser*
2005	425(b)	628(a)	Not applied
2006	509(b)	788(a)	836(a)
2007	578(c)	801(b)	1,085(a)
2008	450(c)	607(b)	1,000(a)

Note:
Yields in the same year followed by the same letter not significantly different (Bonferroni, 5%).

The recommended blanket fertiliser rate of 300 g/stand/year NPK (0:23:19) and 200 g/stand 'N-fertiliser' with 15% N was applied. The yield effect of fertiliser was significant in two out of three years. The results demonstrated the possibility of doubling farmers' yields, from the current low level, ranging from 250 to 600 kg beans/ha, by a combination of improved maintenance, pest control and fertiliser. There was considerable variation of these effects across areas and farmers, however, and applying fertiliser would mean losing money for many farmers when cocoa prices were low.

Crop-livestock integration

A medium-term on-farm trial was carried out with six farmers in three villages in the NGS of Nigeria. The objective was to test the synergistic effects of the simultaneous application of different component technologies, in particular crop rotation, crop residue management, livestock manure production and fertiliser application (BNMS-II, 2003; Franke et al., 2010). The trial was expected to answer two technical questions: (i) what quantities of manure can be produced *in situ* from residues of cereal-legume rotations in zero-grazing systems, and what is the nutrient content of this manure? (ii) does the application of the manure in combination with synthetic fertiliser lead to increased crop production compared to the use of synthetic fertiliser alone?

The trial was complex and it was therefore carried out by the farmers 'with close monitoring and advice from researchers and technicians'. There were two rotations, maize-maize and maize-legume, with the two phases of the latter planted in each year, plus 'farmer's practice' (Table 6.14). Half the farmers planted groundnuts in the maize-legume rotation, half soybean. In treatments 2 and 3 the crop residue was fed to goats. All cropping practices except in treatment 1 were 'best bet' technologies (improved varieties, recommended densities, combinations of fertiliser and manure, produced from crop residue adding up

Table 6.14 Treatments in an on-farm test on crop-livestock interaction in the NGS (BNMS-II, 2003)

Treatment	Year 1	Year 2	Stover/manure management
1	Farmer practice	Farmer practice	Variable
2	Maize	Maize	Fed to animals[b]
3a	Groundnut or soybean[a]	Maize	Removed and sold
3b	Maize	Groundnut or soybean[a]	Removed and sold
4a	Groundnut or soybean[a]	Maize	Fed to animals[b]
4b	Maize	Groundnut or soybean[a]	Fed to animals[b]

Notes:
a Half of the farmers planted groundnuts and half soybeans.
b Manure returned to field.

to the same nutrient content, manure storage). In treatment 1, farmers planted maize, in 11% of the cases with an associated crop, probably cowpeas.

The trials, which ran for two years, were carried out twice, starting in 2002 and 2003. Treatments 2–4 received approximately the same moderate amounts of nutrients, as fertiliser only or as fertiliser + manure (treatment 4), while farmers on average applied about one-third less fertiliser in treatment 1. Table 6.15 shows the yields of maize and legumes, averaged over sites and years.

From the economic analysis and farmers' own assessment it was clear that on average treatment 2, maize followed by maize with 'improved practices', was not an attractive option. Furthermore, the rotation with soybean did not out-perform maize-maize, nor did manure application do any better than applying the same amount of nutrients as fertiliser (treatment 3 versus 4). This leaves the rotation of maize and groundnuts as the only attractive option, due to higher maize yields in the rotation and a much higher market price of groundnuts than soybean. The analysis did not take improved livestock production into account, which would probably make the combination of groundnut-maize with manure the best performing option.

Disaggregation of the results by village showed the same pattern in two villages, while in the third, Ikuzeh, the yields of farmers' maize and soybean were extremely low, also leaving groundnut-maize as the best option, irrespective of the production and application of manure. The low yield of maize was most likely caused by very low fertiliser use by the farmers there, compared with the other villages. Possibly, farmers do not plant much maize there because of low yield expectation. The most distinctive soil property in Ikuzeh was the very low P-content of 2.6 ppm, compared with 7.3 and 14 ppm in the other villages.

The authors concluded that there was 'no large impact' of integrated crop-livestock production on the agronomic and economic performance of crop

Table 6.15 Crop yields and maize equivalent yields in the trial of Table 6.14 (BNMS-II, 2003; Franke et al., 2010). No statistics are reported

Treat-ment	Rotation	Manure	Mean yields (t/ha)			M.e./cycle[a]
			Maize	Groundnut	Soybean	
1	Maize-maize	-	1.71	NA[b]	NA	3.42
2	Maize-maize	+	2.20	NA	NA	4.40
3	Groundnut-maize	-	2.93	1.36	NA	7.44
	Soybean-maize	-	2.45	NA	1.09	4.29
4	Groundnut-maize	+	2.78	1.38	NA	7.36
	Soybean-maize	+	2.20	NA	1.13	4.10

Notes:
a Maize equivalent (M.e.) per cycle of two crops, with legumes counted according to their market price relative to maize.
b 'NA' = 'Not applicable'.

farming (in fact there was none!). The increasing adoption of integrated crop-livestock production in the Guinea savannah was supposed to be driven by factors other than those assessed in the test, such as animal traction.

Coffee-banana intercropping

The productivity and profitability of coffee-banana intercropping relative to their sole crops was evaluated in 152 farmer fields in the Arabica-growing (Mt. Elgon) and Robusta-growing (south and west) regions of Uganda (van Asten et al., 2011b). Data were collected from field observations and through structured farmer interviews. In both the Arabica and Robusta-growing regions, coffee yields in the sole and intercropping systems were similar. In the Arabica region, banana yields in intercrops were significantly higher (P < 0.05) than in mono-crops (20.2 against 14.8 t/ha/year), while they were significantly lower in the Robusta region at 8.9 against 15.0 t/ha/year (Figure 6.17). The marginal rate of return of adding banana to mono-cropped coffee was 911% and 200% in Arabica and Robusta growing regions, respectively. Fluctuations in coffee prices are not likely to affect the acceptability of intercrops when compared with coffee mono-crops in both regions, but an increase in wage rates by 100% can make intercropping unacceptable in the Robusta growing region, because the marginal rate of return will then become too small for farmers to adopt the technology. This study showed that coffee-banana intercropping is much more beneficial than banana or coffee mono-cropping and that agricultural intensification of food and cash crops in African smallholder systems should not solely depend on the mono-crop pathway.

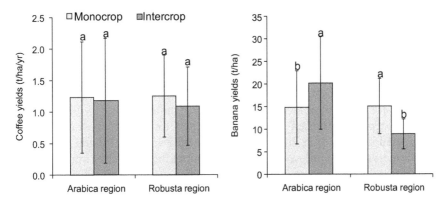

Figure 6.17 Average yields of mono-cropped and intercropped coffee (green beans, t//ha/yr) and banana (fresh fruit, t/ha/yr) in Arabica and Robusta growing regions in Uganda (adapted from van Asten et al., 2011)

Error bars are standard deviations of the individual yields. Treatments with different letters within a site are significantly different at the 0.05 level

6.3.3 Technology validation by farmers

The three major programs of this episode, BNMS-II, PROSAB and CIALCA all positioned themselves at the development end of the research-development continuum (Figure 6.1), with participatory on-farm technology testing as a pivotal activity. Most field research was now being conducted under farmers' conditions, with some back-up research in various research stations. It usually involved a sequence of researcher-managed technology development trials, and on-farm validation or demonstration trials, as exemplified by the 'mother and daughter' trials in PROSAB. Apart from the number of treatments, they also differed in the degree of researcher involvement. The trials controlled by the researchers where technology development was the primary objective were discussed in Sections 6.3.1 and 6.3.2. Here we look at trials which were intended to validate technology performance by farmers under their conditions.

A range of potential technologies were tested by farmers to alleviate constraints or exploit opportunities identified by participatory methods. Some of them came out of the projects' own technology development work, discussed earlier; other technologies were put together from the results obtained by other scientists. The trials served two purposes: (i) validation of the performance of technology under farmer conditions and (ii) familiarisation of the farmers as a first step towards adoption, if the technology turned out to be viable. In principle, these trials-cum-demonstrations were managed by the farmers themselves, although the real extent of farmer management is not always clear.

Methodology

Research projects were now scattered across Africa and involved numerous local and international partners, all with their own predilection for specific approaches and methods and little guidance or coordination from the institute's scientific leadership. A variety of approaches were used in the validation trials, with different levels of farmer involvement in the choice of treatments and trial management, and little methodological harmonisation across projects. Nevertheless, some commonalities can be gleaned from the publications, viz. single replication per farm, stepwise inclusion of a small number of technologies and the use of statistical methods to explain variability.

The stepwise approach as used by CIALCA for the adoption of ISFM is illustrated in Figure 6.18. Improved crop varieties were the first and easiest step for farmers to adopt, followed by different options for soil fertility improvement, starting with fertiliser and ultimately cumulating in integrated soil fertility management, choosing components consistent with the local conditions and the farmers' objectives. As the farmers' knowledge increased more complex interventions could be considered. This was also reflected in some of the on-farm trials where these different steps formed the treatments.

An aspect which often remains unclear is the degree of farmer management. The use of the term 'demonstration' and the experimental details published suggest more than trivial supervision by the research and/or development partners.

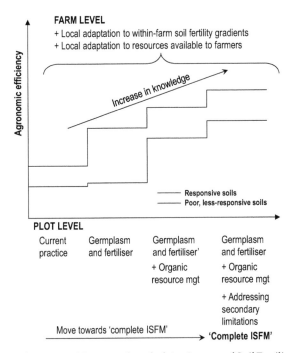

Figure 6.18 Revised conceptual framework underlying Integrated Soil Fertility Management (ISFM), adapted from the original version, presented by Vanlauwe et al. (2010)

The current version distinguished plot from farm-level 'local adaptation' interventions (adapted from Vanlauwe et al., 2015)

Technology validation by farmers

FARMER TECHNOLOGY VALIDATION IN THE GREAT LAKES AREA

'Demonstration trials' were carried out with farmer groups in the CIALCA program to test and demonstrate the effect of various improved technologies in cassava-legume intercropping, viz. different legume species, improved crop varieties, fertiliser rates and planting densities (CIALCA, 2008). A first simple test demonstrated the effect of replacing groundnuts by soybean, which produced much larger amounts of biomass, and were therefore assumed to make a greater contribution to soil fertility as had been found elsewhere in SSA.

More substantive tests with stepwise inclusion of various technologies were to allow farmers to observe the technologies' cumulative effect. These were essentially researcher-controlled trials in farmers' fields, with extensive participatory design, monitoring and evaluation. The sequence of (conventional) technologies considered for inclusion in the trials was:

1   Improved cassava varieties;
2   Choice of legume species (beans, groundnuts, soybean);

3   Cassava planting pattern (1 × 1 or 2 × 0.5 m); and
4   Fertiliser and/or green manure.

Green manure was only tested in Bas-Congo, where land pressure was still low and green manures could be available in reasonable quantities. No results were published of these tests.

On-farm demonstrations were carried out with combinations of local and improved maize varieties with and without fertiliser, simulating the first step in the adoption process. The best variety yielded a very high average of 5 t/ha without fertiliser and close to 6 t/ha with modest fertiliser (60:13:25 kg/ha NPK) across four sites (Vanlauwe et al., 2013).

As a second step towards ISFM the effect of maize, climbing beans (CB) or soybean (SB) on the following maize crop was demonstrated at a low fertiliser level (100 kg/ha of 17:17:17). The results are shown in Figure 6.19, with yields which were probably closer to those obtained by farmers than in the previous demonstration. The advantage of growing a legume on the yield of the following maize stands out, especially for climbing beans.

Erosion control had been analysed as a potentially important innovation for some areas. Technologies for erosion control, including contour planting and minimum tillage had therefore been tested on-station in South-Kivu (DR Congo), where erosion was a serious problem on very steep slopes. Some of these technologies were subsequently tested on-farm in essentially researcher-managed trials, but no results have been published.

FARMER TECHNOLOGY VALIDATION IN THE SAVANNAH AREAS OF WEST AFRICA

In the BNMS-II project, farmer validation of technology was part of the 'demonstration trials', which followed directly on the technology development

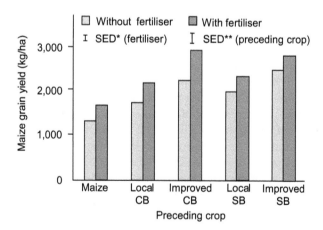

Figure 6.19 Effect of preceding crop on maize yield with and without fertiliser (adapted from Pypers et al., in prep.)

'CB' refers to 'Climbing beans', 'SB' to 'Soybean', and 'SED' to 'standard error of the difference'

process, also mostly carried out on-farm (Sections 6.3.1 and 6.3.2). The distinction between farmer validation and demonstration thus becomes artificial. On-farm validation of the research findings on legume-maize rotation with fertiliser and manure in the NGS is therefore discussed later, in Section 6.4.3 (Dissemination and M&E).

The PROSAB project conducted so-called daughter trials with farmers, the mother trials being researcher-managed tests with a greater range of treatment levels for the selection of 'best-bet' combinations (Kamara et al., 2009). Technologies tested and demonstrated in the daughter trials were varieties of maize, soybean, groundnuts and NERICA (New Rice for Africa) upland and lowland rice, as well as legume-cereal rotation and appropriate fertiliser rates. The effects on *Striga* were an important consideration for farmers when choosing technology combinations. No data on yield and factors affecting yield were collected in the daughter trials, the evaluation being based on farmer assessment of the technologies' performance.

Technology performance under farmer management

The on-farm validation trials discussed here were conducted under supervision by project staff, in some cases with very high yield. The question remains how the technology would perform under full farmer management. Few quantitative data have so far been published on this, but data on the effect of di-ammonium phosphate on beans collected in farmers' fields (Pypers et al., in preparation) illustrate the kind of variation to be expected (Figure 6.20). The boundary line suggests the possible trend of agronomic efficiency of the fertiliser, with low efficiency in the poorest and the most fertile fields and highest efficiency in between. The wide scatter below the boundary line represents the (depressing) effects of a combination of causes, from soil-to-management-related factors, which will have a strong influence of the profitability of fertiliser.

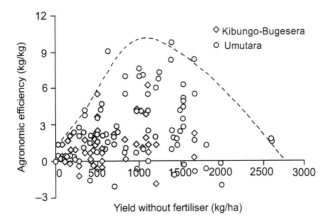

Figure 6.20 Scatter of the agronomic efficiency of fertiliser (di-ammonium phosphate) as a function of yield without fertiliser (adapted from Vanlauwe et al., 2013)

Monitoring adoption

The first genuine proof for the adoptability of new technology is when farmers, who have been involved in its on-farm testing, adopt the technology voluntarily and on their own account. As long as the researchers and development workers are still involved, other factors and expectations may stimulate farmers to continue using it. It is only when that is no longer the case that real adoption can occur, and that is when it should be assessed. The second, and much stronger proof, is when neighbours and other farmers in the community and in neighbouring communities start adopting. This process happens naturally with improved crop varieties, but so far no systematic monitoring has been reported for this kind of adoption of soil and soil fertility management technology, perhaps with the exception of alley cropping (Whittome et al., 1995).

6.4 Technology delivery and dissemination

Technology delivery and dissemination is now treated as an integral part of the research process, not as a separate activity to be taken up by development and extension organisations, once the technology has been validated. In principle, the entire process is jointly planned and carried out by all partners, whereby each partner's specific role is of course different according to the phase of the process.

6.4.1 Databases, digests and guidelines

Little effort was devoted during this episode to the consolidation of the results of previous soil and land use research, even though the institute was reinforcing its role as a development partner and would need digested research results to formulate 'best-bet' recommendations. The 2007 EPMR expressed concern about the absence of systematic efforts to make the body of knowledge generated in earlier decades accessible. For example:

> Research data from the first 25 years […] were summarised as scholarly publications […] but this is not in a form that the fertilizer industry or policy makers could easily use […]. Needed is a comprehensive summary documenting the fertiliser responses […] supplemented by on-farm studies and 'best-bet' technologies in fertility maintenance and measures of soil organic matter (SOM) regeneration.

or

> Alley cropping has largely been abandoned because of poor adoption, though a comprehensive analysis of the causes of this failure was not made available to the EPMR [= did not exist?].

Other areas where the EPMR thought that consolidation of previous research was needed were leguminous cover crops, animal manures and crop residues.

In fact, an inventory of the results of 25 years of research on alley cropping and cover crops had been published in 2004, analysing their performance for restoration or maintenance of fertility (Hauser et al., 2004), but this publication was not intended for practical application.

Results from more recent research were summarised by the BNMS-II project in the form of 'Guidelines for Balanced Nutrient Management' and an extension guide entitled 'Growing Soybean and Maize in Rotation', but it is not clear whether these documents are still available[15].

There is no evidence that the 2007 EPMR observations led to more systematic efforts to assemble past results into a useable form for practical farming. Only CIALCA, with its typical 'downstream' objectives, did put together packages of recommendations around major crops in the Great Lakes area, based on research results from various institutes, including IITA (Vanlauwe pers. comm.). A range of technical sheets on nutrient deficiencies, crop management and plant density for highland bananas was developed as learning materials for 'training of trainers'.

6.4.2. Decision support

IITA supported the Grameen Foundation in Uganda on developing mobile phone apps to deliver recommendations to farmers based on using sms-guided decision-tree questions to tailor the recommendations to their needs. It is not very clear to what extent these apps have really found their way to the farmer (van Asten, pers. comm.).

No further activities have been reported for this episode on the development of Decision Support tools.

6.4.3 Dissemination and monitoring and evaluation

Technology dissemination is the final phase in the research-development continuum. In the historical arrangement of things, it was considered outside the realm of research and the task of the extension service. The failure of the old model was one of the stimuli to reform the process and associate research with the entire research-development process, starting in the 1980s with FSR and culminating in today's R-for-D. During the present episode, research was still grappling with its role, in particular at the transition from on-farm testing to actual adoption by farmers. Once a technology was considered viable, different methods were tried to disseminate it more widely to farmers, some quite similar to old school extension methods, others more innovative. The following methods were experimented with:

- On-farm demonstrations;
- Field days;
- Contact farmers and farmer groups;
- Feed-back meetings;
- Innovation platforms; and
- Farmer field schools.

Here we will review the technologies which were considered ready for dissemination, while dissemination will be looked at in the next section.

The early 2000s marked the eclipse of technologies once considered most promising: alley cropping and short fallows with herbaceous legumes. We will look at their adoption status and the basis for their abandonment in Section 6.5.2 (Technology uptake and impact). Other green manure technologies which had shown promise were the use of the major fallow species *Chromolaena* and *Tithonia*. However, since their performance and profitability were not assured they were not actively promoted.

The results of earlier fertiliser testing programs across SSA had not been translated into practical recommendations, so that locally adapted nutrient packages based on those results could not be put together (CGIAR, 2008).

Production packages were put together for dissemination by the various projects, based on their own research and results obtained elsewhere, with emphasis on the improvement of nutrient use efficiency, by a combination of biological and chemical components.

Starting in 2000, improved system components were demonstrated in farmers' fields in northern Nigeria by BNMS-II in collaboration with the institute of Agricultural Research (IAR) and SG2000. The demonstrations included three packages:

- Continuous maize with sufficient fertiliser (SG-2000 package);
- Continuous maize with less fertiliser, combined with manure; and
- Soybean-maize rotation, with a reduced amount of fertiliser applied to the maize.

Maize yields with manure were similar to those in the soybean-maize rotation and SG2000 technology and significantly higher than in the Farmer Practice. The variability among farmers was large. In the next few years, the differences with Farmer Practice became smaller, which was taken to indicate adoption of the improved practices. Availability of manure, however, was a constraint (BNMS-II, 2008).

As of 2006, a large number of demonstrations with legume-cereal rotation were established in three states in northern Nigeria, again in collaboration with SG2000, using maize, sorghum or millet and cowpea, groundnut or soybean, depending on the preference in the area. Recommended crop varieties were used if available, while management and cultural practices were applied as recommended by SG2000.

PROSAB promoted crop production technologies in north-east Nigeria through their 'daughter trials' and through demonstrations with farmer groups, field days and exchange visits. The demonstrations were concerned with (Kamara et al., 2009):

- Micro-dosing of fertiliser and burying of fertiliser to ensure maximum use and avoid volatilisation and rain wash;

- Cereal-legume rotation to improve soil fertility and control *Striga* infestation of cereal and legume; and
- Adequate plant population and plant spacing to maximise land use and increase productivity.

CIALCA made a distinction between pilot sites, where technologies were developed and validated, and satellite sites where proven technologies were disseminated by development partners. Several technology packages for improved crop management were promoted, with emphasis on maintenance and improvement of soil fertility (CIALCA, 2013):

- Improved maize-legume rotation and intercropping;
- Improved cassava-legume intercropping;
- Banana-beans intercropping;
- Coffee-banana intercropping; and
- Combined manure/compost and fertiliser application.

Local partners and beneficiaries chose technologies to promote according to their own needs. In North Kivu (DR Congo) and Burundi, banana technologies were most important, while the focus in Bas-Congo was on legumes and the management of legume intercropping with maize and cassava. CIALCA's preferred dissemination channel were farmer groups, but it was left to the development partners to choose the dissemination methods. Some of them also experimented on their own initiative with the use of farmer-managed trials as a means to expose farmers to new technologies. A stepwise design was typically used, starting from farmers' practice, and incremental addition of improved varieties, crop arrangement and density, and fertiliser inputs. Development partners collected data through trained lead farmers and discussed the results within their groups of farmers. Some farmers continued to apply or modify certain aspects of the technologies, based on interviews, but this was never verified, let alone quantified in the field (Pypers, pers. comm.).

6.5 Outputs and impact

6.5.1 Highlights of research outputs

Although the continent-wide collection and interpretation of soil data was no longer an objective, some of the work in this episode did contribute to the database on the characteristics and nutrient status of West and Central African soils. In particular, the field research sponsored by the BNMS-II project in Nigeria and Togo included systematic soil data collection in the project's areas of activity. All the projects also paid attention to macro- and micronutrient deficiencies in the savannah areas and used new methods based on tissue analysis to diagnose such deficiencies in various crops. However, as pointed out earlier, an underlying framework was missing to link experimental results to a geographical grid showing major soil groupings with a reasonable degree of detail, i.e. beyond the level of the Great Groups in Soil Taxonomy, or single prefix qualifiers in WRB[16].

On the other hand, there was a profusion of new analytical tools being used in several locations and projects to relate measured plant and soil variable to actual and potential yield:

- DRIS (Diagnosis and Recommendation Integrated System);
- CND (Compositional Nutrient Diagnosis);
- Boundary line analysis; and
- QUEFTS (Quantitative Evaluation of the Fertility of Tropical Soils).

DRIS, CND and boundary line analysis creatively exploited 'big data': large data sets collected in field trials and farmers' fields and analysed to detect relationships between crop yield and plant-, field- and farmer-related variables. Further adaptation of these tools to different crops and growing conditions will be needed before they can become suitable for wider application.

The emphasis during this episode was on the development of methodology for the increasingly important on-farm research activities, including the above techniques to assess nutrient deficiencies in farmers' fields. The analysis of variability and yield gap analysis across farms was to provide a basis for site-specific plant nutrient recommendations.

The technologies actually tested were of a conventional nature, the novelty being that they were evaluated in a whole farm context, rather than in the isolation of research plots.

Station research in the Great Lakes area showed that the short-term yield effects of erosion control were negative, making it difficult for farmers to adopt. In colonial times the 'remedy' had been to impose slope protection; today's program renounced the introduction of direct erosion control measures on-farm.

Testing of amendments for the extremely unfertile soils in South Kivu showed some effect of P and N, but they could not break through the apparent impediments to higher yield. Poor genetic material of maize was, however, suspected to be an important factor in the poor test results.

Most work was done on identifying the most profitable combinations of organic and inorganic sources of nutrients and crop management practices to optimise farm productivity in humid and sub-humid West Africa, the Great Lakes area of Central Africa and mid-altitude East Africa. Amendments for P and secondary and micronutrient deficiencies thereby often emerged as important elements. Furthermore, there was increasing emphasis on grain legumes as part of balanced cropping systems and their ability to extract tightly-bound P, whereby important species and varietal differences were found. The positive rotational effect of *Mucuna* on P nutrition of maize was confirmed, but no further work was done on herbaceous legumes.

The synergy between organic matter and fertiliser-N at relatively low maize yield levels was confirmed by work in Benin and Togo and could in some cases be attributed to improved availability of moisture. A similar synergism was not found for cassava in Bas-Congo, nor did manure application do any better than applying the same amount of nutrients as fertiliser in northern Nigeria.

The yield variability in existing low yield cocoa orchards is poorly understood and the profitability of fertiliser use is not assured when fertiliser prices are high. More research is needed to relate the effect of shade on the response to fertiliser to the soils' fertility status.

In the Great Lakes area, it was found that coffee-banana intercropping is much more beneficial than banana or coffee mono-cropping.

6.5.2 Technology uptake and impact

Information on the adoption of soil and soil fertility technology remained scarce, contrary to improved varieties, the adoption of which is easier to verify. Reported adoption in most cases referred to farmers knowing about or having been exposed to technologies and NGOs and other development partners having taken up technologies in their own extension programmes (BNMS-II, 2008; CIALCA, 2013; Ellis-Jones et al., 2009). Adoption and impact studies invariably relied on interviews, rather than on actual field-level observations, which is the only really reliable method to verify genuine adoption.

In many cases, farmers complained about high cost and poor availability of fertiliser and manure, as they had always done, the difference being that the researchers now developed initiatives together with other stakeholders to improve the situation through stakeholder platforms, in which the parties involved in input supply participated. The success of such initiatives to improve the input supply situation will be a determinant for the future success of the efforts towards intensified production.

6.6 Emerging trends

Major international events occurred in Africa during this episode, which strongly affected the direction international research was taking. The first was the Abuja Fertilizer Summit in 2006, where the Heads of State of African countries met to sketch a strategy for a significant increase in fertiliser use, from 8 to 50 kg of fertiliser nutrients per hectare, as a necessary condition to boost agricultural production and a prerequisite for alleviating rural poverty and accommodating the food needs of a growing population. This was followed in 2007 by the establishment of the Alliance for a Green Revolution in Africa (AGRA) upon the initiative of Kofi Annan. An African Green Revolution, while taking lessons from the Green Revolutions in Latin America and Asia, would have to be uniquely African by recognising the continent's great diversity of landscapes, soils, climates, cultures and economic status (Annan, 2008), as well as the complexity of subsistence farming. AGRA was supported by the Rockefeller and the Bill & Melinda Gates Foundation (BMGF) Foundations, the latter having recently become engaged in agricultural development. A Soil

Health Program was established within AGRA that would support the Africa-wide dissemination of ISFM practices. The vision about the role of international research underwent a most significant change, inspired by these events and led to the adoption of the R-for-D approach. The primary goal was no longer just to generate prototype technology, but rather to contribute directly to improved livelihoods by bringing technical expertise to bear on a multi-stakeholder effort. This change affected soil and land use research more strongly than crop improvement, whose methods remained relevant, even with changed breeding criteria and early exposure of breeding material to the real farm.

A draw-back inherent in the R-for-D approach, in particular the multi-donor variety, is the inevitable fragmentation of research activities, spread out over a wide area. There are different 'centres of inspiration' associated with different projects and donors and central scientific leadership has been lacking. This poses the challenge of maintaining the coherence of the Institute's activities and approaches, a challenge that will grow with its geographical spread, as happened with the numerous 'outreach projects' of the 1980s (Chapters 3 and 4).

The characterisation studies of the previous decade delineated large agro-ecological zones with similar characteristics and research needs and identified representative pilot areas where research was to be carried out. This approach had been abandoned and during the present episode regional studies were carried out by various projects, to characterise the conditions in their more limited mandate areas and identify priority constraints and opportunities for applied research. In some programs, a certain degree of over-kill was apparent, with large and costly socioeconomic surveys to arrive at conclusions which could have been obtained with more informal methods.

In respect of soils and soil fertility, classical soil survey methods and associated mapping were largely abandoned. As a result, many of the trials carried out all over the continent suffered from incomplete knowledge of the soil conditions and their intrinsic deficiencies and potential, making the interpretation of trial results often spurious and hard to extrapolate. The need remains, however, to capitalise on the large amount of soil data collected in previous years.

Classical soil survey had been replaced by a combination of soil, tissue and yield sampling to explain crop responses to nutrients and improved technology. DRIS, CND, Boundary Line analysis and QUEFTS were thereby used as analytical tools. Another tool has recently been advocated for the analysis of soil properties, viz. Near-Infrared Reflectance Spectroscopy (NIRS). With time, the alternative methods may become powerful tools for the identification of and mapping of nutrient contents and deficiencies, but more research will be needed to assess their potential. In future those tools, with precise instructions for their use and indications of their potential and limitations, may become an important contribution to soil and soil fertility research in SSA.

Dynamic modelling has had only a minor history at the institute, with some activities around CERES-Maize and the Rothamsted Carbon model, which ended with the departure of their protagonists (Chapter 5). During this episode, there was some modelling activity with the static QUEFTS model and the dynamic LINTUL-1 model from Wageningen University which was adapted for bananas by two PhD students working in the Great Lakes area.

The 2011 External Review of the Natural Resource (NRM) Management research areas recommended a fresh start with system modelling as a way to bring together the 'dispersed NRM activities' (Lynam and Carberry, 2011; Merckx, 2011). Crop modelling would also be useful in association with on-farm experiments and the other 'new' analytical tools, to assess the effects of climate and soil conditions on potential crop growth, estimate yield gaps and alert researchers to unidentified limiting factors in case those gaps are large. The External Review suggested collaboration with CSIRO (Commonwealth Scientific and Industrial Research Organisation)[17], which would bring in yet another actor in the institute's modelling story, after Rothamsted (Carbon model), the University of Florida (CERES 'Crop Environment Resource Synthesis' Maize) and Wageningen University (LINTUL). In order to avoid confusion and make progress, it would be wise to carefully consider the options and choose one principal modelling partner for the years to come.

The importance of P for intensified cropping systems, particularly in the moist savannah zones, received much attention. Earlier results about the synergistic effects of legumes-cereals rotation on P-availability in P-limited soils were partially confirmed, while the effects of rock phosphate remained confusing. Pypers et al. (2007) stated that: 'high rainfall and low soil pH are absolute prerequisites for [rock phosphate] utilization', which had been the earlier soil scientists' opinion all along, but which was contradicted by the results of the study by Abaidoo et al. (2007), where the largest RP effect was found in the driest site (Shika). At most it confirms another conclusion, viz. that 'beneficial effects of [rock phosphate] application combined with adapted legume-cereal rotation systems are site-specific' as had been shown by Vanlauwe et al. (2000). Participants in the closing conference of the ACIAR project, which had focused on improving P-availability in cropping systems, went even further in stating that the benefits of using RP on non-acid soils were not clear, RP was not readily available to farmers in the open market and that therefore there should be reduced emphasis on work that included RP. In view of the expected increase in P-deficiency with intensified cropping, this conclusion may be ill-advised and should be carefully reconsidered based on the results so far.

The profitability and adoptability of a rotation system with *Mucuna* or other non-edible legumes turned out to be highly dubious and it was thought that 'a dual-purpose grain legume (e.g. soybean), providing edible beans and a direct profit to farmers [...] deserved further investigation' (Pypers et al., 2007), a conclusion which strongly affected research in this episode, with its emphasis on grain legumes as a rotational dual purpose crop, especially soybean.

Alley cropping had by now disappeared from the research agenda, but in some people's minds there could still be a niche for the technology under some conditions (e.g. Dennis Shannon, pers. comm.). What these conditions would be had been examined earlier by Dvorak (1996) and Whittome et al. (1995), but the 2007 EPMR Panel felt that a more thorough assessment would be needed for proper closure of the alley cropping story.

Considerable research efforts were devoted to the tree component (coffee) in the cropping systems in the Great Lakes area, but the work on tree crops in the lowland forest areas dwindled, in spite of their suitability for perennial crop production. This reflected a significant shift away from the humid forest zone compared with earlier years.

The enthusiasm to go and work directly with the farmers and skip on-station research as much as possible was refreshing and the creative use of various new analytical techniques for the interpretation of large amounts of data collected in farmers' fields is admirable. These analyses were very helpful in explaining farmer variability.

As regards experimentation, few truly farmer-managed tests were carried out: most of the experiments continuing to be under researcher control. Even at the dissemination stage technology was *demonstrated* to farmers, rather than *handed over* to them to try in their own way, with observation by the researchers and developers rather than control. Control implies great effort and expense in time and money, which appears often quite unnecessary, when the technology is straightforward and within farmers' competence, as was often the case. The methodology of on-farm experimentation would need to be revisited whereby the experiences of the 1980s would be useful. The use of Hildebrand's 'adaptability analysis' combined with Boundary Line analysis would form an excellent tool for the examination of the variation of nutrient responses and other treatment effects across farmers.

Farmer variability attracted increasing attention, as a consequence of the further transfer of research activities from the station to the real farm. In particular, the analysis of the large variations in nutrient use efficiencies, which had also been apparent in research from earlier years, would be crucial in the search for location-specific soil fertility technologies.

The pressure to show significant impact was rising during this episode, as exemplified by the impact study conducted at the end of the CIALCA project, which stated that 'Impact assessment […] is a requirement to establish accountability for public funds spent […], as well as derive lessons [for] higher impacts and more efficient processes [in the future]' (CIALCA, 2013).

At the same time, it was far from clear what should be considered as significant impact and how it could be measured. The weakest criterion was the often mentioned 'awareness' among intended beneficiaries of a project's activities and the technologies it promoted, the strongest being a measurable increase in productivity in the target zone. The most commonly used methods were stakeholder workshops, focus group discussions and questionnaire-based surveys, covering large numbers of respondents. The conclusions from such studies in respect of the impact of agronomic technology has often been that high awareness was generated, but that it was too early to find measurable impact in terms of increased productivity. In the case of CIALCA, for example:

a couple of years after massive dissemination and out-scaling [...] many adopters might still be at the beginning of their learning curve and hence observed impacts as yet are likely to underestimate full benefits of the technologies. A follow-up study after 2 to 5 years is therefore recommended.'

This, of course, is a valid proviso, but it is unlikely that such follow-up studies will be carried out after a project ends. It also points to the inability of many or most assessment studies to measure genuine adoption in the early stages of on-farm technology testing and dissemination, in spite of their great effort and cost. What is needed is more down-to-earth direct observation of what happens after a technology has been tested or demonstrated, starting with the farmers who were involved and possible copiers. In other words: track the technology, rather than trying to measure its necessarily minuscule effect on regional productivity in its infancy, which is impossible.

Notes

1 Now called '*Conseil à l'Exploitation Familiale*' (CEF) (Paul Kleene, pers. comm.)
2 See Chapter 8 for the trends in restricted and unrestricted funding.
3 ASB ('Alternatives to Slash and Burn'), 1994–2007, an inter-centre collaborative project at the Humid Forest Station, Cameroon, led by ICRAF and funded, amongst others, by the Global Environment Facility; STCP ('Sustainable Tree Crops Program'), 2000–2012, supported by USAID and the World Cocoa Foundation, emphasising cocoa-based tree crop systems; BNMS-II ('Balanced Nutrient Management Systems'), 2002–2006, funded by the Belgian Government; CIALCA ('Consortium for Improving Agriculture-based Livelihoods in Central Africa'), 2006–2013, an inter-agency project in the Great Lakes area, with participation by Bioversity International, IITA, the Tropical Soil Biology and Fertility Institute of the International Center for Tropical Agriculture (TSBF-CIAT), and national research and development partners, in Rwanda, Burundi and DR-Congo, funded by the Belgian Government; 'Improving Phosphorus Availability in Cropping Systems in Sub-Saharan Africa', 2001–2004, with ACIAR and CSIRO Plant Industry, funded by the Australian Government; and N2Africa ('Putting N fixation to work for smallholder farmers in sub-Saharan Africa'), with Wageningen University, the International Livestock Research Institute (ILRI) and research and development organisations in several countries all over Sub-Saharan Africa, and funded by the Bill & Melinda Gates Foundation (BMGF), a new donor, supporting agricultural development in Africa, who appeared on stage during this episode, and was going to be very influential in coming years.
4 PROSAB ('Promoting Sustainable Agriculture in Borno State, Nigeria'), 2003–2009, with participation by ILRI and national research and development organisations based in Borno State and funded by the Canadian International Development Agency (CIDA); and SSA-CP ('Sub-Saharan Africa Challenge Program'), 2008–2013, an inter-Centre collaborative project, with IITA in the lead, funded by a group of donors including DFID through the Forum for Agricultural Research in Africa (FARA).
5 Mandate areas are defined as areas with similar agro-ecological conditions and poverty profiles that have nonetheless relatively good access to large urban markets. The number of people living in each mandate area can vary between 300,000 and 1,200,000 (CIALCA, 2013).

6 Based on critical values for table bananas from literature, for a first orientation.
7 A second improved variety included in the trial in Kenya has been left out in the table.
8 The methodological study preceded this field study but was published later.
9 Surprisingly, the paper's abstract states that the prediction was even better than for Ntungamo, contrary to the main text.
10 The authors postulated a linear relationship beyond 1500 mm, without a clear justification.
11 The range of K-contents given in the paper for the boundary line analysis of Kibungo do not correspond with those shown in the table with physical soil data.
12 The Mg content of all the soils was above 0.48 cmol/kg (Table 6.9) which probably also exceeds the threshold for this element.
13 The Mehlich-3 test gives results comparable with Bray-1.
14 No further studies on the fate of *Mucuna* were carried out in Benin after 2002.
15 They are not mentioned in the IITA on-line bibliography (biblio.iita.org).
16 The authors of the Soil Atlas of Africa recommend this level for small scale soil maps (< 1:1 million) (Jones et al., 2013).
17 CSIRO has been associated for a long time with the CERES modelling group in the USA.

References

Abaidoo, R.C., N. Sanginga, J.A. Okogun, G.O. Kolawole, B.K. Tossah and J. Diels, 2007. Genotypic variation of soybean for phosphorus use efficiency and their contribution of N and P to subsequent maize crops in three ecological zones of West Africa. In: B. Badu-Apraku et al. (eds), *Demand-Driven Technologies for Sustainable Maize Production in West and Central Africa,* WECAMAN/IITA, Ibadan.

ACIAR/IITA, 2005. *Improving phosphorus availability in cropping systems in sub-Saharan Africa.* Final Project Report. ACIAR Canberra/IITA Ibadan.

Ahenkorah, Y., 1981. Influence of environment on growth and production of the cacao tree: soils and nutrition. Paper presented at the International Cocoa Research Conference, Douala, Cameroun, 4–12 Nov 1979.

Annan, K.A., 2008. Forging a Uniquely African Green Revolution. Address by Mr. Kofi A. Annan, Chairman of AGRA, Salzburg Global Seminars, Austria, 30 April 2008.

Asadu, Robert and Felix Nweke, 1999. *Soils of Arable Crop Fields in Sub-Saharan Africa: Focus on Cassava-growing Areas.* COSCA Working Paper No. 18. IITA, Ibadan.

Assiri, A.A., E.A. Kacou, F.A. Assi, K.S. Ekra, K.F. Dji, J.Y. Couloud and A.R. Yapo, 2012. Rentabilité économique des techniques de réhabilitation et de replantation des vieux vergers de cacaoyers (*Theobroma cacao* L.) en Côte d'Ivoire. *Journal of Animal & Plant Sciences*, 14, 1939–1951.

Black, C.A. 1993. Nutrient supplies and crop yields: Response curve. In: C.A. Black (ed.), *Soil Fertility Evaluation and Control.* CRC Press, Boca Raton, FL.

BNMS-II, 2003. Achieving development impact and environmental enhancement through adoption of balanced nutrient management systems by farmers in the west African savanna. *Annual Report 2002.* IITA, KULeuven, DGDC, Ibadan.

BNMS-II, 2005. Achieving development impact and environmental enhancement through adoption of balanced nutrient management systems by farmers in the west African savanna. *Annual Report 2004.* IITA, KULeuven, DGDC, Ibadan.

BNMS-II, 2006. Achieving development impact and environmental enhancement through adoption of balanced nutrient management systems by farmers in the west African savanna. *Annual Report 2005*. IITA, KULeuven, DGDC, Ibadan.

BNMS-II, 2007. Achieving development impact and environmental enhancement through adoption of balanced nutrient management systems by farmers in the west African savanna. *Annual Report 2006*. IITA, KULeuven, DGDC, Ibadan.

BNMS-II, 2008. Achieving development impact and environmental enhancement through adoption of balanced nutrient management systems by farmers in the west African savanna. *Final Report January 2002–December 2007*. IITA, KULeuven, DGDC, Ibadan.

CGIAR Science Council, 2008. *Report of the 6th External Program and Management Review of the International Institute of Tropical Agriculture (IITA)*. CGIAR Science Council Secretariat Rome, Italy.

CIALCA, 2008. *Progress Report December 2006–December 2007*. BioDiversity/IITA/TSBF-CIAT, Cali, Colombia.

CIALCA, 2011. *Technical Progress Report No. 7, January–December 2010*. BioDiversity/IITA/TSBF-CIAT, Cali, Colombia.

CIALCA, 2013. *Assessing the Impact of CIALCA Technologies on Crop Productivity and Poverty in the Great Lakes Region of Burundi, the Democratic Republic of Congo (DR- Congo) and Rwanda*. BioDiversity/IITA/TSBF-CIAT, Cali, Colombia.

Defoer, T and A. Budelman (Eds), 2000. *Managing Soil Fertility in the Tropics. PLAR and Resource Flow Analysis in Practice. Case studies from Benin, Ethiopia, Kenya, Mali and Tanzania*. Royal Tropical Insitute (KIT), Amsterdam, Netherlands.

Diels, J., B. Vanlauwe, M.K. van der Meersch, N. Sanginga and R. Merckx, 2004. Long-term soil organic carbon dynamics in a subhumid tropical climate: ^{13}C data in mixed C_3/C_4 cropping and modeling with ROTHC. *Soil Biology & Biochemistry*, 36, 1739–1750.

Dvorak, K.A., 1996. *Adoption Potential of Alley Cropping*. Resource and Crop Management Research Monograph No. 23. IITA, Ibadan

Ellis-Jones, J., P. Amaza and T. Abdoulaye, 2009. An early assessment of adoption and impact of PROSAB's activities. PROSAB Stakeholder Conference for Sharing Experiences 1st and 2nd September 2009, Maiduguri, Borno State, Nigeria.

Faure, G., P. Kleene and S. Ouedraogo, 1998. Le conseil de gestion aux agriculteurs dans la zone cotonnière du Burkina Faso: une approche renouvelée de la vulgarisation agricole. *Études et Recherches sur les Systèmes Agraires et le Développement*, 31, 81–92.

Fermont, A.M., P.J.A. van Asten, P. Tittonell, M.T. van Wijk and K.E. Giller, 2009a. Closing the cassava yield gap: An analysis from smallholder farms in East Africa. *Field Crops Research*, 112, 24–36.

Fermont, A.M., P. Tittonell, Y. Baguma, P. Ntawuruhunga and K.E. Giller, 2009b. Towards understanding factors that govern fertilizer response in cassava: lessons from East Africa. *Nutrient Cycling in Agroecosytems*, 86, 133–151.

Franke, A.C, E.D. Berkhout, E.N.O. Iwuafor, G. Nziguheba, G. Dercon, I. Vanderplas and J. Diels, 2010. Does crop-livestock integration lead to improved crop roduction in the savanna of west Africa? *Experimental Agriculture*, 46, 239–455.

Hartmann, P., 2004. *An Approach to Hunger and Poverty Reduction for Subsaharan Africa*. IITA, Ibadan.

Hauser, S., C. Nolte and R.J. Carsky, 2004. What role can planted fallows play in the humid and sub-humid zone of West and Central Africa? *Nutrient Cycling in Agroecosystems*, 76, 297–318.

IITA, 2003. *Annual Report 2002*. IITA, Ibadan.

IITA, 2006. *Annual Report 2005*. IITA, Ibadan.

ISPC, 2010, *Report of the Second External Review of the Sub-Saharan Africa Challenge Program (SSA-CP)*. CGIAR, Washington, DC.

Iwuafor, E.N.O., K. Aihou, J.S. Jaryum, B. Vanlauwe, J. Diels, N. Sanginga, O. Lyasse, J. Deckers and R. Merckx, 2000. On-farm evaluation of the contribution of sole and mixed applications of organic matter and urea to maize grain production in the savanna. Presentation at the BNMS Symposium, Oct. 2000, Cotonou.

Janssen, B.H. and F.C.T. Guiking, 1990. Modelling the response of crops to fertilizers. In: M.L. van Beisichem (ed.), *Plant Nutrition: Physiology and Applications*. Kluwer, Dordrecht.

Janssen, B.H, F.C.T. Guiking, D. van der Eijk, E.M.A. Smaling, J. Wolf and H. van Reuler, 1990. A system for quantitative evaluation of the fertility of tropical soils (QUEFTS). *Geoderma*, 46, 299–318.

Jones, A., H. Breuning-Madsen, M. Brossard, A. Dampha, J. Deckers, O. Dewitte, T. Gallali, S. Hallett, R. Jones, M. Kilasara, P. Le Roux, E. Micheli, L. Montanarella, O. Spaargaren, L. Thiombiano, E. Van Ranst, M. Yemefack, and R. Zougmoré (eds), 2013. *Soil Atlas of Africa*. European Commission, Publications Office of the European Union, Luxembourg.

Kamara, A.Y., R. Abaidoo, J. Kwari and L. Omoigui, 2007. Influence of phosphorus application on growth and yield of soybean genotypes in the tropical savannas of northeast Nigeria. *Archives of Agronomy and Soil Science*, 53, 539–552.

Kamara, A.Y., J. Kwari, F. Ekeleme, L. Omoigui and R. Abaidoo, 2008. Effect of phosphorus application and soybean cultivar on grain and dry matter yield of subsequent maize in the tropical savannas of north-eastern Nigeria. *African Journal of Biotechnology*, 7, 2593–2599.

Kamara, A.Y., F. Ekeleme, L. Omoigui, I.A. Teli, Bassi, M.B. Mallah, I.Y. Dugje, J. Ellis-Jones and A. Tegbaru, 2009. Technological interventions to improve crop productivity in Borno State: PROSAB. PROSAB Stakeholder Conference for Sharing Experiences 1–2 September 2009, Maiduguri, Borno State, Nigeria.

Kamara, A.Y., F. Ekeleme, J.D. Kwari, L.O. Omoigui and D. Chikoye, 2011. Phosphorus effects on growth and yield of groundnut varieties in the tropical savannas of northeast Nigeria. *Journal of Tropical Agriculture*, 49, 25–30.

Kissel, E., P. van Asten, R. Swennen, J. Lorenzen and S.C. Carpentier, 2015. Transpiration efficiency versus growth: Exploring the banana biodiversity for drought tolerance. *Scientia Horticulturae*, 185, 175–182.

Kleene, P., B. Sanogo and G. Vierstra, 1989. A partir de Fonsébougou. Présentation, objectifs et méthodologie du "Volet Fonsébougou" (1977–1987). *Systèmes de Production Rurale au Mali : Volume 1*. IER, Bamako and KIT, Amsterdam.

Kwari, J.D., A.Y. Kamara, F. Ekeleme and L. Omoigui, 2009. Relation of yields of soybeans and maize to sulphur, zinc, and copper status of soils under intensifying cropping systems in the tropical savannas of north-east Nigeria. *Journal of Food, Agriculture & Environment*, 7, 129–133.

Kwari, J.D., A.Y. Kamara, F. Ekeleme and L. Omoigui, 2011. Soil fertility variability in relation to the yields of maize and soybean under intensifying cropping systems in the tropical savannahs of northeastern Nigeria. In: A. Bationo, B. Waswa, J.M. Okeyo,

F. Maina and J.M. Kihara (eds.), *Innovations as Key to the Green Revolution in Africa: Exploring the Scientific Facts.* Springer Science + Business Media, Heidelberg.

Lynam, John and Peter Carberry, 2011. *IITA CCER Natural Resources Management.* IITA, Ibadan.

Manyong, V.M, V.A. Houndékon, P.C. Sanginga, P. Vissoh and A.N. Honlonkou, 1999. *Mucuna Fallow Diffusion in Southern Benin.* IITA, Ibadan.

Merckx, R. 2011. Mission report to the International Institute for Tropical Agriculture, Ibadan, Nigeria, 21/03/2011–26/03/2011, as a member of the Board of Trustees, participating in the CCER of NRM research at IITA.

Ndabamenye, T., P.J.A. Van Asten, N. Vanhoudt, G. Blomme, R. Swennen, J.G. Annandale and R.O. Barnard, 2012. Ecological characteristics influence farmer selection of on-farm plant density and bunch mass of low input East African highland banana (*Musa* spp.) cropping systems. *Field Crops Research*, 135, 126–136.

Ndabamenye, T., P.J.A. van Asten, G. Blomme, B. Vanlauwe, B. Uzayisenga, J.G. Annandale and R.O. Barnard, 2013. Nutrient imbalance and yield limiting factors of low input East African highland banana (*Musa* spp. AAA-EA) cropping systems. *Field Crops Research*, 147, 687–678.

Nwoke O.C., B. Vanlauwe, J. Diels, N. Sanginga, O. Osonubi and R. Merckx R., 2003. Assessment of labile phosphorus fractions and adsorption characteristics in relation to soil properties of West African savanna soils. *Agriculture, Ecosystems and Environment*, 100, 285–294.

Nwoke O.C., B. Vanlauwe, J. Diels, N. Sanginga and O. Osonubi, 2004. The distribution of phosphorus fractions and desorption characteristics of some soils in the moist savanna zone of West Africa. *Nutrient Cycling in Agroecosystems*, 69, 127–141.

Nyombi, Kenneth, 2010. Understanding growth of East Africa highland banana: experiments and simulation. Doctoral Thesis, Wageningen University, The Netherlands.

Nyombi, K., P.J.A. van Asten, M. Corbeels, G. Taulya, P.A. Leffelaar and K.E. Giller, 2010. Mineral fertilizer response and nutrient use efficiencies of East African highland banana (*Musa* spp., AAA-EAHB, cv. Kisansa). *Field Crops Research*, 117, 38–50.

Nziguheba, G., B.K. Tossah, J. Diels, A.C. Franke, K. Aihou, E.N.O. Iwuafor, C. Nwoke and R. Merckx, 2009. Assessment of nutrient deficiencies in maize in nutrient omission trials and long-term field experiments in the west African Savanna. *Plant and Soil*, 314, 143–157.

Pypers P., M. Huybrighs, J. Diels, R. Abaidoo, E. Smolders and R. Merckx, 2007. Does the enhanced P acquisition by maize following legumes in a rotation result from improved soil P availability? *Soil Biology and Biochemistry*, 39, 2555–2566.

Pypers P., W. Bimponda, J.P. Lodi-Lama, B. Lele, R. Mulumba, C. Kachaka, P. Boeckx, R. Merckx and B. Vanlauwe, 2012. Combining mineral fertilizer and green manure for increased, profitable cassava production. *Agronomy Journal*, 104,178–187.

Pypers, P., E. Vandamme, J.-M. Sanginga, T. Tshisindad, M. Walangululu, R. Merckx and B. Vanlauwe, undated. K and Mg deficiencies corroborate farmers' knowledge of soil fertility in the highlands of Sud-Kivu, Democratic Republic of Congo. Unpublished manuscript.

Taulya, G. 2013. East African highland bananas (*Musa* spp. AAA-EA) 'worry' more about potassium deficiency than drought stress. *Field Crops Research*, 151, 45–55.

Taulya, G., 2015. Ky'osimba Onaanya: Understanding productivity of East African highland banana. PhD Thesis, Wageningen University.

van Asten, P.A.J., A.M. Fermont and G. Taulya, 2011a. Drought is a major yield loss factor for rainfed East African highland banana. *Agricultural Water Management*, 98, 541–552.

van Asten, P.J.A., L.W.I. Wairegi, D. Mukasa and N.O. Uringi, 2011b. Agronomic and economic benefits of coffee–banana intercropping in Uganda's smallholder farming systems. *Agricultural Systems*, 104, 326–334.

Vanlauwe, B., C. Nwoke, J. Diels, P. Sanginga, R. Carsky, J. Deckers and R. Merckx, 2000. Utilization of rock phosphate by crops on a representative toposequence in the Northern Guinea savanna zone of Nigeria: response by *Mucuna pruriens, Lablab purpureus*, and maize. *Soil Biology and Biochemistry*, 32, 2063–2077.

Vanlauwe, B., K. Aihou, S. Aman, E.N.O. Iwuafor, B.K. Tossah, J. Diels, N. Sanginga, O. Lyasse, R. Merckx and J. Deckers, 2001. Maize yield as affected by organic inputs and urea in the west African moist savanna. *Agronomy Journal*, 93, 1191–1199.

Vanlauwe, B., A. Bationo, J. Chianu, K.E. Giller, R. Merckx, U. Mokwunye, O. Ohiokpehai, P. Pypers, R. Tabo, K.D. Shepherd, E.M.A. Smaling, P.L. Woomer and N. Sanginga, 2010. Integrated soil fertility management: Operational definition and consequences for implementation and dissemination. *Outlook on Agriculture*, 39, 17–24.

Vanlauwe, B., P. Pypers, E. Birachi, M. Nyagaya, B. van Schagen, J. Huising, E. Ouma, G. Blomme and P.J.A. van Asten, 2013. Integrated soil fertility management in Central Africa: Experiences of the Consortium for Improving Agriculture-based Livelihoods in Central Africa (CIALCA). In: Clair H. Hershey and Paul Neate (eds). *Eco-Efficiency: From Vision to Reality*. CIAT Publication No. 381. Cali, Colombia.

Versteeg, M.N. and V. Koudokpon, 1993. Participative farmer testing of four low external input technologies to address soil fertility decline in Mono département (Benin). *Agricultural Systems*, 42, 265–276.

Wairegi, L.W.I. and P.J.A. van Asten, 2010. The agronomic and economic benefits of fertilizer and mulch use in highland banana systems in Uganda. *Agricultural Systems*, 103, 543–550.

Wairegi, L.W.I. and P.J.A. van Asten, 2011. Norms for multivariate diagnosis of nutrient imbalance in the East African Highland bananas (*Musa* spp. AAA). *Journal of Plant Nutrition*, 34, 1543–1472.

Whittome, M.P.B., D.S.C. Spencer and T. Bayliss-Smith, 1995. IITA and ILCA on-farm alley farming research: lessons for extension workers. In: B.T. Kang, A.O. Osiname and A. Larbi (eds), *Alley Farming Reserarch and Development*. Proceedings of the International Conference on Alley Farming, IITA, Ibadan, 1992, 423–435.

7 Addressing farmers' own soil fertility challenges

2012–today

7.1 Scope, approaches and partnerships

This chapter covers the most recent years, characterised by an entirely new set-up of soil and soil fertility research and an increased influence of African institutions on its direction and of various stakeholders on its content. Although some of the work treated here is a continuation of research initiated earlier, an important part consists of new initiatives, which are still in the process of development. More space than in earlier chapters will therefore be devoted to new concepts. This chapter is about 'work in progress', building on and extending experiences gathered during several years of decentralised Research-for-Development.

In 2012, the CGIAR and its donors developed an entirely new strategy, structure and *modus operandi* for its network of international Centres. International research would henceforth be organised around several grand themes, called Consortium Research Programs (CRPs), in which different centres would participate according to their capabilities. Different CRPs focused on major crop-based systems as well as on themes cutting across systems and continents, viz. the 'Humidtropics', the 'Forests, Trees and Agroforestry', or the 'Water, Land and Ecosystems' CRPs. One of the key concepts adopted for the entire system was that its adaptive research would be embedded in the development effort by associating itself with capable national and international partners, in a synergistic association, and would carry out its work in close collaboration with farmers, an approach known as Research for Development (R-for-D). The principles of this approach had been implemented over several years of research on soil fertility management, which had brought home the very variable production environment of smallholder farmers, affecting the performance and the uptake of improved soil management practices (Chapter 6).

This reorientation took place at a time when African political and development organisations had become much more vocal in expressing their vision on the needs for rapid agricultural development as a first step towards the overall economic development of the continent (Chapter 6, Emerging Trends).

One of the new actors supporting this development was the Bill & Melinda Gates Foundation (BMGF) which had become a major investor in soil and soil fertility management in Africa since 2006, through the CGIAR and the Alliance for a Green Revolution in Africa (AGRA). By requiring the direct engagement of development and private sector partners, BMGF funding enhances the conditions for the uptake of best practices developed by research, thus embedding research in an integrated development context. Also, there is a growing awareness that farmers have to deal with different value chains affecting their livelihood, stressing the need for a systems approach even when focusing on specific commodities.

The overriding aim, which has driven IITA's soil and soil fertility research throughout its lifetime, has been to strive for Sustainable Intensification (SI), a term recently coined, which is attracting a lot of interest. The differences between successive periods have been in the interpretation of this aim and the approaches to achieve it. Today, the goals of Sustainable Intensification are three-fold: (i) increase productivity per unit of land, labour or capital, (ii) restore or maintain other critical soil and land-based ecosystem services and (iii) ensure resilience to shocks, including climate change. In practice, this means that soil and land are managed so that yields are increased, soil carbon is stabilised or built up, biological diversity is maintained, and negative nutrient balances, soil erosion and environmental pollution, resulting from the inefficient use of external inputs, are avoided. Sustainable Intensification requires access to quality inputs and labour-reducing devices and to capital at crucial times to pay for them and profitable markets for farm produce will thereby be essential. Ultimately, it will be the profitability of intensification that determines whether or not smallholders will take it up – its sustainability will not necessarily be *their* immediate concern. Appropriate incentives, including payments for environmental services, can also enable investments in sustainable intensification.

The decentralisation process at IITA, which had started in the mid-1990s and gained momentum in the 2000s, came to its conclusion under IITA's recent Refreshed Strategy for 2012–2020 (www.iita.org), resulting in the establishment of four multi-disciplinary, hub-based teams, covering West, Central, East and Southern Africa. Each team covers research activities ranging from germplasm improvement and seed systems, to crop productivity, soil and soil fertility management and pest and disease management. There are no longer separate research divisions or programs, and agronomists and soil fertility specialists work with other disciplines at the country or hub level, all of them making complementary contributions to the multidisciplinary R-for-D process. In order to effectively play that role, soil fertility specialists need to be familiar with market access and value addition issues in addition to concepts of their own discipline such as nutrient use efficiency or soil organic carbon fractions.

A corollary of the changing role of soil research has been an increased need for cooperation with Advanced Research Institutes (ARIs), to ensure continued availability and use of state-of-the-art tools and approaches in the R-for-D programs. This has resulted in a greater level of complementarity, with IITA and partners raising issues and questions through their on-the-ground activities, for which ARIs may be able to provide or jointly develop the necessary answers.

Previously, target areas were chosen on the basis of different sets of criteria. In the early days, these were related to the major West African ecologies and soil classes (Chapters 1–4), in later years modified and extended through the AEZ (agro-ecozone)/Benchmark approach (Chapter 5). More recently, when the AEZ approach turned out to be unwieldy and inconsistent with IITA's size and funding position, intervention areas were chosen on the basis of the need for intensification of smallholder farming and the availability of donor funding (Chapter 6), with agro-ecological representativeness, though still important, no longer the sole criterion.

With the arrival of the Humidtropics CRP at the start of the present episode, four Action Areas were demarcated in humid and sub-humid sub-Saharan Africa (SSA) based on (i) the need for intensification of smallholder farming systems, equated with relatively high population densities and lack of fallow land, (ii) existing partnerships across the R-for-D continuum and (iii) agro-ecological representativeness that captures all relevant diversity. Within these relatively large Action Areas, Action Sites were identified covering gradients of population density, access to markets and status of land degradation. Within Action Sites, the actual research was carried out in representative Field Sites, building on earlier results of 'proof-of-concept' science. Each of the geographic layers had a formal partnership arrangement, with specific roles and responsibilities for the research and development partners. Stakeholder platforms identified priority issues for research, which included soil fertility decline and land degradation. Under-representation in the platforms of organisations with large-scale dissemination objectives, however, is resulting in less than the desired dissemination of good soil and soil fertility management practices through the Humidtropics CRP. The continuation of Humidtropics through agri-food system CRPs beyond 2017 is likely going to improve on this situation.

An alternative strategy for the choice of intervention areas was to seek association with (and funding by) major agricultural development programs supported by BMGF and USAID (United States Agency for International Development), carried out in areas chosen by those organisations. In the case of BMGF, for example, the target countries in 2015 were Nigeria, Tanzania and Ethiopia. Usually, priority crops, cropping systems or value chains were also decided upon during project formulation. Within these 'boundaries', partnerships were set up with National Agricultural Research Systems (NARS), extension organisations, non-governmental organisations (NGOs) and private companies, and the research processes are embedded in the participating projects' dissemination networks. For the Humidtropics strategy research, themes were identified through participatory approaches within pre-set target areas, but for the donor-driven projects, partnerships and target areas were identified within the projects' intervention areas, for pre-set crops or cropping systems. Which of the above approaches generates most impact cannot yet be determined.

In the traditional R-for-D continuum, even where technology is developed in a farmer-participatory manner, the adoption process is often hampered by several factors. The technology may not be of highest priority to the farming households or to development entities catering for their needs, a high proportion of farmers may not actually reap the anticipated benefits, or the adaptation needed for particular groups of smallholders or particular conditions is not made. As a result 'scaling' does not take place and the main output is papers in scientific journals, with the technology remaining 'products on-the-shelf'; the fate of many or most soil and soil fertility technologies of the past.

Alternative models place research within the context of a development initiative, most often led by (non-)governmental development organisations, thereby reaching the large number of farming households or farmer associations which the development partner works with, proposing solutions that are relevant to farming communities. Such an approach, referred to as R-for-D, has important practical advantages: research shares target areas and priorities with the development partners and integrates these into its research program, and it directly engages the farming households and associations assisted by the development partners, which will eventually enhance adoption of profitable innovations, addressing farmers' real needs (Paul et al, 2014). There are also methodological advantages, such as embedding the variation in performance of innovations across the target environment in the research process (Vanlauwe et al., 2016) and generating new research questions shared by farmers and development partners (Giller et al., 2013). Obviously, placing research in development also creates a real demand for the products and solutions to be developed.

After initial problem diagnosis and, whenever necessary, a technology design or assessment phase under controlled conditions, the adaptive research cycle is implemented, consisting of a number of logical steps, leading to eventual adoption or rejection by the farmers (Figure 7.1).

Diagnostic activities are usually carried out at the start of the adaptive research cycle, to learn about the major problems faced by the target population, the potentially existing solutions and the overall environmental conditions in which improved technology must fit. This can include 'fast and cheap' Participatory Rural Appraisals (PRA), a more costly formal baseline study or specific studies, e.g. on soil fertility conditions.

Demonstration trials consist of 'best-bet' technologies that have been derived from earlier research or experiences in other areas with similar conditions. Such technologies could include improved agronomic practices (e.g. adapted cassava (*Manihot esculenta*) spacing to enable a higher population of intercropped legumes, improved nutrient management practices (e.g. micro-doses of fertiliser in combination with locally available organic resources), or improved cropping systems (e.g., the MBILI – 'Managing Beneficial Interactions in Legume Intercrops' – intercropping arrangement to allow soybean (*Glycine max*)-maize (*Zea mays*) intercropping). All demonstrations normally integrate improved germplasm. Farmers assisted by extension officers and researchers chose suitable technologies to solve specific, agronomic problems or meet market demand. The demonstrations always include a best local practice as control and each site serves as a replication. Demonstrations are co-managed between extension staff and farmer groups, and planning meetings, mid-season field days and exchange

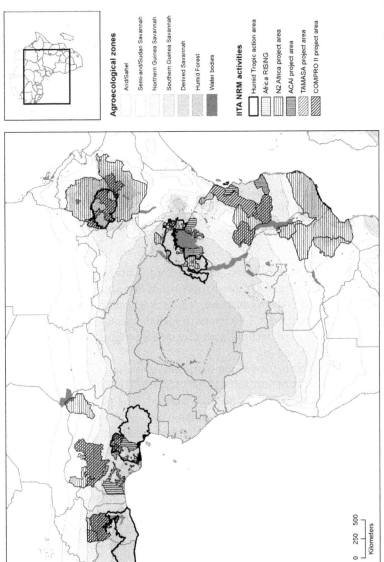

Agroecological zones

- Arid/Sahel
- Semi-arid/Sudan Savannah
- Northern Guinea Savannah
- Southern Guinea Savannah
- Derived Savannah
- Humid Forest
- Water bodies

IITA NRM activities

- Humid Tropic action area
- Africa RISING
- N2 Africa project area
- ACAI project area
- TAMASA project area
- COMPRO II project area

0   250   500

Kilometers

Figure 7.1 Map of the target areas for soil and soil fertility management research during the period covered by this chapter. The Action Areas of the Humidtropics are shown as well as the project areas of various initatives (N2Africa, TAMASA, ACAI, COMPRO, all supported by BMGF, and AfricaRISING, supported by USAID) having an important soil fertility management scaling component

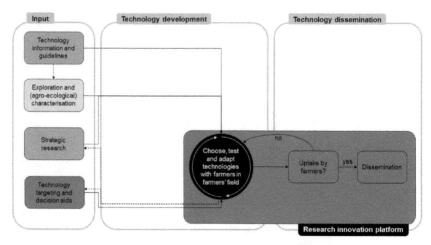

Figure 7.2 IITA's technology development and dissemination model, 2012–2015; towards multi-stakeholder research for development through on-farm technology development and testing and technology dissemination

Shading indicates relative emphasis on the various components with darker ones indicating a relatively higher emphasis

visits and a post-harvest evaluation are organised. Data are collected in the demonstrations by farmers under the supervision of extension personnel, but at a subset of sites (called *focal demonstrations*), researchers collect the data to create a high quality reference dataset. Observations include agronomic and economic performance of the tested technologies and gender-segregated participatory evaluation of the technologies.

Based on their performance at the demonstration sites, a few of the most promising technologies are transformed into adaptation packages, consisting of a maximum of 2–3 technologies and local practice. The packages themselves contain the needed seed, inputs and information tools to allow the household to install the package on an area of about 10 × 10 m per treatment, with assistance of NGO facilitators and members of the technical committees of the participating farmers associations. The participating households record all important farm activities, observations on yields and pests/diseases, with assistance of the technical committee member, and visits to the adaptation site. The number of adaptation trials will vary from 200 to 1,000 per site per season, depending on its goal. Such trials may be accompanied by soil and plant sampling from a sub-sample of sites to have a better understanding of plant-soil relationships underlying the performance of a specific technology.

After the demonstration trials, a number of households will be expected to continue using part or all of the demonstrated technologies on their own, provided their performance is satisfactory and the necessary inputs are available. The impact of those technologies will then be observed against the local practice in an adjacent or nearby area of the household's farm. Direct observation of adoption will be a crucial step in the research cycle, providing evidence about the technologies' performance or failure under farmers' conditions. There is only limited experience so far with the implementation of this step in the process.

7.2 Characterisation of soils, farms and farming systems

Characterisation remains essential for agricultural research at any level. At the more strategic level, it analyses and maps broad soil-, management- and household-related variability as a necessary basis for the development of location- and farm type-specific recommendations. In its more adaptive form, it diagnoses the target production systems in more detail, to identify their constraints and the opportunities for improvement.

The elaboration of farmer and household typologies is now seen as a necessary tool in the process of technology targeting, by classifying farming households according to their constraints and opportunities and their need for improved technologies. There has been a historic debate since the Farming Systems Research (FSR) years on the usefulness of detailed socioeconomic household surveys, as compared with quick and dirty methods. That debate is still relevant in view of the amount of energy and resources invested in system and household surveys today.

7.2.1 Household and farm enterprise studies

In yet another round of detailed surveying, with echoes of the AEZ days (Chapters 4 and 5), the development of a household typology was undertaken in 2012 at three contrasting sites in Burkina Faso, Ghana and Senegal, identified in 2010 by the CGIAR as Climate Change benchmark sites. The purpose was to define household profiles and assess the possible impacts of agricultural practices adapted to climate change on household food security and food self-sufficiency in West Africa. The actual locations for the study were chosen jointly with NARS, NGOs, government agents and farmers' organisations, using criteria such as poverty levels, vulnerability to climate change, key biophysical, climatic and agro-ecological gradients, agricultural production systems and partnerships (Douxchamps et al., 2016).

A survey was conducted using high resolution satellite images, transect drives and focused group discussions with local experts and key informants. Four farm types were identified and described as subsistence, diversified, extensive or intensive farming. Type I (subsistence farming) cropped small pieces of land mainly with staples. They had low market orientation and high dependence on off-farm income but adopted soil and water conservation practices more intensively than any other farm type. Their food security frequency was only 30%. Type II (diversified farming) farms also exploited small pieces of land, but they had more diversified farm enterprises and a higher market orientation. They earned more income from cattle, unlike Type I who earned more from non-ruminants like poultry. Type II farms used more fertilisers and had higher productivity and frequency of food security (40%) than Type I. Type III (extensive farming) focused on staple food crops, as did Type I, and had low market orientation. However, they worked on larger pieces of land, compensating for their low productivity and had higher frequency of food security (55%). Type IV (intensive farming) had both high market orientation and diversified enterprises and a higher frequency of food security (59%) than the others. It was concluded that high input options requiring substantial cash outlay were suitable for Types II and IV due to their high market orientation. Labour-saving soil and water conservation practices and low-cost soil fertility inputs would be suitable for Type I and Type III farms.

This socioeconomic characterisation was complemented with soil characterisation using novel techniques, viz. soil sampling and analysis methods developed by the Africa Soil Information System (AfSIS, Vågen et al., 2010), Fourier-Transform Near Infrared Reflectance (FT-NIR), X-Ray Power Diffraction (XRPD) and Total X-Ray Fluorescence (TXRF) Spectroscopy. Field data are collected from sample farms to develop detailed plot profiles specific to a cropping system. The data are compared with legacy data from what is currently the most comprehensive interactive soil profile database for Africa (ISRIC WISE version 3.1). It is envisaged that the output from this characterisation work will guide soil fertility input dealers in targeting specific products to specific geographical locations in the region.

7.2.2 Characterisation of soils and soil variability

Physical characterisation has been a recurrent theme throughout the life of the Institute, but with only limited consolidation of methods. Current characterisation may be seen as a fresh start, using modern insights, methods and tools.

At a large scale, AfSIS, funded by BMGF, provides important information on attainable yields and yield gaps, as well as on the scope for improvement of non-responsive soils. This knowledge is crucial to effectively address production constraints in the 'highly variable agricultural landscapes in SSA' and develop site-specific recommendations for Integrated Soil Fertility Management (ISFM) practices to improve or restore soil productivity. AfSIS operates in 60 randomly selected sentinel sites of 10 km × 10 km each, which are surveyed in detail, and based on which soil functional properties and land degradation status are mapped for the SSA region. Within these areas, very strict schemes for soil sampling and diagnostic trials are used in representative locations to ensure that the information is representative/unbiased and allows for extrapolation across a wider area. Trials are laid out so that they are distant from each other, cover contrasting soil types and various positions in the landscape, in fields with a homogeneous cropping history[1]. Information thus generated will be essential for extending research results to other areas.

The CIALCA[2] (Consortium for Improving Agriculture-based Livelihoods in Central Africa) project combines knowledge from soil maps and diagnostic trials to understand spatial variation in nutrient requirements using GIS and geo-spatial modelling techniques, to fine-tune recommendations for input use and ISFM technology options, maximising the chance of response and minimising the risks.

7.2.3 Mapping nutrient deficiencies

Another technique, introduced at IITA during the previous decade, was the characterisation of soil nutrient status through the analysis of tissue nutrient content, using Compositional Nutrient Diagnosis (CND) and Diagnosis, Recommendation Integrated System (DRIS) norms (Chapter 6). Indicator leaf samples of Robusta and Arabica coffee plants from 164 plots of on-farm fertiliser trials in Uganda were analysed to derive the CND and DRIS norms. Of the plots sampled, 75 had received N fertiliser ranging from 156 to 392 kg N/ha/year. The rest of the plots did not receive any nutrient inputs. Indicator leaves from eight to ten coffee trees per plot were analysed for nutrient content using standard methods. The norms derived from these analyses were used to identify nutrient limitations to coffee productivity in Uganda.

There were variations in the most limiting nutrients across regions (Figure 7.3), but in general, N was the most limiting nutrient in all coffee producing regions in Uganda. Phosphorus was most deficient in central and eastern Uganda but not in western Uganda and West Nile, where the exchangeable bases were frequently deficient (Wang et al., 2015).

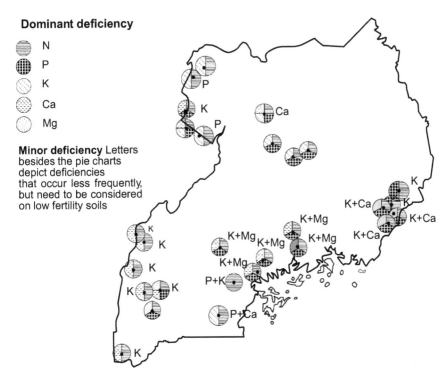

Figure 7.3 Nutrient deficiencies in major coffee growing areas of Uganda (adapted from Ochola et al., 2016)

7.3 Technology development

The principles of technology development have changed considerably, from mostly technology-driven in the first few decades, based on what scientists perceived farmers would need to increase their system's productivity, to the current demand-driven perspective, providing support to decision-making by farmers rather than offering ready-made prescriptions. Even so, a selection of options is necessary for farmers to choose from, while the challenge remains on how to put such a collection together. One possibility is to rely on what is already available, but usually some technology development will be needed to address the specific needs in a given area and, in the case of an international institute, also in a range of conditions occurring in its mandated zone. Research-for-development thus consists of a mix of strategic 'International Public Good'-related research on appropriate crop and soil management and adaptive research with farmer communities, whereby the results of the more strategic research will feed into the more adaptive work. An example is the development of site-specific nutrient management using the Nutrient Expert tool based on the validated QUEFTS (Quantitative Evaluation of the Fertility of Tropical Soils model). An important dimension of this and other soil fertility work is the integration of variability, which should allow the development of village- or household-specific recommendations (Vanlauwe et al., 2016).

7.3.1 Soil and soil fertility management

Soil fertility management by various means has, of course, throughout been the mainstay of research towards SI, ranging from different techniques for soil conservation, straightforward nutrient response research, the use of soil – improving 'auxiliary' species in combination with fertiliser and appropriate crop rotation and association. During the last two decades, with a significant shift of research from the station to the real farm, the emphasis has also shifted to the stepwise introduction of technologies combining efficient fertiliser use, improved crop varieties and N-fixing, dual purpose grain legumes. Boosting the yield of those legumes is thereby seen as key towards the improvement of an entire cropping system, and inoculation of legumes with efficient *Rhizobium* strains as one important method to accomplish that.

Fertiliser management for cassava

For farmers to adopt fertiliser technologies, a decision support system (DSS) is needed that accounts for the complexity of fertiliser response and provides recommendations that are specific to the farmer's production environment, as well as the risk – the likelihood for a non-profitable response – associated with an investment in fertiliser. The 'African Cassava Agronomy Initiative' (ACAI) is a new project that started in 2016 and aims to develop such a fertiliser recommendation system for cassava production in Tanzania and Nigeria (Figure 7.4).

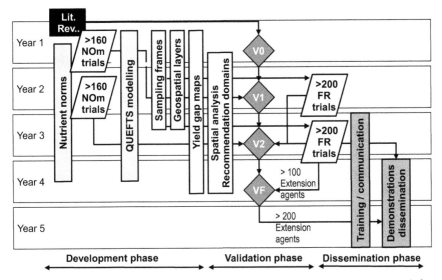

Figure 7.4 Proposed work flow for the development of a decision support tool for site-specific fertiliser recommendations, being developed in the context of a recently approved cassava agronomy initiative

'QUEFTS' means 'Quantitative Evaluation of the Fertility of Tropical Soils'; 'NOm' means 'Nutrient omission trials'; 'Lit. rev.' means 'Literature review; 'FR' means 'FR' means 'Fertiliser recommendations' and 'V0, V1, V2, and VF' refer to versions 0, 1, 2, and the final version of the fertiliser recommendation decision support tool

It started off with a set of nutrient omission trials, established in locations that were chosen to cover the variation in soil and climate conditions in the target areas in the two countries. These trials are primarily meant to determine major deficiencies, as well as to calibrate and validate the QUEFTS modelling framework for cassava. QUEFTS predicts nutrient uptake from soil nutrient reserves and applied fertiliser and the conversion into harvestable produce. The results of these nutrient omission trials and modelling output will then guide a subsequent set of trials to determine fertiliser response curves, which will be laid out following a similar unbiased sampling scheme. The results of these trials will be used to develop a prototype DSS to be validated in the field, involving simple trials with a large number (thousands) of farmers, comparing the recommendation by the DSS to a control and a conventional blanket fertiliser recommendation. Development organisations and private sector fertiliser companies will be involved to help farmers to use the DSS and gather feedback. The data collected along with the trials will then be used to further improve the DSS. A final version of the DSS will aim to provide a recommendation that is in at least 75% of the cases more profitable than the blanket recommendation. Along with the development of this DSS, a tool will be developed to recommend best fertiliser blends for the fertiliser production industry, addressing nutrient deficiencies in the target areas while allowing flexibility for end-users to apply fertiliser according to their specific production objectives.

Legume nutrient management

Dual purpose legumes as components of intensified cropping systems have gradually assumed prominence in R-for-D, especially in savannah areas. The N2Africa project (www.n2africa.org) supports research on legume technology, aimed at a combination of improvement of soil fertility through N-fixation and improving the diet and income of farming families.

On-farm trials at 3,000 sites across Africa measured the yield response of soybean, common beans (*Phaseolus vulgaris*), groundnut (*Arachis hypogaea*) and cowpea (*Vigna unguiculata*) to the application of P fertilisers, showing yield increases of 5–60% to P application depending on the crop and location (Woomer et al., 2014). At many sites application of inoculants in conjunction with P fertilisers improved soybean and bean yield in over 65% of the sites (e.g. Figure 7.5).

A further program of on-farm activities is undertaken in 11 countries across SSA (Ghana, Nigeria, Ethiopia, Uganda, Tanzania, Kenya, Rwanda, Democratic Republic of Congo (DR Congo), Zimbabwe, Malawi and Mozambique), strengthening the role of legumes in the farmers' production systems. A wide range of technology options to overcome various production constraints was demonstrated, diagnostic experiments were carried out to identify biotic and abiotic constraints and adaptive research was initiated to determine how to bridge the important legume yield gaps in the major farming systems.

Baskets of best-bet technologies are demonstrated on a real-life scale (10 × 10 m) in several countries with selected grain legume species, according to the area,

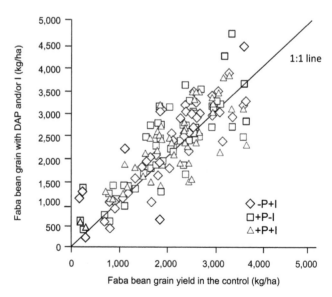

Figure 7.5 Yield increments of faba bean with fertiliser-P, with and without inoculation in Ethiopia, 2014

'DAP' means 'di-ammounium phosphate' and 'I' means 'Inoculant' (of *Rhizobium*). At the 1:1 line treatment yields are equal to control (adapted from Ampadu-Boakye et al., 2015)

in coordination with farmer associations (Vanlauwe et al., 2014). In addition to exposing farmers to a choice of technological options, these demonstrations also offer an opportunity for evaluating their performance. In each country, 25–150 demonstrations trials, co-designed and carried out each year with farmer groups or associations around the themes of variety, inputs and management. A subset of 20–50 *focal demonstrations* is used to collect data of high technical quality.

Large yield variations were found for all the major grain legumes, with yields in control plots often varying from less than 0.5 t/ha to more than 2 t/ha, and strongest responses to improved technology often occurring at control yields of around 1 t/ha. As an example, Figure 7.6 shows the effect of the combination of fertiliser-P and inoculant applied to soybean yield. Yield differences are caused by differences in crop management (time of planting, time of weeding, plant spacing, etc) as well as differences in soil fertility. The findings feed into the agronomic research trials, aimed at closing the yield gaps by reducing yield variability and attain yields of 1.5 t/ha of grain.

Agronomic trials are used to investigate the causes of the yield gaps identified in the on-farm demonstrations (Ampadu-Boakye et al., 2015). Single replicate experiments are conducted with the priority legumes for each area covering the range of land types and soil fertility conditions. Test treatments address the most likely deficiencies at the various sites, Measurements are made on yield, nutrient content of plant tissue, nodulation, rainfall, soil fertility parameters and pest and diseases. These data are used to diagnose the biophysical factors that constrain legume yields, other than P and inoculants.

In northern Ghana, where soils are low in organic matter and nutrient content, application of organic fertiliser increased yields of cowpeas more than the application of P+K+Ca+Zn+B (Ampadu-Boakye et al., 2015). The

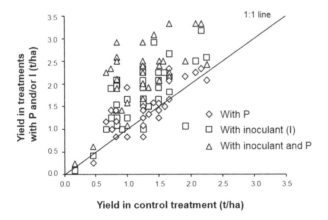

Figure 7.6 Soybean yield with fertiliser-P, *Rhizobium* inoculant, or both inputs against control in demonstration sites in South-Kivu Province, Democratic Republic of Congo
At the 1:1 line treatment yields are equal to control (adapted from Ampadu-Boakye et al., 2015)

combined application of mineral and organic fertilisers was most promising for reducing the yield gap in most locations.

In northern Tanzania, bean yields were significantly increased by P compared to the control and inoculant-only treatment (Figure 7.7) (Bressers, 2014). The additional increments due to K and N or inoculant were not significant at the 0.05 probability level, as shown by the same letter above the bars.

In Uganda, groundnuts responded significantly to triple supper phosphate (TSP) and gypsum (Figure 7.8), but the yields were low (grain yield is around

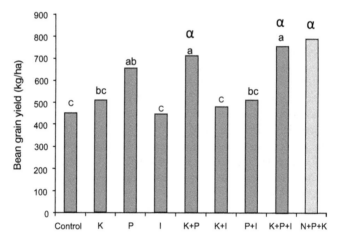

Figure 7.7 Bean grain yield as affected by different nutrient (P, K, or a combination of P and K) treatments and inoculation (I) in Northern Tanzania (adapted from Bressers, 2014)

The grey bar on the right is the positive control (with fertiliser N applied). Bars with different small letters are significantly different at the 5% level. Bars labelled with 'α' are not significantly different from the positive control treatment (N+P+K)

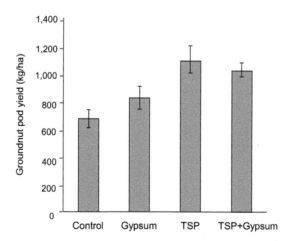

Figure 7.8 Groundnut pod yield response to gypsum and triple super phosphate (TSP) in Northern Uganda (n=10) (Ebanyat et al, Unpublished data)

70% of pod yield), indicating that other soil constraints limited the response (Ampadu-Boakye et al., 2015). These are to be characterised in order to identify appropriate nutrient management packages to close the yield gap of groundnut, from less than 600 to considerably above 1,000 kg/ha. Furthermore, the average results presented here do not differentiate between conditions and types of farmers, which will be necessary in order to formulate specific recommendations for different conditions and different categories of farmers, one of the key objectives of today's R-for-D.

Soil microbiology, including rhizobiology

The importance attached to legumes as components in ISFM is based on the findings that stimulating the production of legumes will result in more biomass production. When incorporated in the soil as crop residue, the biomass will, in turn, enhance the yield of a following fertilised non-leguminous crop by synergy between fertiliser and organic matter. Effective *Rhizobium* strains will enhance legume production and specificity of legume-*Rhizobium* symbiosis gives ample opportunity to select *Rhizobium* × legume combinations to maximise N-fixation. Once effective strains have been identified, they can be impregnated into appropriate carrier material as inoculants and applied to legume seeds at planting. Rhizobiology, or the study of the biology and use of *Rhizobium*, is carried out using a common research approach to ensure consistency which allows meta-analysis across the N2Africa core countries. The work includes screening of national collections of *Rhizobium* strains for symbiotic effectiveness. Exploration for additional collections for common bean, soybean, chickpea (*Cicer arietinum*), groundnut, cowpeas and faba bean (*Vicia faba*) is underway in collaboration with national institutions (Dianda et al., 2014). In five countries more than 1000 new *Rhizobium* isolates were identified in 2015, some of which are being advanced to the screening stage for effectiveness in the screen house, resulting in 32 candidate elite strains. Several of these will be transferred to IITA where a collection of promising strains and several worldwide reference strains for the target crops are being established for sharing among the participating countries. To date, the sharing program has started to reach different countries, such as Malawi, Tanzania, Uganda and Ghana.

Increases in yield of at least 10% above current practice (clearly visible to farmers) are a threshold for identification of better inoculant technologies. This threshold should be reached consistently before N2Africa will actively promote the new technology with smallholder farmers.

Most of the rhizobiology has so far been conducted on soybean, which is relatively specific in its requirement for *Rhizobium* and responds strongly to inoculation. Current evidence indicates that common bean responds to inoculation in some but not all soils. Given the relatively low cost of inoculants, it is probably worth recommending inoculation of common bean. Although several companies are selling *Rhizobium* inoculants for groundnut, there is little evidence that this is worthwhile and we do not recommend inoculation of groundnut. Recent evidence indicates that in Ghana cowpea can respond strongly to inoculation with new strains from Brazil. This work is still at the experimental stage and needs wider confirmation.

To date no strong and consistent interaction between soybean variety and *Rhizobium* strain has been found to warrant selection of particular strains for particular crop varieties. This means that only one inoculant strain needs to be used in commercial preparations, but it does not preclude the examination of inoculants containing more than one strain. The 'Nodumax' plant, at IITA Ibadan, Nigeria produces and markets quality inoculant for soybean in Nigeria. Batches of Nodumax inoculant are produced, packaged and commercialised across the country, with labelling that conforms to international standards.

Bio-fertilisers are defined as an input containing living micro-organisms as the main active ingredient that is used to improve plant nutrient availability[3]. Examples are formulations for seed or soil inoculation with *Rhizobium*, *Azotobacter*, *Azolla*, blue-green algae and P-solubilising bacteria. Their production costs are low compared to inorganic fertilisers and they can be cost effective when properly used (Jefwa et al., 2014). So far, most bio-fertilisers, other than *Rhizobium*-based products, have not been found to be effective as demonstrated by the 'Commercial Products – COMPRO' project (www.compro2.org) project, aiming at institutionalising an appropriate regulatory environment for such products. In a study conducted in Ethiopia, Kenya and Nigeria, for instance, Jefwa et al. (2014) demonstrated that over 40% of the marketed bio-fertilisers were of poor quality and that certain label claims were not substantiated. Quality control is essential to reduce the proliferation of poor quality products in the marketplace and to protect the farmer against fraud. The COMPRO project has built the capacity of competent national partners for effective regulations so that registered products contain the appropriate active ingredients, are efficacious and safe, when used as directed, and use proper labelling to prevent misleading statements about product performance. The regulatory bodies in the pilot countries have started registering products using the regulatory framework developed and high quality bio-fertilisers have recently been registered in the pilot countries as a result. Demonstrated cost-effective use of high quality bio-fertilisers would result in improved trust by relevant stakeholders and consequently promotion and adoption of the technology.

One of the limitations of bio-fertiliser, even effective ones, is their site- and crop-specificity. Large-scale testing is therefore required to develop site- and crop-specific recommendations for use. A large-scale testing campaign with *Rhizobium* inoculants, including commercial and native strains, was implemented, together with national partners in Ethiopia, Ghana, Kenya, Nigeria, Tanzania and Uganda, to develop site-specific directions for product formulation and use. Results have confirmed the spatial variability of soybean response to *Rhizobium* inoculants in all the pilot countries as shown for Nigeria in Figure 7.9 (Okuku, unpublished data). Research is currently focusing on the improvement of legume responses, either through identification of native strains, or through soil amendment, appropriate to specific sites. In countries

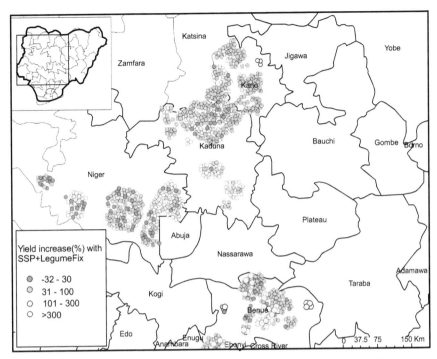

Figure 7.9 Spatial variability of soybean responses to 'LegumeFix' (a Rhizobium inoculant) plus phosphorus fertiliser ('SSP' means 'Single super phosphate') in four states of Nigeria (adapted from Okuku, unpublished data)

like Ethiopia, the investigation led to identification of native strains adapted to local conditions for legumes such as faba bean, soybean, chickpea, lentil (*Lens culinaris*) and field peas (*Pisum sativum*). Ongoing research is also assessing soil amendments, such as application of organic matter, use of starter N in low fertility soils or liming in acid soils, to improve legume response to *Rhizobium* inoculants. Preliminary results have shown improvement of crop response to *Rhizobium* inoculants when appropriate soil amendments are applied.

Soil fertility management in cocoa

Thus far, little soil fertility work had been done by IITA on perennial crops in general and cocoa (*Theobroma cacao*) and coffee (*Coffea sp.*) in particular. In view of their importance as smallholder crops, some activities were undertaken in the previous episode in Cameroon and Ghana and continued in the present episode with cocoa in Ghana.

Cocoa is demanding in its soil nutrient requirements and the crop is usually found on the heavier soils in West Africa. Ahenkorah (1981) provided threshold values for several nutrients necessary for cocoa in Ghana (last column of Table 7.1). A baseline soil analysis conducted by Asare et al. (2016) in four locations

Table 7.1 Mean and standard deviation (between brackets) of soil chemical properties within 0–30 cm on cocoa farms in four districts in Ghana, from a total of 112 samples (Asare et al., 2016)

Chemical characteristics	Districts				Thresholds in Ghana[a]
	Amansie West	Atwima Nwabiagya	Wassa Amenfi West	Sefwi Wiawso	
pH	5.94 (0.04)	5.64 (0.06)	4.98 (0.02)	6.01 (0.09)	5.6–7.2
Total N (%)	0.12 (0.005)	0.14 (0.005)	0.12 (0.008)	0.13 (0.004)	0.09
Organic C (%)	1.14 (0.05)	1.36 (0.04)	1.05 (0.05)	1.12 (0.05)	2.03
Available P (mg/kg)	1.66 (0.23)	5.27 (0.29)	7.17 (0.49)	8.28 (0.61)	20
Exchangeable K (cmol$_c$/kg)	0.25 (0.01)	0.43 (0.24)	0.28 (0.007)	0.33 (0.01)	0.25
Exchangeable Mg (cmol$_c$/kg)	1.94 (0.06)	1.44 (0.12)	0.66 (0.06)	1.53 (0.09)	1.33
Exchangeable Ca (cmol$_c$/kg)	6.75 (0.23)	6.36 (0.30)	2.84 (0.11)	7.08 (0.40)	7.5
Available Zn (mg/kg)	2.73 (0.11)	2.65 (0.12)	2.16 (0.06)	5.70 (0.31)	1.33
Available Cu (mg/kg)	2.70 (0.22)	3.62 (0.14)	0.91 (0.07)	3.01 (0.18)	1.33
Available Fe (mg/kg)	19.48 (0.36)	27.23 (0.63)	34.78 (0.66)	22.49 (0.23)	1.33

Source: Adapted from Ahenkorah (1981).

in the cocoa-growing belt shows limiting levels of P, while other nutrients were mostly above their threshold levels (Table 7.1). High yield has been found to be associated with N-application (Afrifa et al., 2009, quoted by Asare et al., 2016), while K-effects has been negligible (Ahenkorah et al., 1982), consistent with the high available-K content of most soils.

Cocoa in West Africa is mainly a smallholder crop grown under low management as part of a subsistence economy in which there is little fertiliser used (Ogunlade et al., 2009). When establishing new plantations farmers rely on nutrients stock in newly cleared forests or fallow lands with deep, well drained red or reddish brown clay soil (Oxic Paleustalf, Egbeda series in south-west Nigeria).

The rarity of fertiliser use has resulted in depletion of nutrient stocks and declining yields, particularly in the older cocoa-growing areas. Affordable and profitable pathways need to be developed to stabilise nutrient balances and sustainably increase cocoa productivity. This will require development of viable fertiliser technologies for cocoa, based on reliable soil suitability maps.

Identification and management of non-responsive soils

In field trials carried out over five decades, nutrient responses have often remained below expectation, resulting in low nutrient use efficiencies. This has often gone unexplained, beyond the rather obvious comment that some unidentified limiting factors were causing this. With the current integration of variability in the Institute's R-for-D and the large increase in coverage and numbers of field tests, there is a heightened awareness that the low responses to 'standard' fertiliser, usually containing only N, P and/or K, in many soils needs attention (e.g. Chapter 6). Such soils are now referred to as 'non-responsive'.

Nutrient response trials were carried out on-farm between 2012 and 2015 in central Tanzania (Tabora), western Kenya (Yala), eastern DR Congo (Walungu) and northern Nigeria (Zaria), covering the gradients in soil texture and topography. The profitability of inorganic inputs over two growing seasons was investigated, using value:cost ratios (VCR). The study found that common NPK fertiliser inputs on maize crops resulted in a loss on investment (VCR < 1) for 32% of fields in the Kenya study area as compared to 55%, 41% and a whopping 83% in DR Congo, Tanzania and Nigeria respectively. On the other hand, a satisfactory profit (VCR > 2) was obtained on 35, 10 and 9% of the farmer fields in Kenya, DR Congo and Tanzania respectively, and none in Nigeria. For soybean, common PK fertilisers resulted in a loss on investment for 67, 50, 56 and 68% of the cases in Kenya, DR Congo, Tanzania and Nigeria respectively, and generated a satisfactory profit on 10, 31 and 4.5% of the fields at the study area in Kenya, Tanzania and Nigeria, and none whatsoever in DR Congo. The differences in prevalence of non-responsiveness were attributed to soil mineralogy and management history. Furthermore, within each location there were important differences in responsiveness according to soil texture, season, soil acidity, phosphorus sorption and nutrient deficiencies, while responsiveness was often lower as non-fertilised yield was lower.

The project also carried out trials where fertiliser was combined with: (i) broadcasting of agricultural lime to correct soil acidity, (ii) addition of sulphur, zinc, boron and/or molybdenum for tackling nutrient deficiencies, (iii) input of organic resources, such as compost or manure to enhance soil microbial functions and (iv) deep tillage up to 30 cm for addressing soil compaction. Lime and secondary and micro-nutrients caused major decreases in the prevalence of non-responsiveness at the study area in Kenya. At the Tanzania site, crop yield and fertiliser responses were significantly increased by secondary and micro-nutrients, whereas deep tillage and manure were required for improving fertiliser efficiencies in the Nigeria site.

This project is expected to contribute important insights about the profitability of common fertilisers at landscape and farm level as well as the need for additional management practices for improving their efficiency. Analyses are ongoing to quantify the effect of soil factors and ameliorating measures, which will be important to target interventions more precisely to farmers' conditions (Roobroeck et al., *forthcoming*).

7.3.2. Cropping systems

Most agronomy and soil fertility research is embedded in the farmers' cropping systems context, even when component technology is tested for specific crops. Research on typical cropping systems issues such as crop densities and crop associations were started in the previous episode and resulted in recommendations added to the 'baskets of options' (Chapter 6). No new cropping systems research *sensu stricto* with annual crops or bananas was started in the last three years. However, many activities aiming at integrating dual purpose grain legumes in maize and cassava-based systems continue to explore the potential contributions of fixed N and organic inputs to companion crops and improve soil fertility conditions.

Cocoa-based systems

Cocoa plantations are created by partial clearing and/or burning of the understorey vegetation, and thinning or completely eliminating the overstorey trees. Some farmers preserve certain mature trees for shade and household uses (Asare, 2005). Food crops such as plantain (*Musa* spp.), cocoyam (*Colocasia esculenta*), maize and cassava are then planted, followed by the cocoa, to provide initial shade to the cocoa seedlings and to provide food and income to the farm family over the following growing season.

During the maturing stages of the cocoa plantation, weeds and seedlings of non-cocoa trees establish rapidly alongside the cocoa seedlings. The open conditions favour a range of species, but it is difficult to predict which trees will emerge from the initial cohort. The intensity of weeding and the thinning of tree seedlings, whereby trees with economic, domestic or environmental value are retained, will also have a significant effect (Asare and Asare, 2008). The emerging species further depend on the species that grow in the surrounding landscape, the trees' regeneration mechanisms and the presence of birds or mammals that play a role in pollination and seed distribution.

In the ensuing years, farmers coppice naturally regenerated tree saplings to grow in tandem with the cocoa and provide essential shade. Many farmers also plant high value tree crops such as *Persea americana, Elaeis guineensis, Carica papaya, Mangifera indica* and *Citrus* spp., and in rare cases valuable timber species (Asare, 2005)[4].

As the cocoa matures into full production, farmers continue to manage trees according to their priorities and the changing conditions within the farm. In a study conducted in Ghana, Asare and Ræbild (2015) reported that as farm sizes increased, density and canopy cover of shade trees decreased (Figure 7.10). The authors also showed that men had a significantly higher tree density and species diversity compared to women.

In another study in Ghana, the effect of differences in soil conditions and the intensity of shade on nutrient response of cocoa were explored, as a first step towards the development of area- or even farmer-specific nutrient

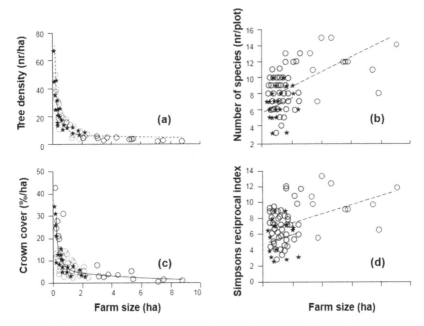

Figure 7.10 The relationship between tree density (a), species richness (b), crown cover (c), the Simpsons reciprocal index (d), and farm size men (open circles and broken line) and women (asterisks and continuous line) (adapted from Asare and Ræbild, 2015)

recommendations, rather than following blanket prescriptions (Asare et al., 2016). The effect of shade received particular attention, because of the unsettled issue of the role of shade in intensive cocoa production, the general opinion, based on station research, being that shade was beneficial in low-input cocoa production, but would put a ceiling to productivity and reduce nutrient response.

The study was carried out on 26 farms in four locations in two cocoa-growing regions, covering two years, with large variations in crop age, shade and soil conditions across farms. Four plots were demarcated in each farm, with factorial combinations of with or without shade and fertiliser. The recommended fertiliser rate of 375 kg/ha/year of NPK (0:18:22 + Ca + Mg + S) was applied. Average yield by location ranged from a high 1,230 to a low 386 kg/ha/year and the mean effect of fertiliser was just 14.5%. The low overall yield in Amansie West (Figure 7.11) was attributed to the soil's exceptionally low available P-status of 1.66 ppm Truog-P. There was a significant negative effect of shade on yield in two locations and no effect in the other. When percentage crown cover was included for the shaded plots, however, there was a tendency for yield to *increase* with shade in both fertilised and unfertilised plots. This effect may actually have been an effect of higher soil fertility in the plots with higher crown cover. No further analyses were carried out to relate the effect of shade on the response to fertiliser to the soils' fertility status, leaving the issue of nutrient response and shade unresolved.

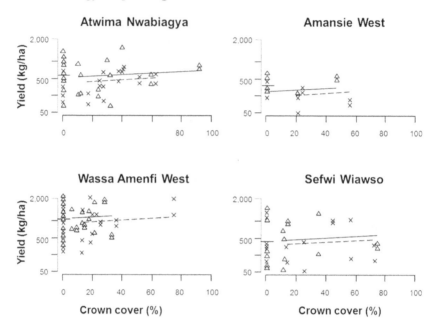

Figure 7.11 Canopy cover and yield of cocoa with (triangles) and without fertiliser (asterisk) in four location in Ghana; drawn lines are 'predicted' yields with (solid) and without fertiliser (dashed) (adapted from Asare et al., 2016)

7.3.3 Technology validation by farmers

Whereas in the days of Farming Systems Research, technology validation by farmers was an activity separate from the technology generation process (Chapters 3 and 4), it has now become integrated into the technology development and testing process itself. Validation with and by farmers occurs throughout the process and is managed by the same scientists. What used to be known as 'farmer-managed on-farm trials' is similar to the 'adoption step' in the current *demonstration-adaptation-adoption cycle* (Section 7.1.4).

7.4 Technology delivery and dissemination

In past years, communication technologies and approaches have developed in all sectors, including agriculture, both in developed and developing countries. The expansion of the internet, mobile phone technologies and interactive radio programmes have created a need for much more 'piecemeal' and 'demand-based' information supply. Much more effort is currently going into tailoring information to demand by translating and carving up science results into easily accessible pieces. This is in line with the transformation of applied research from the generation of single prescriptive solutions to context- and end-user specific decision-support.

7.4.1 Databases, technology digests and guidelines

Examples of the kind of information which is being generated:

- Short briefs (1–4 pagers) on technologies and tools that farmers can use, such as (i) nutrient diagnostic pictures and related site-specific management options, (ii) ISFM, (iii) intercropping options, such as cassava + legume, banana + legume, banana + coffee, (iv) methods for reduced tillage and soil cover mulching; and
- Mobile phone apps such as SMS-based decision trees (e.g. work with Grameen Foundation) to provide advice snippets based on needs and site.

These short information guides, together with video training material, are also used to strengthen training-of-trainers. Examples are CIALCA training videos (www.cialca.org) and supporting initiatives. The videos are also put online, to allow dissemination beyond the original project areas and partners. Training-of-trainers approaches are further supported by comprehensive training manuals on improved cropping systems management by the Africa Soil Health Consortium (ASHC; http://africasoilhealth.cabi.org). For example, the banana-coffee intercropping guide[5] not only advises on appropriate plant spacing options, but also provides information on the system's nutrient requirements and nutrient management options.

7.4.2 Decision support and systems modelling

In earlier years, there was only limited experience at IITA with the development of DSS. Examples are the LEXSYS database package for the choice of herbaceous legumes (Chapter 4) and an early decision scheme for the choice of suitable tillage systems (Chapter 3). With today's much greater direct involvement in the development process the need for DSS tools has been increasingly felt, to be used by development workers, assisted by researchers, to help farmers choose the most appropriate technologies for their circumstances.

Customised decision support (DS) applications are being designed for specific technologies as requested by the partners (referred to as 'use cases'). As a first step, a prototype is built from existing information, which is then populated with data, and improved through a number of iterations. The necessary data will be collected from socio-economic surveys, agronomic trials and geo-spatial information sources (crop maps, soil constraint maps and historical and near-real time weather information). These data, in combination with crop modelling output, will be used to provide site-specific recommendations. A variety of formats could be considered for the DS tools, such as a software tool to be used on laptop, a smartphone application, an interactive web-based map, a paper booklet with decision rules or a simple set of rules of thumb, as is found most appropriate.

To foster wide-scale use of the DS tools, on-farm demonstrations and technology tests will be carried out, facilitated by extension agents, highlighting the performance of agronomic practices as suggested by those tools, in comparison with current practices, and associated field days and exchange visits are held.

Wide-scale technology dissemination

The 'African Cassava Agronomy Initiative' (www.cassavamatters.org) intends to build a knowledge base on cassava agronomy and make it widely accessible through Decision Support (DS) applications. The project works on site-specific fertiliser recommendations and customised fertiliser blends for cassava, best planting practices, optimised intercropping practices, staggered planting for smooth supply to the processing industry and practices for increased root starch content. The DS applications will target development organisations, the private sector and last-mile development agents to reach considerable numbers of farming households.

The DS tools will have to be validated against existing recommendations. For example, site-specific fertiliser blends proposed for cassava by the DS application on the basis of limiting nutrient assessments will be compared to standard NPK fertilisers in a multi-locational field trial campaign. The DS tool is expected to provide recommendations that ensure considerably higher profits while minimising the risk for applying fertiliser under non-responsive conditions.

The project will facilitate the use of the DS applications within the dissemination networks of development partners and collect feedback to further improve the 'packaging' of the applications.

The TAMASA ('Taking Maize Agronomy to Scale') (http://research.ipni.net) project supports the dissemination of improved maize technology in Ethiopia, Nigeria and Tanzania through the development, validation and wide-scale application of DS tools. The tools to be built were chosen through extensive consultation with potential users, and once developed will be scaled-out in collaboration with development partners. Service providers are evidently very keen on tools for fertiliser recommendation and variety choice. Their interest stems from the quest for site-specific nutrient management and the need to match maize varieties with suitable soil and climatic conditions. These tools are now under development for northern Nigeria.

A Fertiliser Recommendation Tool[6] (Figure 7.12) is being developed, based on the existing Nutrient Expert (NE) application of the International Plant Nutrition Institute (IPNI)[7]. Its core is the QUEFTS package (Janssen et al., 1990). In order to calibrate and validate NE, a large number of representatively located nutrient omission trials were established in 2015 to obtain calibration coefficients for the initial version. The calibration process will be repeated in new locations in 2016 and the recalibrated version-2 will be validated with partners in 2017. A second Variety Tool is being developed on the basis of the CERES-Maize simulation model, following a similar three-year process, which involves growing over 20 popular maize varieties at four locations to derive genetic coefficients for tool calibration, with irrigation during the dry season.

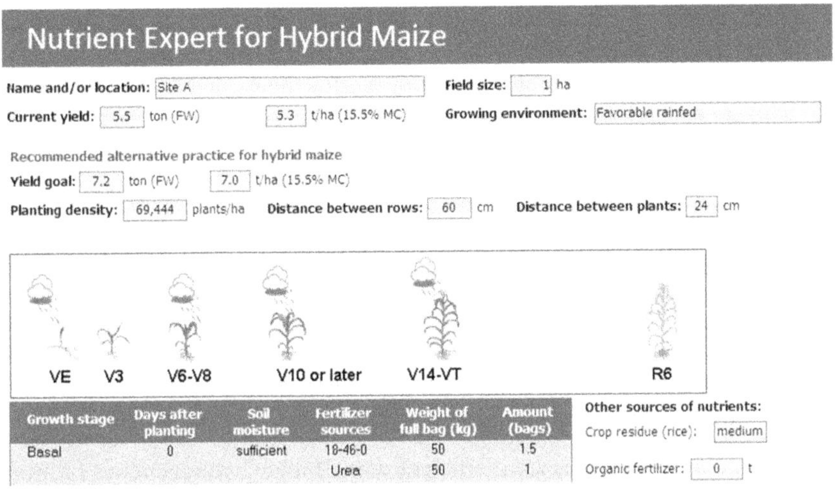

Nutrient Expert for Hybrid Maize

Name and/or location: Site A   Field size: 1 ha

Current yield: 5.5 ton (FW)   5.3 t/ha (15.5% MC)   Growing environment: Favorable rainfed

Recommended alternative practice for hybrid maize

Yield goal: 7.2 ton (FW)   7.0 t/ha (15.5% MC)

Planting density: 69,444 plants/ha   Distance between rows: 60 cm   Distance between plants: 24 cm

	VE	V3	V6-V8	V10 or later	V14-VT	R6

Growth stage	Days after planting	Soil moisture	Fertilizer sources	Weight of full bag (kg)	Amount (bags)
Basal	0	sufficient	18-46-0	50	1.5
			Urea	50	1

Other sources of nutrients:

Crop residue (rice): medium

Organic fertilizer: 0 t

Figure 7.12 Screenshot of the Nutrient Expert for maize (http://software.ipni.net/article/nutrient-expert)

Strengthening the role of multipurpose legumes

A project entitled 'LegumeCHOICE' was initiated in 2014 in the Humidtropics action sites in Kenya, Ethiopia and DR Congo. Its goal is to realise 'the underexploited potential of multi-purpose legumes towards improved livelihoods and a better environment in crop-livestock systems in East & Central Africa'. The project's rationale, motivated by the conviction that the potential of legumes is often much greater than their actual role, is to identify and exploit niches for multi-purpose legumes in existing systems, thereby building on past experiences with legume intensification. By their smart integration legumes are expected to provide the farmer with a choice of potential contributions that best match their aspirations: protein-rich food and fodder, income, fuel and improved soil fertility.

The project provides knowledge and tools to farmers and development partners to make rational decisions about integration or strengthening the role of legumes, based on their specific conditions and needs. One of those tools being developed is *LegumeCHOICE*, a DS framework for system intensification, by fitting appropriate legumes into niches in the farmers' production systems. The tool condenses all available knowledge into an operational form, to be used in the context of demand-driven development efforts. It consists of four components:

1 Qualitative diagnosis of the farming system focusing on the elements of relevance to legume integration, including identification of constraints and possible niches for legume integration;
2 A more detailed context assessment considering key constraints to legume use (e.g. land availability, inputs supply);

3   A participatory community needs assessment in relation to the functions that legumes might offer; and

4   Screening of a long list of legume options for their suitability in relation to constraints and expected legume functions, derived from the participatory analyses.

The output is a short list of promising legume options for the target community. They are discussed with communities, who will then grow them for evaluation. The feedback from this evaluation is used to refine the tool.

The tool development is still in its initial stage and it is too early to present concrete results.

Decision support for farm level management

Blanket recommendations are inefficient in the highly heterogeneous farming environment of SSA. There is therefore need for a tool that allows rapid assessment of the likely effect of off-the-shelf technologies on household nutrition and income and on ecosystem health, both in the short and the long run. The FARMSIM ('FArm-scale Resource Management SIMulator') model, developed at Wageningen University, is such a tool (www.africanuances.nl). It simulates the effects of management interventions in mixed crop-livestock farms, and can be used in support of ex-ante analyses of management decisions or production techniques.

FARMSIM is composed of three sub-models. The LIVSIM sub-model simulates production (meat, milk, progeny and manure) of individual animals in the farm's herd, based on genetic potential, feed requirements and feed availability. The HEAPSIM sub-model simulates the nutrient cycling of manure produced by the livestock and keeps track of collection and storage. Manure in the heap can then be allocated to the fields in the next season. Finally, FIELD calculates crop dry matter production per season, as determined by radiation and temperature and influenced by water and nutrient availability, for the different fields on a farm. Crop residues can either be left in the field or used as livestock feed. Fodder crops (e.g. Napier grass) can be chosen and used as feed for the livestock. FIELD integrates the QUEFTS routines (Chapter 6) as a module to relate soil nutrient content to crop growth.

The FARMSIM model is being tested in the Great Lakes region of Africa, but in its present configuration it can simulate only a limited range of crops out of those commonly found there. Currently, routines are therefore built into the model for important crops in the region. Furthermore, data from the nutrient omission trial conducted at Ntungamo (south-western Uganda) were used to calibrate QUEFTS for East African highland bananas (Chapter 6; Nyombi et al., 2010). The calibrated QUEFTS model reasonably predicted East African highland banana yields as measured in another nutrient omission trials, at Kawanda ($R^2 = 0.57$). This marked the first step towards extending FARMSIM to simulate banana-based farming systems.

A graphic user interphase is now being developed for FARMSIM, to promote its wider use by extension workers, advising farmers on best-bet interventions, in particular in support of decision making on integrated soil fertility management. First results are expected to be published by the end of 2017.

7.4.3 Dissemination and M&E

Classical dissemination pipelines (CGIAR-NARS-Public extension), never very effective in the first place, perhaps with the exception of improved crop varieties, have virtually collapsed over the past two decades since the structural reforms of the late 1980s and early 1990s under the initiative of the World Bank and the International Monetary Fund. They have been replaced by a complex tapestry of partnerships between national and international research institutions, NGOs, the remnants of public agricultural extension services, farmer organisations and private sector actors. Their operations are in most cases highly dependent on donors, who are increasingly asking for more investment in delivery and demanding proof of impact. The CGIAR, in general, and IITA, in particular, are therefore making great efforts to define the role of research in the dissemination process and devise effective ways to examine whether dissemination activities result in uptake and impact of the technologies delivered.

Dissemination

For technology dissemination to be successful, three conditions must be met, namely (i) that research responds to clearly articulated demands and needs from the intended users, (ii) that information on the technology is appropriately packaged and passed through existing networks (see Section 7.4.1) and (iii) that there is an 'enabling environment', providing the services needed for farmers to adopt the technology.

An example of the latter is the need for sustained availability of inputs for the adoption of legume technology promoted by N2Africa, for example, certified seed of improved varieties, inoculants and legume fertilisers, which remains a challenge. Research is therefore embedded in the development process, as reflected in the term 'Research-in-Development', and operates in a demand-driven mode, targeting technology options for which development partners have expressed a need and are keen to cooperate, and promoting the development of stable input and output chains. Different business models are now being considered to obtain such chains, for example: producer collectives, out-grower schemes and input supplier-driven schemes. Information on the functioning of such models elsewhere is being sought and partnerships are established with existing value chain projects with similar objectives.

Monitoring and Evaluation

All projects continue to struggle in building effective Monitoring and Evaluation (M&E) methods into their activities, both to keep track of adoption and to learn from successes and failures and adjust operations accordingly. There are two components of the M&E framework: (i) project M&E that routinely collects data on indicators for the project's progress and (ii) impact assessment to evaluate the effects of the project on farmers' productivity and livelihood. Implementing these components creates learning loops, from development to research and back to development, assigning a central role to M&E in the dissemination, adaptation and adoption of technologies.

M&E takes place at different levels. At the highest level, it involves assessment of the number of development actors that have gained access to the knowledge generated by the project, the number of policies or strategic investment plans that have changed due to project activities, and the performance of multi-stakeholder networks and platforms in capturing stakeholder needs and expectations with respect to knowledge.

At the user level, various approaches are used to monitor uptake of innovations by target smallholders and to verify that all assumptions for impact are met, whereby the development partners are always fully involved. M&E is based on regular measurement of indicators elaborated in the projects' Logical Framework, some of the instruments allowing to track achievements in real-time. The information is used to amend project strategies and approaches if required (i.e. the learning component in ME&L).

M&E tools are embedded in all dissemination activities and used by extension agents to gather feedback on the performance of the recommendations suggested by DS tools and other applications. These applications have a module to allow users to provide feedback, whereby the frequency and intensity of use is recorded to continuously improve the DS applications. Partners are facilitated to carry out these activities through their own dissemination networks. As an example, TAMASA has trained extension agents working with partners on the use of smartphone-based tools for field data collection, and researchers from Bayero University on the use of eBee drones for low altitude geospatial data collection. Plans are underway to train NARS staff on improved methods for yield survey based.

The N2Africa project has probably advanced most in implementing a structured M&E system. An ICT (Information Communication Technology)-based system has been adopted by the project, using tablets for both routine and case study data collection, based on pre-defined indicators. Programmed data collection tools are used on android systems (tablets and smartphones) in all the 11 countries. Such data is aggregated on a centralised system and linked to an online summary analysis tool ('Shiny' application) for easy access and reporting. This allows for real-time data analysis and timely feedback, allowing technology packages to be redefined on a seasonal basis and integration of the findings for updating project interventions. The data collection processes with timelines are developed together with partners and integrated in their existing structures. Feedback is brought into end-of-season evaluations and annual planning workshops with all major partners and stakeholders. Finally, external evaluations are planned at selected project locations representing the different agro-ecological zones and participating countries.

Although the principles and objectives of M&E are well established, the most effective methods are still being worked out and tested, whereby 'real time' monitoring of adoption, in particular during the fully farmer-managed phases of the adoption process, must receive greater attention. The new tools being applied now can help to quickly adjust programs to the reality of the field, provided relevant observations are made and the right kind of questions are asked.

7.5 Outputs and impact

7.5.1 Highlights of research outputs

Several of the research programs described in this chapter started during the last five years and are still largely 'work in progress'. Some of the results obtained to date are summarised here.

Progress is being made in several areas in establishing nutrient norms and quantifying and mapping nutrient deficiencies and yield gaps of a range of crops, including perennials (coffee, cocoa, bananas), using modern analytical techniques. The data are the basis for technology targeting and provide inputs for site-specific nutrient management recommendations, which are being generated using the Nutrient Expert tool. Progress is also made in identifying the causes for non-responsiveness of soils in various areas and developing ameliorating interventions.

Wide-ranging R-for-D activities around dual purpose grain legumes in many countries are showing the prevalence of response to *Rhizobium* inoculation, often combined with and enhanced by P-application. Yields of all grain legumes vary widely, mostly between 500 and 1500 kg/ha, and research is carried out to identify the factors causing the yield gaps as a basis for site-specific recommendations to close the gap.

Soybeans have been found to respond often to inoculation, while common beans respond in some, but not in all soils. There is little evidence for a response of groundnuts and inoculation is not recommended. Crop response to *Rhizobium* inoculants can be improved when appropriate soil amendments are applied. Quality inoculant for soybean is produced commercially and marketed by the NoduMax plant, at IITA Ibadan, Nigeria. Collections of *Rhizobium* strains for major grain legumes are assembled and screened.

P was also often found limiting for cocoa in the Ghana cocoa belt and site-specific nutrient recommendations must be developed, for which reliable soil suitability maps are thought to be needed. The controversial effect of shade on the nutrient response of cocoa is being elucidated.

Common NPK fertiliser inputs on maize in non-responsive soils are likely to result in a loss on investment. Remedies varied, with lime, secondary and micro-nutrients, deep tillage and manure alleviating non-responsiveness, depending on the region. Further analyses of causal factors are needed to target interventions to farmers' conditions.

Several technology guidelines and DSS are under development in support of maize production, integration of legumes and mixed crop-livestock farming.

7.5.2 Uptake and impact

Adoption and impact surveys have been conducted or are planned for most projects implemented during this period, using methodology which varied between projects. There is a need to critically review the different methods, come to an agreement about the most meaningful and effective way to conduct impact studies and lay down the agreed methods in a methodological write-up.

N2Africa

For N2Africa impact surveys were conducted in 2013 among farmers who had participated in N2Africa dissemination trials carried out between 2009/2010 and 2012, with the following findings (van den Brand, 2016):

- Farmers reported a change in legume area ranging from a decrease of 0.05 ha per farm in Zimbabwe, to a 0.37 ha increase in Nigeria, with an average increase of 0.10 ha across all surveyed countries; and
- The proportion of farmers cultivating legumes at the time of the baseline survey varied from 47% in Mozambique to 90% in Rwanda, while at the time of the impact survey they had gone up to 91% in Nigeria and 100% in DRC, Rwanda and Mozambique.

More detailed analysis of the extent that farmers, who had received legume packages, continued to use the technology, showed that:

- In Ghana, Kenya, Malawi and Rwanda, the proportion of farmers cultivating soybeans increased from 49% to 90% after receiving a soybean package, in DRC their proportion decreased from 57% to 44%, while in Mozambique and Nigeria there was little change;
- The number of farmers cultivating groundnut in Ghana, common bean in Kenya and Rwanda and cowpea in Malawi decreased after having received the specific legume package; and
- In all countries except DRC the proportion of farmers that used P-fertiliser and/or inoculants increased after having received a package, compared with the start of the project.

No explanation was given for those cases where technology adoption had actually declined.

COMPRO

An assessment under the COMPRO project of whether farmer participation in technology demonstrations increased their knowledge and the uptake of those technologies was conducted in Kenya by comparing participating with non-participating farmers in different areas (Laajaj and Macours, 2016).

The overall knowledge of technologies increased by up to 30% for the treatment group compared to control group after three growing seasons, while there was a wide variation in the level of uptake of some practices, such as the use of crop-specific fertilisers. The proportion of farmers using mavuno, a locally manufactured compound fertiliser, after the demonstration, was 9%, compared to 0.9% in the control area. Fifty percent of farmers in the demonstration group were willing to pay for mavuno compared to 8% in the control group, while 70% of the latter were willing to pay for the commonly known di-ammonium phosphate, compared with 30% of the former, indicating a shift in appreciation due to participation in the demonstration (Figure 7.13). Other technologies, such as the use of P fertiliser specific to legumes and the use of inoculants showed a similar trend.

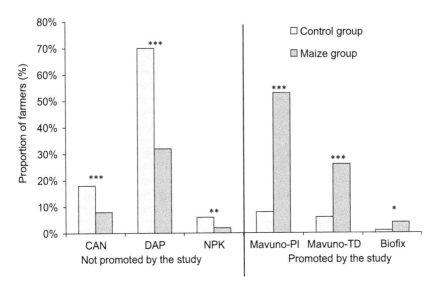

Figure 7.13 Farmers' willingness to pay for selected (bio-)fertiliser products in control and maize groups in Western Kenya (adapted from Laajaj and Macours, 2016). 'CAN' means 'calcium ammonium nitrate', 'DAP' means 'di-ammounium phosphate', 'PI' means 'planting initiation', and 'TD' means 'top-dressing'. NPK refers to commonly available NPK fertiliser, mavuno is a multi-nutrient fertiliser produced in Kenya, and Biofix is a *Rhizobium* inoculant also produced in Kenya

* = significant differences at 10% probability level; ** = significant differences at 5% probability level; *** = significant differences at 1% probability level

CIALCA

From an impact survey conducted in 2014, overall the yields of target crops (cassava, banana, soybean) had increased compared to before the project. Adoption of technologies varied between sites. Whereas the recommended spacing of banana and the rotation of maize with high biomass producing climbing beans showed high adoption rates of 28% and 26% respectively in Rwanda, there was much less adoption in DR Congo (5% and 18%). The use of fresh decomposed manure was the most successful technology in DR Congo, with 28% adoption. Intercropping of coffee with banana did not take off in any country. Farmers in Burundi and Rwanda perceived that 41% of their plots have increased soil fertility in 2014 compared to before the project, whereas a decrease was perceived in respectively 13% and 31% of the plots. In DR Congo there was a perceived increase of soil fertility in 31% of the plots and a decrease in 40% (CIALCA, preliminary results).

7.6 Emerging trends

There has been an upsurge in characterisation activities, which stems from the realisation that research findings must be extrapolated within and beyond the confines of project territory, for which biophysical and socioeconomic characterisation are essential. This has, of course, been a concern throughout 50 years of research on soils and soil fertility, but the methodology has changed considerably during the last 10 years. Today the emphasis is on new analytical methods for mapping nutrient deficiencies, whereby nutrient content of plant tissue plays an important role. Also, new techniques are being tried for *in-situ* measurement of soil nutrient content. The question is whether there remains a place for the 'traditional' survey-based soil mapping, which represents a huge resource from many decades of soil research. ISRIC for one is exploring whether soil survey units can be brought into correspondence with QUEFTS-based soil fertility classes (Rutunga et al., 2010). A similar attempt was undertaken earlier for Soil Taxonomy and CERES-Maize (Ritchie and Crum, 1989).

Another issue, attracting much attention, is the variation at a smaller scale, within short distances, both in biophysical and in socioeconomic conditions and farm management. New statistical tools are being used to capture such variation and use it to categorise farmers and technologies as a basis for better technology targeting. Also, rather than pushing one-fits-all technologies, farmers are presented with 'baskets' of best practices to allow them to decide on what best suits their resource endowments and production objectives, an approach similar to the farm counselling methodology developed in Mali in the 1990s (e.g. Defoer and Budelman, 2000).

A specific area of research is the diagnosis and rehabilitation of non-responsive soils or soils on which crops do not respond to commonly available fertiliser. Earlier studies in the Walungu area of South Kivu remained largely inconclusive (Chapter 6) and a broader 'dedicated' project began during this episode. There was also renewed interest in developing soil fertility management recommendations for cash crops (coffee and cocoa) due to the higher prices for these commodities.

In continuation of a trend which started in the previous decade, research is now to a large extent associated with or even embedded in projects aiming at agricultural development. This trend is strongly supported by a variety of donor organisations, which set high targets as to the effects that are expected from the application of research findings on the livelihoods of large numbers of farming families. This trend has important implications for the process by which research findings move into farmers' fields. One of them is the need for decision support tools which can assist development workers and farmers to choose technologies suitable for their conditions. There is increasing emphasis on the development of such tools, but current technical capabilities to do so are still limited. The Institute therefore intends to build in-house IT capability in the near future. Since the challenges are similar across the different projects, coordinated choices will have to be made about the kind of tools to be built and the programming techniques to be used. Association with advanced institutions will be needed to build up the expertise.

Another implication is the need to know how technologies are faring once they have passed into farmers' hands. The measurement of technology adoption and impact is therefore acquiring growing importance. More attention needs to be devoted to the improvement of the measurement methodology itself, especially for early adoption, which is difficult to capture with the customary large-scale interview-based surveys. The current projects are, in fact, paying attention to early adoption through their *demonstration-adaptation-adoption* scheme and it would be useful to monitor how this approach affects technology uptake.

There is an increased engagement of governments in agricultural development, through Agricultural Transformation Agendas in various countries, promoted by the African Development Bank. This will have an impact on the use of fertiliser which is seen to have become more common according to recent LSMS data for six countries in Africa (Sheahan and Barrett, 2014). Also, there is increasing engagement by the private sector in the agro-input supply and value addition chains. All of this will impact on the potential adoption of external inputs for soil fertility improvement by smallholders.

Notes

1 For publications, see the AfSIS website http://africasoils.net/
2 Now part of the Humidtropics CRP.
3 Local production facilities are piloted by the business incubation platform of IITA in Ibadan (Nigeria), the Menagesha Biotech PLC in Ethiopia, MEA Ltd in Kenya, Makerere University (Uganda) and Sokoine University of Agriculture (Tanzania).
4 In south-west Nigeria *Cola nitida* is often planted in cocoa orchards and naturally seeded timber trees such as *Obeche (Triplochiton scleroxylon)* and *Iroko (Milicia excelsa)* are commonly found there (Mutsaers, 2007, p. 607).
5 http://africasoilhealth.cabi.org/wpcms/wp-content/uploads/2016/03/102-Banana-coffee-English-colour-low-res.pdf.
6 Bayero University, the main TAMASA partner responsible for the conduct of field research activities in Nigeria agreed to host the Fertilizer Recommendation

Tool once fully developed at the end of the Project period, while the Institute for Agricultural Research (IAR) in Samaru, Nigeria, will host the Variety Tool since the Institute has the national mandate for maize improvement in the country.

7 http://software.ipni.net/article/nutrient-expert.

References

Afrifa, A., K. Ofori Frimpong, S. Acquaye, D. Snoeck and M. Abekoe, 2009. Soil nutrient management strategy required for sustainable and competitive cocoa production in Ghana. Paper presented at the 16th International Cocoa Conference, Bali.

Ahenkorah, Y., 1981. Influence of environment on growth and production of the cacao tree: soils and nutrition. Paper presented at the International Cocoa Research Conference, Douala, Cameroun, 4–12 Nov 1979.

Ahenkorah, Y., B. Halm, M. Appiah and G. Akrofi, 1982. Fertilizer use on cacao rehabilitation projects in Ghana. Proceedings of the 8th International Cocoa Research Conference, Cartagena, Colombia, 18–23 October 1981, Cocoa Producers' Alliance.

Ampadu-Boakye, T., M. Stadler, G. van den Brand, F. Kanampiu et al., 2015. *N2Africa Annual Report 2015, Phase II, Year 2*. N2Africa.org, Wageningen, The Netherlands.

Asare, R., 2005. *Cocoa Agroforests in West Africa: A Look at Activities on Preferred Trees in the Farming Systems*. The Danish Centre for Forest, Landscape and Planning (KVL), Horsholm.

Asare, R. and R.A. Asare, 2008. *A Participatory Approach for Tree Diversification in Cocoa Farms: Ghanaian Farmers' Experience*. STCP Working Paper Series 9 (Version September 2008). International Institute of Tropical Agriculture, Accra, Ghana.

Asare, R., R.A. Asare, W.A. Asante, B. Markussen and A. Ræbild, 2016. Influences of shade trees and fertilization on on-farm yields of cocoa in Ghana. *Experimental Agriculture*. Available online at https://www.cambridge.org/core/services/aop-cambridge-core/content/view/S0014479716000466, doi: 10.1017/S0014479716000466

Asare, R., and A. Ræbild, 2015. Tree diversity and canopy cover in cocoa systems in Ghana. *New Forest*, 47, 287–302

Bressers, E., 2014. Nutrient deficiencies and soil fertility constraints for common bean (*Phaseolus vulgaris* L.) production in the Usambara Mountains. MSc. thesis, Plant Production Systems, Wageningen University.

Defoer, T. and A. Budelman (eds.), 2000. *Managing soil fertility. A Resource Guide for Participatory Learning and Action Research*. Royal Tropical Institute, Amsterdam.

Dianda, M., K.E. Giller and P.L. Woomer, 2014. *Rhizobiology Master Plan*, Version 1.1, 29 November 2014, N2Africa.org, Wageningen, The Netherlands

Douxchamps, S., M.T. Van Wijk, S. Silvestri, A.S. Moussa, C. Quiros, N.Y.B. Ndour, S. Buah, L. Somé, M. Herrero, P. Kristjanson, M. Ouedraogo, P.K. Thornton, P. Van Asten, R. Zougmoré and M.C. Rufino, 2016. Linking agricultural adaptation strategies, food security and vulnerability: evidence from West Africa. *Regional Environmental Change*, 16, 1305–1317.

Giller, K., A. Franke, R. Abaidoo, F. Baijukya, A. Bala, S. Boahen, K. Dashiell, S. Kantengwa, J.M. Sanginga, N. Sanginga, A. Simmons, A. Turner, J. De Wolf, P. Woomer and B. Vanlauwe, 2013. N2Africa: putting nitrogen fixation to work for smallholder farmers in Africa. In: B. Vanlauwe, P. Van Asten, P. and G. Blomme (eds), *Agro-Ecological Intensification of Agricultural Systems in the African Highlands*. Routledge, London.

Janssen, B.H, F.C.T. Guiking, D. van der Eijk, E.M.A. Smaling, J. Wolf and H. van Reuler, 1990. A system for quantitative evaluation of the fertility of tropical soils (QUEFTS). *Geoderma*, 46, 299–318.

Jefwa, J.M., P. Pypers, M. Jemo, M. Thuita, E. Mutegi, M.A. Laditi, A. Faye, A. Kavoo,

W. Munyahali, L. Herrmann, M. Atieno, J.R. Okalebo, A. Yusuf, A. Ibrahim, K.W. Ndung'u-Magiroi, A. Asrat, D. Muletta, C. Ncho, M. Kamaa and D. Lesueur, 2014. Do commercial biological and chemical products increase crop yields and economic returns under smallholder farmer conditions? In: B., Vanlauwe, P. van Asten and G., Blomme (eds), *Challenges and Opportunities for Agricultural Intensification of the Humid Highland Systems of sub-Saharan Africa*. Springer International, Heidelberg.

Laajaj, R. and K. Macours, 2016. Farmers' experimentation and learning about inputs and practices. Unpublished COMPRO survey data.

Mutsaers, H.J.W., 2007. *Peasants, Farmers and Scientists. A chronicle of tropical agricultural science in the twentieth century*. Springer, Dordrecht, The Netherlands.

Nyombi, K., P.J.A. van Asten, M. Corbeels, G. Taulya, P.A. Leffelaar and K.E. Giller, 2010. Mineral fertilizer response and nutrient use efficiencies of East African highland banana (*Musa* spp., AAA-EAHB, cv. Kisansa). *Field Crops Research*, 117, 38–50

Ochola, D., P. van Asten, and S. Mbowa. 2016. "The land that feeds us by sun and fertile soils": Should we be guided by our national anthem? Poster Presentation at the National Validation Workshop for the Regulatory Impact Assessment on the National Fertilizer Policy. 16 Apr 2016 Protea Hotel, Kampala Uganda. DOI: 10.13140/RG.2.1.4017.2409.

Ogunlade, M., K. Oluyole and P. Aikpokpodion, 2009. An evaluation of the level of fertilizer utilization for cocoa production in Nigeria. *Journal of Human Ecology*, 25, 175–178.

Paul, B.K., P. Pypers, J.M. Sanginga, F. Bafunyembaka and B. Vanlauwe, 2014. ISFM adaptation trials: farmer-to-farmer facilitation, farmer-led data collection, technology learning and uptake. In: Vanlauwe B, P. van Asten and G. Blomme (Eds). *Challenges and Opportunities for Agricultural Intensification of the Humid Highland Systems of Sub-Saharan Africa*, Springer, Switzerland.

Ritchie, J.T. and J. Crum, 1989. Converting soil survey characterisation data into IBSNAT crop model input. In: A. Bouma and J.K. Bregt (eds), *Land Qualities in Space and Time*. Proceedings of the ISSS Symposium, Wageningen. Backhuys, Leiden.

Roobroeck, D., G. Nziguheba, B. Vanlauwe and C. Palm, forthcoming. Fertilizer use in maize and soybean cropping on smallholder farms in sub-Saharan Africa: Responses in productivity and returns under heterogeneous conditions.

Rutunga, V., B.H. Janssen and P.S. Bindraban, 2017. *Use of QUEFTS-calculated maize yields for the establishment of relationships between soil fertility classes and soil classification units in southern Africa* (submitted to Geoderma).

Sheahan, M. and C.B. Barrett, 2014. *Understanding the Agricultural Input Landscape in Sub-Saharan Africa: Recent Plot, Household, and Community-level Evidence*. Policy Research working paper no. WPS 7014. World Bank Group, Washington, DC.

Vågen, T.-G., K.D. Shepherd, M.G. Walsh, L.A. Winowiecki, L. Tamene Desta and J.E. Tondoh, 2010. AfSIS Technical Specifications - Soil Health Surveillance. Africa 76 pp. World Agroforestry Centre, Nairobi, Kenya.

van den Brand, G., 2016. *Synthesis report of the N2Africa Early Impact Survey 2016*.

Vanlauwe, B., K. Giller and J. Van Heerwaarden, 2014. *Agronomy Plan*, Version 1.1, 29 November 2014. N2Africa.org, Wageningen, The Netherlands.

Vanlauwe, B. R. Coe and K.E. Giller, 2016. Beyond averages: new approaches to understand heterogeneity and risk of technology success or failure in smallholder farming. *Experimental Agriculture*. doi: 10.1017/S0014479716000193.

Wang, N., L. Jassogne, P.J.A. van Asten, D. Mukasa, I. Wanyama, G. Kagezi and K.E. Giller (2015). Evaluating coffee yield gaps and important biotic, abiotic, and management factors limiting coffee production in Uganda. *European Journal of Agronomy*, 63, 1–11.

Woomer, P.L., J. Huising and K.E. Giller, 2014. *N2Africa Final Report of the First Phase 2009–2013*. N2Africa.org, Wageningen, The Netherlands.

8 Looking back and moving forward

8.1 Introduction

The story told in this book covers nearly a half century of soil, land use and soil fertility research in and for sub-Saharan Africa (SSA). In this final chapter, we bring together its main features and its evolution over the years, drawing conclusions, assimilating the implications and laying out prospects for the future.

But first this: if we were to choose a single lesson learned in these years it would be that research can only make genuine contributions to farmers' productivity if it is carried out in close association with them, testing technologies as much as possible under their management and in their own fields. There is no substitute for exposing technology to farmer management as early as possible, as a reality check for researchers' assumptions; failing to do so will ultimately be penalised when it turns out that some essential elements standing in the way of adoption have been overlooked.

In the following paragraphs, we will scrutinise these and other lessons emerging from the work described in the preceding chapters, which hopefully can help current and future researchers to avoid the pitfalls of the past and increase their chances to contribute to the improvement of agricultural production in Africa.

This final chapter is sub-divided in the same manner as the 'subject matter chapters' 1–7, into six major sections:

1 Scope, approaches and partnerships;
2 Characterisation of soils, farms and farming systems;
3 Technology development;
4 Technology delivery and dissemination;
5 Impact; and
6 Emerging trends.

Under each of these headings a number of 'story lines' are presented, which run through the successive episodes in different guises, and which characterise the evolution, achievements and failures of the research on soils, soil fertility and land use.

This chapter ends with the lessons the authors have drawn from this book's story, which will hopefully be useful for the future quest towards improved soil and soil fertility management, for the benefit of the African smallholder farmer.

8.2 Scope, approaches and partnerships

8.2.1 'Strategic research', 'Research for Development' and balancing acts

Soil research during the first two decades can unhesitatingly be termed 'strategic'. Characterisation of soils and farming environments was to provide a basis for systematic targeting of technology, while technology development was to generate different prototypes broadly adapted to specific classes of soils and environments. These two components, characterisation and the development of prototype technology, were in fulfilment of the original goal of international research, of generating information for and providing innovative technological components to the National Agricultural Research Systems (NARS), for further adaptation and eventual transfer to the farmer through the extension service.

IITA's soil and land use research of the early years, however, rather than aiming at the improvement of existing cropping systems, implicitly targeted a hypothetical new type of farmer who would integrate the research findings into an efficient, intensified, semi-mechanised operation. This explains the emphasis on erosion, land clearing, zero tillage, mulching systems and intensified permanent cropping in the period from 1968 to the mid-1980s (Chapters 1–3), topics which had minor relevance for most existing farming systems in the humid area, the then target zone of the Institute. Adoption of technologies such as no-tillage, alley cropping, short herbaceous fallows and live mulch, although in principle 'system neutral', would also have required a significant overhaul of the existing farmers' systems, an unlikely choice for risk-averse farmers, even should the new systems be highly performant.

The earlier research did address real technical challenges for intensified production, but it was essentially supply-driven, because it did not respond directly to demands from existing producers. During the mid-1980s, however, an On-Farm Research (OFR) capability was created, following the international Farming Systems Research (FSR) movement. Initially, it was more successful in stimulating NARS in Nigeria and Côte d'Ivoire to carry out demand-driven on-farm research than inducing IITA's own soil and land use researchers to do so, but by the late 1980s the latter started to move more and more of their researcher-controlled trials to farmers' fields, especially in the savannah areas.

As from about 1995, a gradual shift occurred towards a more demand-driven approach, whereby research associated itself increasingly with development partners, and addressing the needs of existing farmers assumed prevalence. This trend strengthened further in the 2000s, culminating in what is now called Research for Development (R-for-D), with research becoming a partner in the development process through joint activities with other stakeholders, adaptive research with and by farmers and participation in development platforms.

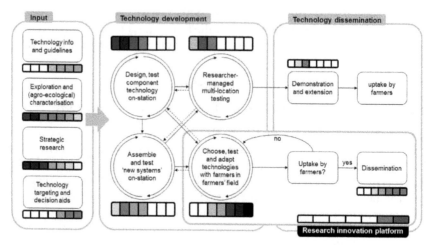

Figure 8.1 Evolution of IITA's technology development and dissemination model, 1967–2015: from technology generation to multi-stakeholder research in development

Shading in the rule bars indicate relative emphasis on the various phases in successive episodes, corresponding with Chapters 1–7, with darker areas indicating relatively more emphasis

Figure 8.1 depicts the changes in research concepts and approaches as they have evolved since 1967.

Today's R-for-D[1] focuses on the stepwise improvement of existing systems through innovations developed or chosen through a participative process, which involves the major stakeholders. This process makes credible claims of enhancing the chances of technology adoption and impact, a promise which is, however, yet to be proven by the significant adoption of soil and soil fertility management practices by farmers.

One of today's challenges is finding a balance between strategic or upstream, and adaptive or downstream research. There is a certain danger of over-shooting into a development-oriented organisation. There is a temptation to assume tasks which belong to NARS and development organisations, to compensate for the latter's shortcomings. International research, however, continues to have a more strategic role to play, studying principles, processes and methods, beyond the strictly local level, even though that role is less prominent now. It should contribute to the elucidation of processes underlying soil fertility and productivity and to the development of novel research methods, 'International Public Goods' (IPG) research for short. The challenge is to combine the two within the R-for-D framework.

8.2.2 The role of research and development partners

The role of partnerships has undergone fundamental changes over the years (Figure 8.2). In the 1970s and the early 1980s three types of partnership were primarily engaged in:

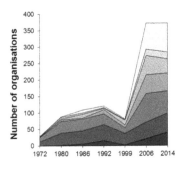 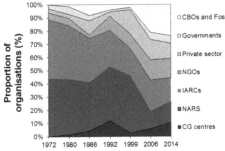

Figure 8.2 Changing partnerships in soil and soil fertility research, showing the approximate number (a) and proportions (b) of stakeholder categories' interaction with the research activities

1   Partnerships with Advanced Research Institutes (ARIs) in mostly western countries, in support of strategic research;
2   Partnerships with NARS, including universities, in Nigeria, with national scientists benefiting from association with international science, and IITA using NARS facilities for field research; and
3   Loose association with Agricultural Development Projects (ADPs) in Nigeria, with some scientists testing land clearing and management techniques at scale.

The opportunities provided by the ADPs were exploited to a limited extent by the Farming Systems Program (FSP) to put its technologies to a reality test with real farmers, in particular alley cropping. Otherwise little use was made of these opportunities, contrary to the Crop Improvement Programs, probably because most FSP-developed technologies were not yet considered ready for that. The dissemination of improved crop varieties is probably more straightforward, provided the varieties perform well under farmers' conditions, and primarily requires effective distribution of seed or planting material. FSP on the contrary needed to deliver understanding of soil and soil fertility management techniques, which were quite different from business as usual.

In the course of the 1980s, a considerable number of so-called outreach projects were contracted with United States Agency for International Development (USAID) and World Bank and staffed by IITA-hired scientists. Projects were established in Burkina Faso, Cameroon, Ghana, Rwanda and Zaire (now Democratic Republic of Congo), in support of the NARS in those countries. Although the emphasis was mostly on improved crop varieties, FSP scientists also collaborated to varying degrees with the scientists in the outreach projects, who adopted various research concepts and technologies, such as the use of auxiliary legumes. The cooperation in most cases lacked structure however, due to insufficient guidance from the FSP. Most of these projects closed down by the early 1990s.

A different kind of partnership was established in the 1980s between a relatively small OFR group at IITA and research groups in several NARS in Nigeria and

Ivory Coast and later also in Cameroon and Zaire, for joint development and promotion of on-farm research activities. These activities led to the creation of the West African Farming Systems Research Network (WAFSRN), in which both Anglophone and Francophone countries participated. These initiatives survived into the early 1990s, when interest by both IITA and donors waned.

An Alley Farming Network for Tropical Africa (AFNETA) was created in 1986 in collaboration with the World Agroforestry Centre (ICRAF), which promoted testing of alley farming by participating NARS. Many trials were carried out all over SSA, almost exclusively under researcher management, until the network was transferred to ICRAF in the mid-1990s.

Around 1990, following the adoption of the agro-ecological zone (AEZ) approach by the CGIAR, IITA initiated the decentralisation of its research, and the delineation of Benchmarks to represent the conditions in the AEZs under its mandate. The nature of collaboration with NARS thereby changed into more formal institutional arrangements to develop joint research programs in the NARS-mandated areas. The Benchmark approach was largely abandoned again ten years later, but research continued in a decentralised mode afterwards, now as R-for-D in many locations across SSA, in partnership with NARS, non-governmental organisations (NGOs), farmer organisations and other stakeholders, mostly with special project funding.

Today, development partners and the research community interact and collaborate through formal and informal platforms, jointly plan and implement programs and assess the results and their implications. Development partners play an increasingly important role in disseminating innovations, collecting feedback on their performance, as well as organising the farmers, improving marketing opportunities and facilitating an enabling environment for the uptake of improved soil and soil fertility management practices.

IITA scientists have also become more closely associated with ARIs again for the development and application of novel research tools (see Section 8.5.4), from which many PhD theses have resulted.

8.2.3 Who sets the research agenda?

The research agenda of an institute is a function of its goals, which for IITA were defined at its inception in 1967 as, freely rendered, increasing agricultural production by Africa's smallholders through the intensification of their production systems. The term 'sustainable' was added explicitly as from the late 1980s, but otherwise the original goal has remained valid until today. The question is how this goal was translated into the research agenda.

In the 1970s, it was the scientists who did the translation, assuming they knew what kind of technologies were needed to bring about intensification. This resulted in a soils and soil fertility research program of long duration which was carried out essentially under controlled conditions. In the end, very little if anything was adopted by the intended beneficiaries, a harsh and costly lesson. From the early 1990s, more attention was paid to the farmers' own needs for technology, which, however, initially did not lead to a major reorientation of the research agenda: it remained mostly defined by the scientists themselves, although the work was increasingly carried out in collaboration with farmers, and in their fields.

It was only during the course of the 2000s that research gradually became more demand-driven, through its association with, or involvement in development projects, often supported or even designed by donors, aiming at the improvement of agricultural production in specific areas. Although even then the influence of the scientists on the agenda remained strong, mechanisms were put in place for other stakeholders, not only to be heard, but to have a strong say in the content and implementation of the agenda, such as innovation or development platforms. In recent years, the role of development partners and partnership platforms in setting the research agenda has further increased, and the influence of scientists has been scaled down accordingly. It is expected that this will increase the likelihood of future technology adoption and impact.

Effective mechanisms to associate all major partners with the development process are a holy grail of today's international development efforts. Innovation Platforms and other collaborative mechanisms have been tried out for some time now, but evidence for their effectiveness remains sketchy. It would be very useful to review the experiences so far, for future guidance.

8.2.4 Geographical focus and strategy

Between 1967 and 1982, the focus of IITA's soil and land use research was on the humid and sub-humid zones of West Africa, with most activities carried out in Nigeria, the home of IITA and harbouring all relevant AEZs. The Benchmark Soils project of the 1970s surveyed soils across these zones, in order to identify representative sites where technology performance on different soil classes could be studied. Experimental research was, however, limited to a small number of sites in humid south-west and south-east Nigeria with predominantly Alfisols and Ultisols and associated soil types[2], mostly at the Ibadan and Onne research stations, and the wetlands in central Nigeria. With Alfisols and Ultisols and their associated soil classes being dominant in the West Africa humid and sub-humid region, these sites were considered broadly representative for the region.

Over the course of the 1980s, more research was started in the Nigerian Guinea savannah. Geographical coverage also expanded through 'outreach projects' in several West and Central African countries, where conditions ranged from humid forest and forest-savannah transition to coastal derived savannah, Guinea savannah and mid-altitude conditions (Rwanda). Support was further provided to NARS for the development of OFR activities in Nigeria, Benin, Ivory Coast and Cameroon. Most of these projects did not develop truly integrated partnerships with IITA in the form of a mutually reinforcing soil and soil fertility research network[3]. The geographical coverage contracted again with the termination of most outreach projects in the early 1990s.

The period 1995–2001 saw increasing decentralisation of research activities, with new projects in the moist savannah of West Africa. The short-lived

AEZ/Benchmark program delineated six zones in West Africa and Cameroon and identified representative benchmark areas in each of them, but research never extended beyond on-going activities in a few of the benchmarks, viz. in Cameroon (humid forest), Nigeria (degraded forest, Northern Guinea savannah) and Benin (derived savannah).

Since the early 2000s, decentralisation continued with increasing momentum, initially in West Africa and the Great Lakes area of Central Africa. Research was no longer bound to benchmarks but rather followed donor priorities with the focus remaining on the humid and sub-humid zones. Since 2010, research networks of all CGIAR Centres have greatly expanded, with IITA having evolved into today's fully decentralised setup, with four regional 'hubs' for West, Central, East and Southern Africa. Each of these hubs has a number of 'stations' in collaboration with NARS, from where 'project-driven' research activities are carried out. Choice of research location is conditioned by a combination of factors. First of all, there are the institute's 'traditional' target systems: cereal-, roots and tubers-, and highland banana-based. Other factors include population density and the need for intensification, the road infrastructure, the effectiveness of input and output markets, on-going partnerships, and last but not least, donors' geographical focus. The extent of donor influence on the research programs can be inferred from the relative weight of unrestricted and restricted funding in the Institute's budget (Figure 8.3).

Today, the IITA research setup is highly decentralised, with small groups of scientists scattered thinly over the continent, a trend also observed for most CGIAR Centres at global level. Maintaining scientific consistency and reconciliation of ideas and methods in such a distributed system requires effective communication and leadership to ensure scientific synergy between the groups.

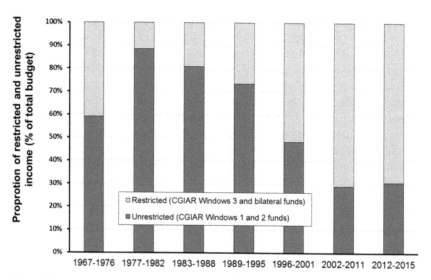

Figure 8.3 Average volume of IITA's annual unrestricted and restricted funding, 1967–2015

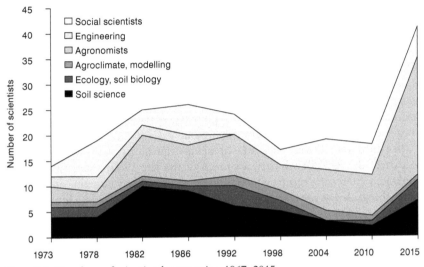

Figure 8.4 Numbers of scientists by expertise, 1967–2015

8.2.5 Scientific disciplines

The disciplines of the internationally recruited senior scientists associated with soils and soil fertility research has changed significantly over the years (Figure 8.4) showing several striking trends: (i) an overall decline in numbers since the peak in the 1980s and a renewed surge in the last five years, (ii) a steady decline of engineering and specialised soil science disciplines (soil physics, chemistry, fertility) from a peak in 1982, and a temporary surge in soil biology between 1990 and 1998, (iii) substitution of soil scientists by systems agronomists from 1998 until today and (iv) the fluctuating representation of social scientists of different denominations, and a fairly strong increase since 2000. These changes reflect the evolution of the Institute's research orientation. It was strongly 'physical' until the late 1980s, and dominated by soil taxonomy, physics, chemistry and fertility. In the 1990s, there was a change in emphasis to the more biological aspects of soil fertility, such as organic matter dynamics and the process of nitrogen fixation, followed by an increasing focus in the last 15 years on Integrated Soil Fertility Management (ISFM) in the context of priority farming systems. The current strong representation of agronomists and social scientists is related to the increasing dominance of Research-in-Development (R-in-D), the management of its collaboration and learning processes, and the importance of dissemination and impact assessment.

8.3 Characterisation of soils, farms and farming systems

Characterisation has been an ever changing kaleidoscope of targets and methods. In the early years, social scientists and agronomists carried out various surveys of farming systems, mostly in Nigeria, to identify leverage points for the development of smallholder agriculture, while the soil scientists conducted soil surveys to select benchmark soils for future systematic technology testing. The two hardly came together and ended in dead alleys, without their results being effectively used or even adequately published.

In a second wave, multi-disciplinary diagnostic surveying came in vogue as an essential component of the FSR approach to research. The OFR group had some success in promoting this approach at national institutes in several countries, but it left IITA itself mostly untouched.

In a third wave, broad systematic surveying of farming systems was undertaken as part of the AEZ/Benchmark approach of the late 1980s and early 1990s. It generated interesting results which influenced technology targeting and, among other things, led to the conclusion that there would only be a very limited niche for alley cropping, if any. The AEZ/Benchmark approach never came to full fruition, probably because of its overly ambitious goals and limited funds.

Today, there is a variety of characterisation activities associated with R-for-D projects across the continent, combining the collection and analysis of socio-economic and physical data as a necessary basis for area- and farm type-specific technology targeting. New survey and analytical methods are being used, especially for the characterisation of soils and soil fertility, focusing more on the properties of the topsoil, which most affect plant growth. Profile-based classification systems and the earlier Fertility Capability Classification (FCC) do provide information on general soil properties, but they are not sufficient to give clear relationships with crop growth and responses. The new topsoil-based methods will need to be consolidated into an effective tool set for future use.

There has been a historic debate since the FSR years on the usefulness of detailed socio-economic characterisation and typology of smallholder farms and households. In the 1970s and 1980s francophone *Recherche-Développement* workers argued for the elaboration of detailed survey-based farm typologies, while 'anglophone' workers usually opted for quick and dirty methods to identify constraints and opportunities shared by different types of farmers (e.g. Fresco, 1984). The effectiveness of either of these approaches has not been systematically assessed, perhaps because the FSR era ended without much to show in terms of technology adoption. IITA sided with the latter approach during the 1980s, but has been turning to the former in the past two decades, carrying out extensive household and livelihood surveys. The effectiveness of such expensive surveys has not been ascertained, but their often rather obvious conclusions raise doubt of their advantage over quick informal approaches.

8.4 Technology development

8.4.1 From new systems to the Sustainable Intensification of existing systems

The evolution of technology development research over the years is illustrated in Figure 8.5, which indicates the approximate periods of development of

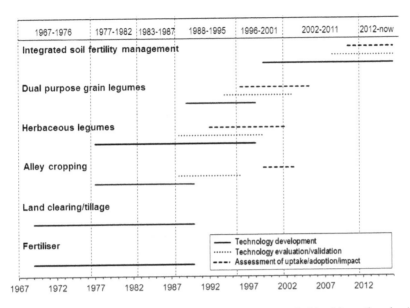

Figure 8.5 Evolution of the technologies and interventions prioritised by soil and soil fertility research initiatives with an indication of the technology development, evaluation/validation, and uptake/adoption/impact phases from 1967 until today

different technologies, their evaluation and validation, and the assessment of uptake, adoption and impact.

In spite of substantial surveying of existing production systems by social scientists in the first two decades, the themes of much of the experimental soil research in those years show that its goal was the replacement of assumedly outdated production systems by new, more productive ones. This is particularly obvious from the work on land clearing, soil erosion, zero tillage, combined with chemical weed control and small- or medium- level mechanisation. In particular, the work on erosion control had no relevance for the existing smallholder farming systems in humid and sub-humid Africa, since erosion was not a major issue. Other technological components such as high levels of external inputs, live mulch, short duration leguminous fallows and alley cropping were more 'system neutral' in that they could, in principle, be introduced into existing systems. Sustainability was clearly a major concern, as testified by the emphasis on soil protection, even though it was about sustainability of an as yet imaginary, or, more positively stated, a prospective future system.

A consequence of the absence of a target system for most novel technologies developed during these years was that there were hardly any opportunities for testing them in the real world of existing farming systems. Only alley cropping was evaluated extensively in farmers' fields and showed serious constraints after more than 15 years of sheltered development. Ultimately, none of the novel technologies found their way to the farm, which, in retrospect, represents a substantial amount of misspent resources.

As from the mid-1980s, attention began to shift towards existing production systems and their capacity for change and sustainable intensification. Alley cropping and planted, leguminous fallows remained on the agenda for some years, but were eventually replaced by technologies which were closer to the farmers' own experience, such as dual purpose legumes. These became part of the tool box of ISFM, the approach which gained prominence in all the Institute's projects from the mid-1990s, as part of its drive to intensify farming in SSA. Its principles are aligned with those of Sustainable Intensification (SI), which encompass (i) increased production per unit of land, (ii) maintenance of essential soil-based ecosystem services and (iii) resilience to shocks, especially climate change.

Research now addressed both strategic and adaptive issues, always in the context of the improvement of existing systems, towards sustainably intensified production, using judicious amounts of external inputs. In the first category were studies on efficiency- and sustainability-related factors, such as improved biological N- fixation, use-efficiency of inputs and maintenance of soil organic matter. Adaptive research developed baskets of options, assembled from different sources including own field research, and delivered in association with NARS and development organisations, thereby blurring the traditional distinction between the tasks of international and national research and extension.

8.4.2 Soil fertility and the role of fertiliser, auxiliary crops and grain legumes

The availability of nutrients in different soils, its changes under intensified cropping and the options for improving nutrient availability have been a consistent concern throughout 50 years of soil and land use research. In the early years, the use of inorganic fertiliser[4] (and liming in the wetter areas), was considered as the primary factor for increased productivity, following the example of the Green Revolution. Some of the more basic research on soil acidity, nutrient leaching, nutrient deficiencies, and perhaps also the soil classification work, made significant contributions to the 'body of knowledge', and much of it still has relevance for today's technology development. A thorough inventory of the results and the accumulated data would be needed, however, for them to become operational. The information presented in the first four chapters of this book can be a starting point for such an inventory.

Scientists soon became aware that fertiliser alone would not result in sustainable intensification. Conservation, crop rotation and the retention of crop residue helped to slacken fertility decline, but in the humid zone they were not enough. After a number of years, yields would still go down and it was necessary to fallow the land to prevent serious degradation, even at the research station (Chapter 3). This realisation ushered in a long period of experimentation with organic methods of soil improvement, through live and dead mulches and herbaceous and woody auxiliary crops, in particular *Mucuna pruriens* as a short duration fallow and alley cropping, integrating cropping and fallowing. The application of fertiliser, however, remained a necessity for an acceptable yield, whereby a synergistic effect between organic inputs and fertiliser was sometimes found (Chapters 5 and 6).

Today, after having given up on the adoptability of auxiliary crops whose single contribution was their anticipated effect on soil fertility, research turned to rotations with grain legumes with dual or triple purposes: edible grain, fodder and a contribution to the N-economy of cropping systems. Fertiliser remains an essential component of ISFM, to generate satisfactory amounts of useful products, including biomass that can be returned to the soil, directly as fresh crop residues, or after passing through animals as livestock feed.

In view of the lack of access or costliness of fertiliser, high nutrient-use efficiency has become paramount as part of SI, contrary to earlier years when there was surprisingly limited attention to this important aspect. In many trials in former years, the relatively low nutrient use-efficiency remained unexplained. It was only by the second half of the 1990s that it received the attention it deserved and many studies have since been undertaken to explain and remedy lower than expected efficiency.

Figure 8.6 illustrates the changing emphasis on different technologies for soil fertility management over the years, through their relative representation in IITA's publications[5].

Much strategic work remains on the analysis and mapping of nutrient availability and the response to applied nutrients in relation to soil type, one of the original, uncompleted tasks of international research. More sophisticated analytical tools are now being introduced for this purpose.

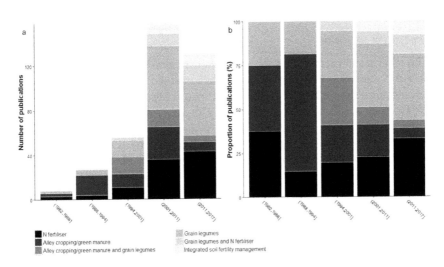

Figure 8.6 Number (a) and percentages (b) of IITA publications covering specific research themes, 1982–2015

8.4.3 From the research station to the farm

The unit of intervention: from plots to farms

When soil and land use research was still mostly an on-station affair, the unit of intervention was the experimental plot. The effects of single or combinations of a small number of factors were studied with single species crops and rotations, while other factors were kept at a fixed level, in most cases quite different from the usual level in farmers' fields. A few on-station studies were carried out simulating a whole-farm situation, such as the 'Unit Farm' test of the 1970s, whereby the simulated farm consisted of a compilation of recommended practices but did not even remotely resemble a real existing farm. At an intermediate level, mixed cropping trials were carried out to study compatibility and possible synergism between species, which added some insights to existing knowledge.

When research started to migrate into farmers' fields, some 'on-farm trials' were simply researchers' experiments carried out in farmers' fields, similar to traditional multi-locational testing. Since the 1980s, more meaningful on-farm trials from an adaptive research perspective were beginning to be carried out, with innovations tested within farmers' usual cropping systems, each farmer becoming an experimental unit. Such trials have gradually increased in importance, until in the course of the 2000s they became the mainstay, with on-station trials only used for precise process or technology screening studies. On-farm research also increasingly addressed the effects of innovations on whole-farm productivity, rather than on single crops or crop combinations. Given that complex innovations will affect a farmer's whole enterprise and the family's wellbeing, in short, their livelihood, the analysis of their adoption or non-adoption needs to consider physical, economic and social factors.

Strategies for site selection

The criteria and methods for site selection have undergone significant changes over the years, though their rationale has remained the same. It dictates that research sites should be representative for an area of significant extension where the findings are likely to apply. In the 1970s representative Benchmark Soils, or rather benchmark toposequences, were chosen in southern Nigeria on the basis of the Soil Taxonomy classification. Soil fertility trials were carried out in several of them, but extrapolation of the results to the wider area for which the benchmarks were representative was never rigorously applied.

In the 1990s, a serious attempt was made to re-introduce a Benchmark concept, with large benchmarks identified in six humid and sub-humid zones, where trial locations would be chosen along transects within benchmarks, covering the variation in environmental conditions and production systems. Again, the approach was never fully implemented and research remained limited to a few locations in five benchmark areas: the Humid Forest Margins (Cameroon), the Degraded Forest (Onne), the Forest-Savannah Transition (Ibadan), the Derived

Savannah (Benin, Togo), the Southern Guinea Savannah (Cote d'Ivoire) and the Northern Guinea Savannah (Kaduna area).

Since then, site selection has increasingly been based on pragmatic considerations and the presence of favourable conditions for the uptake of improved soil and soil fertility management practices, important among them being the availability of donor funding, as much as possible within major intervention zones (see also Section 8.2.4). Geographical variation continued to be accounted for by arranging experiments along transects covering as much as possible of the variation occurring in the intervention area.

Accounting for variation at field, farm and agro-ecology level

Variation appears in different guises at different levels. Apart from large-scale ecological differences, variation was initially recognised mainly at the field level, in the form of considerably different soil types, even at short distances along toposequences. The interaction of technologies with these variations was accounted for by replicating the tests on representative benchmark soils. Other sources of variation were eliminated as much as possible by uniform management in researcher-managed trials. Even then, unidentified factors often led to smaller than expected treatment effects, expressed, for example, in low N-use efficiency in many trials.

Once research went on-farm, scientists became aware of the need to account for the effect of farmer-related factors, rather than trying to simply isolate or even eliminate them. With the first wave of farmer-managed trials in the 1980s, Hildebrand's adaptability analysis (Hildebrand, 1984) was used with some success to relate treatment effects to farmers' overall yield level, but that left the question 'what caused the differences in those overall yield levels?' unanswered. The statistical methods used at the time were not sufficiently discriminating to link them to (combinations of) soil-, climate- and management-related sources of variation.

With the advent of the agro-ecological approach in the 1990s, research was 'stratified' for ecological conditions in the benchmarks. This allowed the identification of broad AEZ-related differences in yield levels and responses to treatments, but it did not help to understand the more intractable but important variation between farms within ecologies.

In the 2000s, research became more directly associated with development, and involved itself in the task of tailoring technologies to the needs of different categories of farmers, in collaboration with NARS and extension organisations. Accounting for differences among fields and farmers as a basis for the formulation of farmer-specific recommendations thereby assumed a new urgency, to ensure maximum returns to investment in soil fertility management. New methods were therefore needed to analyse how variation in soil fertility conditions and farmer management affect crop response to improved technologies. Some existing, hitherto unused methods are showing promise but they are still very much in the evaluation stage (see Section 8.4.4).

8.4.4 Advanced research tools and the changing role of Advanced Research Institutes

There has always been a search for advanced research tools to strengthen strategic soil and soil fertility research. Also, with the move into farmers' fields, the need for analytical tools more powerful than standard statistics was increasingly felt. Scientists have therefore gradually increased their collaboration with ARIs for the joint development of novel approaches for which IITA lacked sufficient in-house manpower and expertise.

Several areas can be distinguished where advanced tools are needed today:

1   Geo-spatial tools to map different features of intervention areas, such as ecological zones, soil types, nutrient deficiencies, crop distribution in support of site selection for technology development and testing;
2   Advanced analytical tools to identify and map nutrient deficiencies in farmers' fields on the basis of nutrient balances in soils and crops;
3   Statistical tools to relate crop yields to measured field characteristics as a tool for targeting technology to farmers with different characteristics;
4   Static and dynamic crop modelling to estimate yield gaps; and
5   Decision Support Systems (DSS) as tools in technology development, evaluation and dissemination.

Links with ARIs have been important to remain conversant with scientific advances elsewhere, but there has been a lack of consistency, because of shifting alliances[6]. It may be necessary to review the field and decide on a limited number of advanced institutions as collaborators in soil and soil fertility research and dynamic modelling for the coming years, fostering relations and maintaining consistency across the Institute's different research groups.

8.4.5 Technology validation by farmers

Technology validation by farmers has now become an integral part of technology development, a shift away from almost exclusively station-based research that has taken time to materialise. In the traditional research-extension model, research ended at the gates of the research institute and its multi-locational testing sites, where extension would take over. In the western world this has worked, as on-station conditions reflect to a great degree those at the farm, while in Africa the discrepancy remains large. There have been various movements to bring researchers into farmers' fields, such as the FSR movement and its off-shoots such as various farmer-participatory approaches and, on the extension side, the 'Training and Visit' and the 'Farmer Field School' approaches, which brought together researchers, extensionists and farmers in learning and exchange sessions. At IITA, a first attempt at OFR was undertaken in the 1980s, initially with simple technologies and essentially under full farmer management, later followed by alley cropping. Since then the role of on-farm testing has gradually increased and it has now become part of the technology development process itself.

Today, the trials in most cases are 'co-managed' by researchers and farmers and fully farmer-managed trials are still rare. Past experiences have shown that even partial researcher management may bias results, because farmers tend to approach the trials differently from their own fields. This also applies to the choice of trial locations, stratified for different classes of fields and farmers. It leads to *a priori* selection of farmers, entailing the risk of lukewarm participation and drop-out. For these and other reasons the methodology of on-farm trials, especially in the 'farmer-validation' phase, is in need of scrutiny, and the production of a new methodological text for on-farm experimentation would be welcome.

8.5 Technology delivery and dissemination

8.5.1 Consolidated results and databases

The accessibility of research findings for potential users has been and continues to be problematic. Up to 1984, the Institute published comprehensive Annual Reports which provided overviews and details of all the work conducted, while more in-depth journal publications contained additional evidence and the context of the wider scientific world. In 1985, the Annual Report changed into a glossy summary, but the FSP published its own detailed Annual Report for most years between 1985 and 1992. Thereafter a few separate fairly detailed project reports were published, until 1998 (Chapter 5), but there were no longer comprehensive publications providing an overview of all soil and soil fertility research. Since then, there has been a variety of special project reports, conference proceedings and books on specific themes, such as Environmental Characterisation and ISFM, but most significant research findings were published in scientific papers in a wide variety of journals. It has therefore become increasingly difficult for the interested outsider to obtain a good picture even of individual research themes, let alone the full range of the IITA's research topics and results. The databases, which were set up over the years (e.g. soils databases, the Resource Information System – RIS, the RCMD trials database) are no longer maintained and may now even be difficult to find or access.

There is need for publications of consolidated results, such as the research monographs published up to the mid-1990s, as well as repositories of all publications in searchable open-access databases, and the restoration of the older soil databases, with consistent meta-data, to optimise the accessibility of past research results.

8.5.2 Formulation of recommendations and the use of Decision Support Tools

During the exploratory years, the issue of recommendations did not arise, research being busy laying the groundwork on which recommendations would later be founded. In the course of the 1980s, recommendations were formulated for land clearing and management, but the target group for such recommendations was barely visible in West Africa. For soil fertility management, a few concrete recommendations with their area of applicability were formulated, including liming of acid soils, but the vast amount of soil fertility studies did not result in precise recommendations, nor were the results adequately collated in comprehensive analytical publications. Later, when the attention turned to auxiliary crops for short fallows, live mulch and alley cropping, more attention was paid to recommending the most suitable species for different soils and zones, in the case of herbaceous legumes with the aid of LEXSYS ('LEgume eXpert SYStem'), a decision support tool (DST).

Only over the past decade, since the Institute has associated itself directly with the development community, has the elaboration of concrete recommendations for soil fertility management received more attention. Rather than developing prototype technologies in the isolation of the research station, to be further adapted and disseminated by national research and extension institutions, the Institute is now assembling baskets of technologies for farmers to choose from, aided by DSTs. The development of the technologies is attuned to farmers' expressed needs, based on their resources, production objectives, risk aversion, etc.

8.5.3 Where does the technology fit?

There are two sides to the question of applicability of improved technology. The first is technical: under which physical and environmental conditions is a technology likely to perform? The other is 'entrepreneurial': for what kind of farming enterprise is this technology suitable and how can the technology be profitably incorporated into the farm?

The assessment of technical applicability requires delineation of the areas where the technology is likely to fit and experimentation in sites which are representative for these areas. Although in recent times, the choice of research areas and sites has increasingly been based on availability of donor funding and only to a limited extent on agro-ecological considerations, this does not mean that the question of applicability to other areas does not arise. It remains a strategic task of international research, which cannot limit itself entirely to 'downstream' research and extension. Different approaches have been used in the past to quantify and map the capabilities and limitations of soils as a basis for an effective extrapolation methodology. The legacy of these attempts is a large amount of partly digested data on soils, which at one time were supposed to be deposited in a database (the RIS), the fate, or even existence of which is unclear. The situation is more favourable in respect of herbaceous legumes and alley cropping, for which treatises have been written about their suitability for different ecologies and the conditions for their adoption. In spite of their rather dramatic failure in smallholder farming, a more comprehensive analysis of their potential niches and the causes of their failure would still be worthwhile. Today, new methods are being developed to characterise and map regional soil resources, in particular likely nutrient deficiencies and simulate the likely performance of new technology under different conditions. They can be expected to become valuable tools for future technology targeting.

Even if a technology is suitable for an area in a technical sense, it must fit well into a farmer's specific production system in order to be adoptable. There are again two aspects to this. The first is the nature of the farmers' production system and their attitude towards farming[7]: a technology may be suitable for one type of enterprise and not for another. The concept of farm typology is being considered as a tool to assess the kind of technology suitable for particular types of farms or, conversely, the types of farm for which a technology may be suitable.

This concept has been used in applied research in francophone Africa since the 1970s, under the name of '*typologie des exploitations agricoles*', and has recently gained attention in IITA's projects. The second aspect is the small-scale variation in conditions within a farm, in particular the nature and variability of its soils, i.e. the fertility gradients within the farm, which affect the technology's performance. The study of this micro-scale variability and its integration in recommendations for ISFM is now one of the research topics on the Institute's agenda.

8.5.4 Principles and approaches of dissemination

Initially, dissemination was not IITA's primary concern; its focus was on the collection of information and the development of principles and prototype technologies. The NARS were expected to take these up and translate them into concrete recommendations for extension to then pass on to farmers. Over time it became clear that this process did not function properly and gradually more attention was paid to dissemination. At first user-oriented guidelines were produced, for example on no-till and alley-cropping, to be used primarily by NARS. Later on the LEXSYS tool was introduced, again mainly targeting NARS, and seed of some legumes was made available to them and to development projects and NGOs for demonstration. Also, two editions were published of practical guidelines for on-farm experimentation.

With the transition to the R-for-D approach over the past decade, there has been more direct involvement in technology dissemination, through various mechanisms, including demonstration plots, in which different aspects of crop and soil management are demonstrated (plant spacing, fertiliser use, different varieties etc.), and through collaboration with actors in input and output markets. The guiding principle thereby is 'responsible scaling', which recognises that recommendations only make sense if favourable conditions exist to enable uptake. Dissemination is increasingly supported by the use of an array of dissemination and media tools, such as radio, extensions materials, video and, more recently, through rapidly improving telephone applications. Scientists have also become acutely aware of the need for proper monitoring, not least because of pressure from donors who wish to see results for their investments. Emphasis on Monitoring and Evaluation (M&E) has been adopted as a priority in all the Consortium Research Programs (CRPs) of the CGIAR.

The R-for-D approach brings research into the realm traditionally occupied by extension and development organisations, making research into a multi-tasking organisation. The question may be raised on whether this is the right direction and whether it will allow research to continue performing its own strategic tasks. That question applies to much of current CG programmes dealing with resource management under the Strategic and Results Framework.

8.5.5 The changing role of monitoring and evaluation

In the past, M&E consisted mostly of the collection of physical data on technology performance. As research was transferred from the station to the farm, data collection was often augmented with information on environmental and socio-economic factors which could explain differences in performance between farmers. More comprehensive M&E is now an integral part of R-for-D in all its phases, with specific roles for all partners, including the farmers. The objectives are to follow the implementation and results of the research activities in 'real time', explain differences in relation to the variation in conditions, and adjust the approach and the technologies to the findings: M&E thereby becomes ME&L, i.e. Monitoring, Evaluation and Learning. Methods and content of ME&L are specified at the start of every project, as part of project planning. Having extensive databases with a wide range of geo-spatially defined response variables and covariates could ultimately allow soil fertility specialists to use 'big data' approaches to fine-tune recommendations.

8.6 Impact and emerging trends

8.6.1 Technology uptake and impact

We can distinguish three levels of uptake and impact monitoring:

1   Observing whether farmers who have been involved in the on-farm testing and validation process continue to use the technology independently; this can only be reliably monitored after completion of the testing, otherwise uptake is confounded by the presence of the promotors;
2   Observing independent uptake by non-targeted farmers in the same community; and
3   Estimating uptake beyond the immediate target to the wider target area.

Separate methodology is needed for the three levels, whereby levels 1 and 2 are best conducted by 'tracking the technology' through physical field observations, not just by interviews.

During the first two decades, there was very little if any monitoring of uptake of soil and soil fertility management technologies, as most technologies under development were not systematically promoted. For land management technologies, practical guides were prepared, but with no obvious users in sight.

In the early 1990s, level 1 and 2 monitoring took place for alley cropping in south-west Nigeria, showing very clearly that the technology was not moving out from the on-farm testing fields to other fields in the same or other farmers' fields. For *Mucuna* as a planted fallow, level 1 and 2 monitoring in Benin showed active adoption for use in controlling *Imperata cylindrica*, while later claims of massive adoption were largely based on secondary indicators, viz. the amount of seed distributed by development organisations and NGOs. Anecdotal evidence indicates that the technology is presently being used at a small scale only, since this artificial market for *Mucuna* seeds collapsed.

So far, reported impact in IITA's R-for-D programs has largely been based on level 3 monitoring through large-scale household and group interviews. At an early stage of adoption this method will not be able to find evidence for burgeoning adoption and level 1 and 2 monitoring should have priority. It will be both more informative and less expensive. Furthermore, it should allow early adjustments of research objectives, methods and content to the reality of uptake or rejection and their causes.

8.6.2 Investments in research by donors and African governments

In the early years, there was regular exchange between donors and international research institutes on the direction research was taking. Major donors were represented on the institute's Boards of Trustees and there were strong personal contacts between scientists and functionaries of donor organisations, themselves often former scientists. Board members would also bring in research institutes from their home countries as collaborators, with funding attached. Impact monitoring was not envisaged, or considered premature; the system was based on confidence.

Beginning in the 1980s, doubt surfaced and donors began to question the effectiveness of research in solving Africa's food production problems. As a result, the willingness to provide unrestricted core funding declined and the balance between core and special project funding gradually changed in favour of the latter (Figure 8.3).

A turning point was reached in the period from 2000 to 2008 when public investment in agricultural R&D increased by over 20%, mainly because of concerns about the surge in global food prices, which peaked in mid-2008 (World Bank – Global Economic Monitor (data until May 2015)). The Alliance for a Green Revolution in Africa (AGRA) was established in 2006, with the belief that investing in agriculture is the surest path to reducing poverty and hunger. AGRA's Soil Health Program, which focuses on ISFM, illustrates the renewed attention for improved soil fertility management as a prerequisite for agricultural development in SSA. During this period the unrestricted funding declined sharply and restricted funding became more prominent.

The Bill and Melinda Gates Foundation (BMGF) started with their investments in agricultural development in 2005. In response to the food crisis, the G8 countries pledged $20 billion to agricultural development at the 2009

Summit in L'Aquila, Italy. Obama's pledge of $3.5 billion became the Feed the Future program that started in 2010. CGIAR funding doubled from 2008 to 2013 to $1 billion, mainly from bilateral and specific project funding. Donor organisations' demand for demonstrable impact of research on development has strongly influenced the way research is conducted, whereby the emphasis is on the dissemination of productivity-enhancing technologies, as well as on value addition by putting enabling conditions in place. This requires stronger collaboration with development and business partners. Donor organisations nowadays set the target areas and crops for which they are willing to provide research funding and insist more than ever on monitoring of the results through systematic M&E schemes. In recent years, the support for agricultural research and development seems to decline again, as other priorities are competing for funding. From 2013 World Bank funds for the CGIAR have declined further from $50 million to $30 million in 2015. Also, the Feed the Future's funding plateaued in 2015 and is set to decline in 2017.

Meanwhile, African governments have become much more proactive, through regional institutions, such as the Comprehensive Africa Agricultural Development Programme (CAADP), which are becoming important actors in setting priorities for development and the research needed to support it. This has only resulted in marginal and variable, country-specific increases in spending, despite various declarations calling for increased investment in agricultural research and development. The Maputo declaration of 2003, wherein heads of state called for an increase in spending on agriculture up to 10% of total public expenditure, for instance, has not marked a turning point. Likewise, since the Abuja declaration of 2006, which called for an increase in fertiliser consumption up to 50 kg of nutrients per ha by 2015, there has been some increase in fertiliser use with a number of countries, such as Nigeria, Ethiopia and Malawi, approaching this target (Sheahan and Barrett, 2014) but overall use is still far from the target. The Malabo declaration of 2010 called for concerted action to double farmers' yields but so far has not induced increased public expenditure on agriculture. For West and Central Africa, we see some overall increase, but it is modest at only half of the targeted 10%. For East Africa, public spending on agriculture reached a peak of 6.8% in 2004 followed by a decline to 4.3% in 2014. For Southern Africa, it increased consistently from 1.3% in 1998 to a maximum of 2.9% in 2007 and then declined again to around 2% in 2013–2014. These are general trends, with some countries actually spending over 10% of their national budget on agriculture. Spending on agricultural research, including research personnel, in SSA has increased slowly but steadily over the past 15 years (Figure 8.7).

In 2011, Africa invested 0.51% of the agricultural GDP in research, below the African Union's target of 1%. Again, there are notable exceptions with some countries spending around 2% on research. Most of it is spent on salaries, and research and development infrastructure and equipment depending heavily on donor aid. This is a worrying trend. International research depends on the national research and extension systems for the delivery and dissemination of

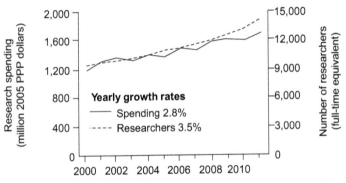

Figure 8.7 Trends in public spending levels in agricultural research and researcher numbers in sub-Saharan Africa, 2000–2011 (adapted from Beintema and Stads, 2014) 'PPP' = 'Purchasing power parity'

their technologies, and the sustainability of R-for-D requires strong national systems. International investment to strengthen NARS is not sufficient to build strong research systems and investment by the countries themselves is needed.

8.6.3 From 'Research-for-Development' to 'Research-in-Development'

The term R-for-D is now gradually being replaced by 'Research-in-Development' (R-in-D), which stresses the need to integrate research within development programmes and contribute directly to their development efforts. Research aligns its priorities with those of the development partner and tests, validates and disseminates successful technologies through the organisation's network. The target areas of research and development are equal and so is the variation in conditions affecting the performance of innovations. This should enhance the likelihood of outputs and outcomes satisfying the goals of both.

8.7 Looking forward

This book started off as a project to document the history of soil, land use and soil fertility research in SSA and the relevance of its results for today. As the story unfolded, it was felt that the subject had to be widened, as the practice of soil research itself had widened as time went on, into what is now called R-in-D, where the quality of the soil and the adequacy of soil fertility is addressed in the context of agricultural development initiatives (Section 8.6.3).

In these final paragraphs, we bring together the views of the authors about the role soil and soil fertility research has to play today, within and as part of the agricultural development effort, against the background of what has and what has not been accomplished in the past 50 years. Rather than dissecting those views into parts, as we have done with the history of soil research itself, we present an integrated vision here, in the form of a narrative, built from the views expressed by the authors in the course of the writing process. We are aware that some of the views are 'stating the obvious' but we hope that the lessons which emerged while writing this book will enrich current reflections on how to carry out soil and soil fertility research which is more effective and more efficient in reaching real impact on smallholder farming.

There is broad agreement that the way forward for agriculture in SSA is through SI, and research has an important role to play on the road towards that goal. Agricultural development is a multi-faceted process, hence the need for firm partnerships between research and other actors right from the start. New methods such as network analysis (Schut et al., 2016) can be used to identify partners who are most influential and most closely associated with the farming community, through whose network technologies can eventually be disseminated. This is one of the cornerstones of what is referred to as 'R-in-D'.

Research must begin at the real farm, through early, rapid characterisation of the target area by cost-effective methods, including informal surveying and big data techniques. The purpose is to gain a preliminary insight in the environmental and socio-economic conditions of smallholder farming and its current farming practices, which is needed for future technology development. The role of particular crops, animals, practices and technologies in the farming system must be understood to address the constraints and interests of the intended clients. Baseline surveys are often used as well, but their cost- and time-effectiveness can often be questioned.

The next step in the process is setting research goals. It has become very clear that impact will depend on the degree of ownership of the research agenda by the partners in the development process and by the end-users: old-school 'top-down' research may have created some scientific insights but very little impact. The technologies to be developed and tested must respond to the end-users' expressed needs and priorities, if they are to stand a chance at sustained adoption. First of all, the perceptions, drives and decision-making of farmers must be known, as they will influence eventual adoption. Next, the perceptions of well-informed, experienced and effective development partners are of great importance and the questions articulated by them must be discussed and agreed upon at the inception of every initiative. A distinctive feature of R-in-D compared with earlier participative approaches is the systematic involvement of development partners. Making their contributions effective will be essential for success. Development partners must enable end users to articulate their needs, which research can then further scrutinise in order to identify best-fit interventions, taking into account other factors that may have synergy with the prioritised needs or minimise trade-offs with other components in the farming system.

Platforms for interaction and integration of the various partners' activities should continue to be used and made more effective, as a tool for setting and prioritising the research agenda with all local stakeholders and defining the role of each, consistent with their competences. Multi-partner development efforts are often slow, however, and improvement of mechanisms such as Innovation Platforms will be needed for more effective cooperation. In order to be effective, these Platforms need 'facilitation' and should address specific issues rather than just broad themes.

There is no guarantee that the demands expressed by farmers and other stakeholders are the real and relevant issues, but the R-in-D approach is being adopted to enhance the likelihood that they are. Whether that is the case, should become clear once the research process gets underway. Apart from the immediate stakeholders, the policies and objectives of donors and national governments, as well as the vision of the research institution itself about its role have to be accommodated in the overall program objectives and strategies.

The actual content of the research will be the combined result of what has been learned through observation and discussion in the field, the capabilities of the research teams and their partners, and the priorities of the intended beneficiaries. An important *a priori* question, which is related to the goals of governments, donors and institutes, is about the broad target population, e.g. should research preferentially focus on the (semi) commercial farmer, as an effective driver of development?

Technologies must be chosen which have shown promise, through ex-ante analyses, to 'fit' within the environmental and socio-economic conditions of the intended users and that can satisfy their needs. Sustainable intensification, the overall goal of soil and soil fertility R-in-D, requires the protection, improvement and maintenance of soil resources, and a sustained supply of the nutrients needed for economically optimum crop yields. Research must develop and help disseminate technologies which allow farmers to use native and applied nutrients to the best effect at the lowest cost. That requires reliable information about the availability of soil nutrients across the range of soil types in the area, the crops' nutrient demands and the response to and profitability of additional nutrients supplied to the soil. Market opportunities and the availability and prices of inputs are thereby important factors. Industry and the private sector must be engaged early to ensure sustainable upscaling and profit-driven demand.

The technical objectives must be translated into precise research themes and into a program of activities to address them. The first theme is the adequate characterisation of the soil resources and their spatial diversity. The diversity is large in African soils, both at a local and regional scale, and its characterisation and mapping are needed to extrapolate the effect of innovations to other locations with similar soil conditions. The question is now, what the principles of characterisation should be. Much existing pedological information remains of great use, but new techniques for soil characterisation and the rapid assessment of specific soil properties should be widely used, for example non-destructive measurement tools for nutrient content of soils and leaves, which can greatly increase the scale and efficiency of research. A comparative analysis of approaches and techniques would be advisable. Also, more attention is needed for farmers' own soil and soil fertility classification.

Research-in-Development will always be a combination of applied and strategic research, whereby the primary question will be how agriculture can be sustainably intensified for as large a number of farmers as possible. For that purpose, technologies need to be identified for farmers' integrated farming systems, not only for single components. They must either be chosen from the Institute's or the partner NARS' technology shelves, or developed specifically for the target farmers and their conditions and tested under farmers' conditions and with their full involvement. Some of the earlier technologies, despite low rates of adoption, may still be relevant due to change in farmers' current conditions. In respect of soil and soil fertility improvement, the emphasis in most research areas is now on appropriate fertiliser formulations in tandem with dual purpose legumes for integration into the farmers' systems. What fertiliser formulation is appropriate depends on the nature of the soil, its native nutrient stock and the response of crops to native and applied nutrients. We must be able to verify which factors determine this response and identify the conditions and locations where a good response may be expected. Nutrient response trials are therefore needed, which can account for the differences in both environmental conditions and farmer management. Focusing on the 'positive deviants' instead of just embracing overall variability is quite novel and can be revealing.

Existing profile-based soil information will be helpful for broad landscape-level stratification, supplemented by modern methods for rapid assessment of fertility differences and gradients within landscapes. Laying out on-farm trials along environmental gradients will help understand technology performance in dependence of those differences. This will be essential for the eventual formulation of site-specific recommendations. A similar reasoning applies to other technologies, such as crop management practices and the incorporation of dual purpose legumes in a system.

Adoptability of technology does not only depend on physical performance of technology but also on socio-economic factors. There is a need for robust methods for integrating the various diversities in site selection for on-farm trials and for later demonstrations of proven technology. Pragmatic site selection holds more promise than sticking to bench mark sites and will allow using development partners' grassroots infrastructure and networks for participatory testing of interventions and extending successful ones. Multivariate analysis can be used to cluster farmers in an agro-ecological zone into farm types on the basis of their biophysical and socioeconomic characteristics. The farm types can then constitute recommendation domains that can be used for on-farm technology validation.

On the technical side, an important parameter for profitable fertiliser use is the efficiency with which nutrients are converted into yield, expressed as nutrient use efficiency (NUE) or fertiliser use efficiency (FUE), a topic which had received insufficient attention in the early years but is essential in the quest for Sustainable Intensification. The use efficiency of a specific nutrient or the absence of response depends on several factors, an important one being the abundance or shortage of other nutrients. Low NUE for one nutrient may then indicate deficiency of others, but there are other factors, such as P-fixation in some soils. Much attention is needed for low NUE and for measures to correct them, through balanced fertiliser applications, agronomic practices or micro-biological measures. Soil organic matter content is also important for soil quality, through soil structure, by buffering for nutrients, and by storing and releasing N and P, and methods are needed for its build-up and maintenance. An important question thereby is whether fertiliser use is a valid entry point for the enhancement of soil organic matter stocks through the production of increased amounts of crop residues. A complementary question is whether research should primarily concentrate on the more fertile soils, or should the reclamation of poor or non-responsive soils or alternative uses be a goal from the beginning? The role of fertiliser will remain important for the intensification of both productive and less productive, impoverished soils. Strategic research remains necessary to address questions that come up during technology development.

Results of on-farm trials must be used for the development of recommendations, whereby advanced tools are needed which can help explain technology performance across highly variable conditions and predict technology performance under the specific conditions of particular farmers or farmer categories. Such tools, including rapid *in situ* nutrient assessment, novel methods for the analysis of variation, nutrient response functions, yield gap analysis and crop growth models, are being tested and their role will increase. It would also be useful to evaluate earlier methods to account for within and between field and farmer variation and compare those with more recent techniques, for relevance to today's needs.

Effective implementation of on-farm research, technology validation, demonstration and dissemination is essential for successful R-in-D. An active network of extension agents, also called nowadays 'last-mile delivery agents' is indispensable for wide-scale agronomic testing and dissemination. They facilitate interactions with farmers and should guide and support them to test and adapt technologies, and collect data on performance of technologies using simple tools in the form of smart devices. Maximising farmer management in on-farm trials is crucial for ownership of the results. Voluntary participation is thereby paramount and should not be hampered by an inflexible site selection scheme. On-farm experimentation should also take into account that farmers are more likely to adopt ISFM options in a stepwise manner, rather than complex assemblies of technology, and the on-farm testing and validation process should mimic this by a stepwise inclusion of ISFM components. Once enough has been learned about technology performance, baskets of options are put together, allowing end-users to choose what best suits their circumstances, adding on additional options as more opportunities unfold. The basket of options must be supported by DSTs, to help extension workers and farmers to choose technology components suitable for their conditions. DSTs for local nutrient management, for example, are being developed, based on new techniques for soil and soil fertility characterisation combined with response patterns from on-farm trials. As long as extensions workers are needed as intermediaries between the farmers and the researcher, the tools are best developed for the extension worker. More attention is needed, however, for the development of DSTs, which are suitable for use by the end-users themselves, not asking technical information that they do not know.

Monitoring, evaluation and impact assessment should be a continuous process. Appropriate and cost-effective methods for assessing adoption and impact should be agreed upon right at the beginning of a project. The process for the assessment of technology adoption is often overlooked, even though there is a general agreement that it is required. Direct monitoring, by physically 'tracking the technology' in the field, should have priority as the only way to verify adoption or rejection at an early stage. Examples are actual fertiliser use by farmers, their yield gaps and FUE. The use of DSTs should also be an integral part of monitoring and evaluation, and feedback should be collected on their performance. Actual findings and feed-back during the implementation of the program will be an important modifying factor for the evolution of research priorities.

Storage and accessibility of previously collected data and research results has been problematic. All data from every trial and every survey should be captured in an institutionally supported open-access database including IITA's soil and nutrient response database since 1968. The analysis of past results could take the form of a series of analytical reports on specific themes, for example nutrient responses in relation to soil types in SSA; soil acidity, Al-toxicity and P- and Zn-response; occurrence of secondary and micronutrient deficiencies; retrospective analysis of the alley cropping (and *Mucuna*) story. Furthermore, an annual or biennial technical publication would be important, documenting all of IITA's on-going soil and soil fertility research at a technical, rather than just anecdotal, level to enhance transparency. Another approach could be the creation of a document repository with well-defined meta-data to allow for queries and effective searching.

Since consistent, shared and reliable methods for on-farm technology testing, validation, dissemination and monitoring adoption are crucial, the publication of a new book of methods for R-in-D would be welcome. It would, for instance, distinguish on-farm trials, adaptation trials, farmer-managed trials, farmer experimentation and demonstrations, to highlight different aspects of technology validation by farmers. The removal of researcher bias is thereby essential. Past research failed to generate adoptable technologies and the question still is whether today's approach will be more successful. The book of methods should therefore also analyse how features of R-in-D enhance its chances of success, compared with the earlier methods, using examples to substantiate the claims.

Advanced tools are increasingly used in all phases of R-in-D. The most suitable tools and modelling systems for soil and soil fertility research for the coming years should be chosen seeking consistency across the Institute. Collaboration with advanced institutions is needed to develop in-house capacity for the informed choice and use of tools and models, or to adapt existing tools to the Institute's requirements.

Covering a large number of sites in response to (donor) opportunities is good, but with many small and widely scattered teams there is a risk of losing the opportunity to learn from each other. Decentralisation needs a strong base of resource persons. Common principles and sharing data across initiatives will accelerate gathering knowledge and increase understanding of soils, farms and farming systems. Various mechanisms for regular technical harmonisation may be considered, such as cross-project meetings around concrete methodological and technical issues.

International soil and soil fertility research has gone through a number of evolutionary cycles in the past 50 years which have brought it to its current configuration: a flexible operation, responding to the needs of today's smallholder farmers and working in close cooperation with NARS and development organisations, to identify those needs and deliver the technologies which will hopefully satisfy them. The logic of this approach, called R-in-D, is compelling, but there are many impediments standing in the way of achieving its ambitious goals, among them the dependency on effective performance of various partners in the process. The coming years should show whether the promises are going to be fulfilled, and the researchers' avowed dedication to genuine and effective monitoring and faithful reporting should clearly bring out both the successes and the failures.

Notes

1 The colloquial term R4D is commonly used in the dedicated literature as short hand for Research-for-Development.
2 The USDA soil classification was mostly used in those years.
3 The exception was the *Recherche Appliquée en Milieu Réel* (RAMR) project in Benin Republic, which had strong professional ties with IITA-Ibadan.
4 In a sense, the term 'organic fertiliser' is a contradiction in terms since Prof Rudy Dudal stated that no organic inputs contain a sufficient amount of readily available plant nutrients to quality as 'fertiliser'; henceforth, we use the terms 'fertiliser' and 'organic inputs'.
5 Meanwhile the numbers of publications have increased steeply, reaching about 160 in the period 2001–2011.
6 An example is dynamic modelling. In the 1990s a member of the University of Florida modelling group was recruited, who introduced the CERES family of models. Next, the Rothamsted soil organic matter model was introduced through the Leuven University collaborative program, and more recently the Wageningen University crop models have been used. The external NRM review of 2011 proposed yet another institution as a partner for dynamic modelling, viz. the Australian CSIRO. A similar picture emerges for soil and soil fertility characterisation, which has travelled from the University of Hawaii and its Benchmark Soils and IBSNAT projects based on the USDA Soil Taxonomy, through the North Carolina State University with its FCC system, then to the adoption of the FAO soil classification system and the semi-empirical QUEFTS method for soil fertility assessment.
7 In some areas, for instance in eastern DR Congo, it is common that smallholder farmers reply 'none' to the question 'which job do you have', while they are gaining their livelihood through farming.

References

Beintema N. and Stads G.J., 2014. *Taking Stock of National Agricultural R&D Capacity in Africa South of the Sahara.* ASTI Synthesis Report, International Food Policy Research Institute, Washington, DC.

Fresco, L., 1984. *Comparing Anglophone and Francophone Approaches to Farming Systems Research and Extension.* FSSP Networking Papers No. 1. University of Florida, Gainesville, FL.

Hildebrand, P.E., 1984. Modified stability analysis of farmer-managed on-farm trials. *Agronomy Journal,* 76, 271–274.

Schut, M., P. J.A.Van Asten, C. Okafor, C. Hicintuka, S. Mapatano, L. Nabahungu, D. Kagabo, P. Muchunguzi, E. Njukwe, P. Dontsop, B. Vanlauwe, 2016. Sustainable intensification of agricultural systems in the Central African Highlands: the need for institutional innovation. *Agricultural Systems,* 145, 165–176.

Sheahan, M. and C.B. Barrett., 2014. *Understanding the Agricultural Input Landscape in Sub-Saharan Africa Recent Plot, Household, and Community-Level Evidence,* Policy Research Working Paper 7014, World Bank, Washington, DC.

Annexe I
Examples of analyses of soils
from various (benchmark) sites

1. Analyses and simple land evaluation of an Egbeda and an Apomu soil at the Ibadan station (Moormann et al., 1975); Chapter 1

Table A1.1 Egbeda (Oxic Paleustalf), IITA Ibadan station

Depth (cm)	Mechanical analysis (%)				pH		Exchangeable cations (cmol/kg)					CEC (cmol/kg)	Organic C (%)	Total N (%)	Bray-I P (mg/kg)
	Gravel	Sand	Silt	Clay	H2O	KCl	Ca	Mg	K	Mn	Al				
0-5	25	67	12	21	6.5	5.9	3.50	1.20	0.39	0.09	0.01	5.45	1.54	0.138	14.9
5-15	20	58	17	26	6.4	5.4	5.20	1.31	0.15	0.11	0.01	6.39	1.50		6.3
15-45	45	49	14	37	6.8	5.7	3.52	1.36	0.12	0.05	0.03	5.26	0.76	0.063	2.6
45-65	40	30	14	56	6.4	5.1	3.05	0.89	0.10	0.02	0.04	4.33	0.28	0.040	0.7
65-95	40	30	6	64	6.0	5.5	2.90	0.75	0.11	0.03	0.06	4.06	0.26	0.044	0.7
95-110	30	20	14	65	5.9	5.4	2.75	0.68	0.10	0.04	0.06	4.49	0.18	0.040	0.7

Suitability for crop production:
- Moderately suitable, increasing limitations, recurrent inputs economical for: maize, cassava, cowpeas
- Moderately suitable, increasing limitations, recurrent inputs not economical for: yams, pigeon pea, soybeans

Table A1.2 Apomu (Psammentic Ustorthent), IITA Ibadan station

Depth (cm)	Mechanical analysis (%)				pH		Exchangeable cations (cmol/kg)					CEC (cmol/kg)	Organic C (%)	Total N (%)	Bray-I P (mg/kg)
	Gravel	Sand	Silt	Clay	H2O	KCl	Ca	Mg	K	Mn	Al				
0-9	2	78	11	10	6.0	5.6	3.95	1.68	0.26	0.13	0.05	6.27	2.48	0.175	12.4
9-36	3	82	7	10	6.3	5.7	1.05	0.46	0.06	0.07	0.02	1.85	0.30	0.031	2.8
36-87	4	87	3	10	6.3	5.6	1.67	0.12	0.04	0.02	0.01	1.04	0.08	0.010	2.6
87-105	5	82	5	13	6.1	5.2	0.80	0.16	0.03	0.01	0.02	1.21	0.05	0.010	3.2
105-178	30	74	5	21	6.0	4.8	1.72	0.38	0.11	0.02	0.02	2.43	0.13	0.019	3.5

Suitability for crop production:
- Moderately suitable, increasing limitations, recurrent inputs economical for: cassava
- Moderately suitable, increasing limitations, recurrent inputs not economical for: maize, cowpeas, pigeon pea
- Marginal, severe limitations for: yams; unsuitable for: soybeans

2. Analyses of two soils in Cross River State, South-East Nigeria (Juo and Moormann, 1981); Chapter 1

Table A1.3 Orhoxic Tropudult (Uyanga toposequence)

Depth (cm)	Mechanical analysis (%)				pH		Exchangeable cations (cmol/kg)					CEC (cmol/kg)	Organic C (%)	Total N (%)	Bray-I P (mg/kg)
	Gravel	Sand	Silt	Clay	H2O	KCl	Ca	Mg	K	Mn	Al				
0-5	0	79	13	8	5.2	4.9	6.49	1.99	0.39		0.24	9.51	3.00	0.19	NA[a]
10-23	19	63	9	28	4.7	3.9	0.63	0.25	0.10	1.82	0.42	3.37	0.70	0.05	NA
23-94	30	59	8	33	4.8	4.0	0.60	0.20	0.07	1.98	0.14	3.13	0.35	0.02	NA
94-117	28	59	9	32	4.9	4.1	0.53	0.28	0.07	1.73	0.27	2.99	0.23	0.02	NA
117-155	22	64	23	13	5.0	4.1	0.53	0.05	0.05	1.57	0.23	2.55	0.54	0.02	NA

Table A1.4 Orhoxic Tropudult (Netim toposequence)

Depth (cm)	Mechanical analysis (%)				pH		Exchangeable cations (cmol/kg)					CEC (cmol/kg)	Organic C (%)	Total N (%)	Bray-I P (mg/kg)
	Gravel	Sand	Silt	Clay	H2O	KCl	Ca	Mg	K	Mn	Al				
0-6	1	62	16	22	3.9	3.5	1.87	0.58	0.24	1.78	1.38	6.07	2.94	0.26	NA[a]
6-13	29	65	13	22	4.0	3.5	0.67	0.16	0.19	2.25	1.36	4.78	1.39	0.12	NA
13-27	60	59	11	30	4.1	3.7	0.50	0.23	0.12	2.27		3.25	1.15	0.08	NA
27-100	62	56	9	35	4.8	3.9	0.37	0.07	0.10	1.72	0.88	3.26	0.83	0.06	NA
100-140	63	54	9	37	4.8	3.9	0.34	0.04	0.07	1.33	0.87	2.77	0.52	0.05	NA

a 'NA' = 'not available'; surprisingly, no data were given for available P-content.
Potential for agricultural use:
- Soil erosion and subsoil acidity are probably the most important factors limiting potential agricultural land use in the area for extensive food crop production.

3. Analyses of 27 hydromorphic soils from the savannah and the forest-savannah transition zone of Nigeria (IITA, 1981); Chapter 2

Table A1.5 Forest-savannah transition zone (17 soils)

Layer	Statistic	Mechanical analysis (%)			pH (H₂O)	Exchangeable cations, me/100g (cmol/kg)			ECEC (cmol/kg)	Organic C (%)	Bray-1 P (mg/kg)
		Sand	Silt	Clay		Ca	Mg	K			
Surface soil	Range	13–89	4–51	4–68	4.7–6.4	0.67–12.8	0.29–7.40	0.03–0.26	1.51–23.60	0.37–1.96	2–14
	Mean	64	21	15	5.6	3.48	1.53	0.13	5.99	0.91	5
Subsurface soil	Range	10–84	5–53	3–70	5.0–6.5	0.47–11.1	0.21–7.89	0.02–0.46	1.06–21.26	0.07–1.36	0.2–4
	Mean	61	18	21	5.8	3.34	1.67	0.11	6.14	0.42	2

Table A1.6 Savannah zone (17 soils)

Layer	Statistic	Mechanical analysis (%)			pH (H₂O)	Exchangeable cations, me/100g (cmol/kg)			ECEC (cmol/kg)	Organic C (%)	Bray-1 P (mg/kg)
		Sand	Silt	Clay		Ca	Mg	K			
Surface soil	Range	4–82	8–51	6–50	4.2–6.1	0.55–7.94	0.20–3.36	0.04–1.92	1.73–11.31	0.29–2.70	2–39
	Mean	42	37	22	5.3	3.50	1.32	0.40	6.26	1.04	9
Subsurface soil	Range	5–89	6–55	5–70	4.6–7.4	0.40–6.84	0.06–3.25	0.04–0.29	0.99–10.61	0.07–0.75	0.4–5
	Mean	43	29	28	5.7	3.34	1.30	0.14	6.13	0.42	2

Table A1.7 Analyses of 2 contrasting Ultisols from Eastern Nigeria used in a liming trial (Friessen et al., 1980); Chapter 2

Soil series (Soil type)	Mechanical analysis (%)		pH		Exchangeable cations (cmol/kg)				Exchangeable acidity (cmol/kg)		Total free Fe₂O₃ (%)	Organic C (%)	Bray-1 P (mg/kg)
	Sand	Clay	H₂O	KCl	Ca	Mg	K	Na	Al	Al+H			
Nkpologu (Ustoxic Paleustult)	64	28	4.72	3.83	0.21	0.06	0.05	0.14	1.88	2.48	3.2	0.63	1.7
Onne (Oxic Paleudult)	73	21	4.32	3.77	0.21	0.08	0.04	0.14	1.79	2.53	1.5	0.98	35

Soil was sampled from 15–40 cm depth, 'in order to approximate the depleted state approached by the surface soil after intensive cropping and also to minimise the complicating effects of organic matter on soil Al chemistry and of organic P mineralization.'

4. Analyses of 4 hydromorphic soils from Sierra Leone (IITA, 1984); Chapter 3

Table A1.8 Typic Tropaquent (BO)

Layer	Mechanical analysis (%)			pH (H₂O)	Exchangeable cations (cmol/kg)				Exch. acidity (cmol/kg)		ECEC (cmol/kg)	Organic C (%)	Total N (%)	Bray-I P (mg/kg)
	Sand	Silt	Clay		Ca	Mg	K	Na	Al	H				
0–15	64	33	3	4.6	0.32	0.08	0.06	0.03	1.16	1.24	2.89	2.4	0.24	11.7
15–35	68	29	3	4.8	0.27	0.07	0.05	0.02	0.96	0.96	2.33	1.2	0.18	8.2
35–60	70	25	5	4.8	0.21	0.42	0.03	0.02	1.09	0.97	2.76	0.6	0.07	9.2
60–80+	70	23	7	4.8	0.21	0.09	0.03	0.03	1.11	1.05	2.52	0.6	0.09	6.0

Table A1.9 Umbric Tropaquult (Gbundapi)

Layer	Mechanical analysis (%)			pH (H₂O)	Exchangeable cations (cmol/kg)				Exch. acidity (cmol/kg)		ECEC (cmol/kg)	Organic C (%)	Total N (%)	Bray-I P (mg/kg)
	Sand	Silt	Clay		Ca	Mg	K	Na	Al	H				
0–23	16	38	47	4.8	0.97	0.49	0.16	0.08	3.20	4.56	9.47	6.8	0.58	5.4
23–32	8	21	71	4.5	0.46	0.35	0.07	0.08	3.04	4.16	8.16	1.7	0.22	4.6
32–55	6	37	57	4.5	0.25	0.29	0.06	0.07	2.42.	3.50	6.59	0.8	0.10	4.2
55–80+	6	63	31	4.5	0.21	0.37	0.06	0.09	2.51	3.30	6.53	0.9	0.11	

Table A1.10 Plinthic Tropaquult (Gbonsamba)

Layer	Mechanical analysis (%)			pH (H₂O)	Exchangeable cations (cmol/kg)				Exch. acidity (cmol/kg)		ECEC (cmol/kg)	Organic C (%)	Total N (%)	Bray-I P (mg/kg)
	Sand	Silt	Clay		Ca	Mg	K	Na	Al	H				
0–20	33	57	11	5.0	0.29	0.09	0.07	0.03	1.94	1.86	4.28	1.6	0.21	7.9
20–40	38	53	9	5.0	0.16	0.06	0.04	0.02	1.88	1.92	3.88	1.6	010	4.5
40–65	40	45	15	4.8	0.19	0.09	0.04	0.02	1.63	1.57	3.56	0.3	0.07	1.9
65–90+	80	17	3	5.0	0.30	0.14	0.05	0.03	2.05	2.35	4.92	0.2	0.05	

Table A1.11 Typic Sulfaquent (Rokpur)

Layer	Mechanical analysis (%)			pH (H$_2$O)	Exchangeable cations (cmol/kg)				Exch. acidity (cmol/kg)		ECEC (cmol/kg)	Organic C (%)	Total N (%)	Bray-1 P (mg/kg)
	Sand	Silt	Clay		Ca	Mg	K	Na	Al	H				
0-15	16	53	31	3.9	1.38	0.61	0.32	2.04	1.56	1.96	2.89	3.8	0.30	4.8
15-50+	38	49	13	3.0	2.29	0.61	0.10	6.52	2.93	11.79	24.43	4.5	0.28	4.5

Annexe II
Soil Taxonomies

Table A2.1 Correlation between Soil Taxonomy and World Reference Base at Soil Order level (Jones et al., 2013).

Soil Taxonomy	WRB	Soil Taxonomy	WRB
Alfisols	Luvisols, Lixisols	Inceptisols	Cambisols, Umbrisols
Ultisols	Acrisols, Alisols	Entisols	Umbrisols
Oxisols	Ferralsols, Nitisols	Fluvents	Fluvisols
		Psamments	Arenosols

Table A2.2 Correlation between Soil Taxonomy and World Reference Base at Sub-Group level for some benchmark soils in Nigeria (Moormann, 1981; Moormann et al., 1974) and Ghana (USAID/IITA, 1983).

Soil Taxonomy	WRB	Soil Taxonomy	WRB
Alfisols		*Inceptisols*	
Oxic Haplustalf	Ferric Luvisol	Aeric Tropaquept	Eutric Gleysol; Dystric Gleysol
Udic Haplustalf	Ferric Luvisol (gleyic)	Vertic (Mollic) Tropaquept	Mollic Gleysol; Eutric Gleysol
Oxic Paleustalf	Ferric Acrisol; Eutric Nitosol; Ferric Luvisol	Oxic Ustropept	Eutric Cambisol; Eutric Cambisol (stony phase)
Rhodic Paleustalf	Ferric Luvisol	Paralithic Ustropept	Chromic Luvisol; Eutric Cambisol
Oxic Plinthustalf	Plinthic Luvisol; Plinthic Luvisol, (pertroferric phase)	Aquic Tropaquept	Dystric Gleysol
Typic Plinthustalf	Plinthic Luvisol		
Typic Plinthaqualf	Plinthic Luvisol		
Oxic Rhodustalf	Chromic Luvisol (stony phase)		

continued…

Table A2.2 continued

Soil Taxonomy	WRB	Soil Taxonomy	WRB
Alfisols		*Inceptisols*	
Udic Rhodustalf	Chromic Luvisol	Tropaquent Eutric Gleysol	Mollic Gleysol
Udic Haplustalf	Orthic Luvisol	Aeric Tropaquent	Eutric Gleysol or Eutric Fluvisol
Aeric Tropaqualf	Gleyic Luvisol	Psammaquentic Tropaquent	Dystric Fluvisol or Dystric Gleysol
Aeric Tropaquent	Dystric Gleysol	Psammaquentic Aeric Tropaquent	Eutric Regosol
		Psammentic Troporthent	Luvic Arenosol
Ultisols		Ultic Ustipsamment	Lihosol
Plinthic Paleustult	Plinthic Acrisol	Lithic Ustorthent	Albic Arenosol
Oxic Plinthudult	Plinthic Acrisol	Psammentic Ustorthent	Eutric Regosol
Orthoxic Tropohumult	Ferric Acrisol	Psammaquentic Ustorthent	Eutric Gleysol
Orthoxic Tropudult	Ferric Acrisol	Vertic Tropaquent	
Plinthic Orthoxic Tropudult	Plinthic Acrisol		
Plinthic Paleustult	Plinthic Acrisol		
Typic Rhodustult	Orthic Acrisol		
Oxic Paleustult	Ferric Acrisol		
Rhodic Paleustult	Dystric Nitisol		

Note

There are distinct differences between the current and the earlier versions of Soil Taxonomy at the Order level.

Index

Page references in *italic* indicate figures and tables. The acronym SSA stands for sub-Saharan Africa, and IITA for International Institute of Tropical Agriculture.

Milton Keynes UK
Ingram Content Group UK Ltd.
UKHW021818071024
449327UK00021B/1343